SpringerBriefs in Architectural Design and Technology

Indexed by SCOPUS

Understanding the complex relationship between design and technology is increasingly critical to the field of Architecture. The *Springer Briefs in Architectural Design and Technology* series provides accessible and comprehensive guides for all aspects of current architectural design relating to advances in technology including material science, material technology, structure and form, environmental strategies, building performance and energy, computer simulation and modeling, digital fabrication, and advanced building processes. The series features leading international experts from academia and practice who provide in-depth knowledge on all aspects of integrating architectural design with technical and environmental building solutions towards the challenges of a better world. Provocative and inspirational, each volume in the Series aims to stimulate theoretical and creative advances and question the outcome of technical innovations as well as the far-reaching social, cultural, and environmental challenges that present themselves to architectural design today. Each brief asks why things are as they are, traces the latest trends and provides penetrating, insightful and in-depth views of current topics of architectural design. *Springer Briefs in Architectural Design and Technology* provides must-have, cutting-edge content that becomes an essential reference for academics, practitioners, and students of Architecture worldwide.

Vikas Chand Sharma · Sandeep Dua ·
Pankaj Kumar Tyagi · Elena Tarsi
Editors

Architectural Vision of Overtourism

Volume 2

Springer

Editors
Vikas Chand Sharma
Liberal Arts and Humanities Cluster,
University Institute of Architecture
Chandigarh University
Mohali, Punjab, India

Pankaj Kumar Tyagi
University Institute of Tourism
and Hospitality Management
Chandigarh University
Mohali, Punjab, India

Sandeep Dua
Department of Architecture
Guru Nanak Dev University
Amritsar, Punjab, India

Elena Tarsi
Department of Architecture
University of Florence
Florence, Italy

ISSN 2199-580X ISSN 2199-5818 (electronic)
SpringerBriefs in Architectural Design and Technology
ISBN 978-981-96-5793-3 ISBN 978-981-96-5794-0 (eBook)
https://doi.org/10.1007/978-981-96-5794-0

This Springer imprint is published by the registered company Springer Nature Singapore Pte Ltd.
The registered company address is: 152 Beach Road, #21-01/04 Gateway East, Singapore 189721, Singapore

Preface

In an era of unprecedented global mobility and digital connectivity, overtourism has emerged as one of the most critical challenges facing destinations worldwide. The phenomenon has evolved from a mere management issue to a complex crisis that threatens the very essence of our cultural heritage sites, historic cities, and local communities. While tourism continues to be a vital economic driver, generating approximately 10% of global GDP and supporting millions of jobs, its unchecked growth has created an unsustainable burden on many destinations. From the crowded streets of Venice to the delicate ecosystems of the Himalayas, from ancient religious monuments to modern urban centers, the impacts of overtourism are increasingly visible and concerning. This crisis demands a fundamental rethinking of how we approach tourism development, particularly from an architectural and urban planning perspective. The intersection of architecture, technology, and sustainable tourism practices offers promising pathways forward, yet also raises critical questions about preservation, authenticity, and the future of travel itself. This book, presents a comprehensive examination of these challenges and opportunities, exploring how architectural interventions and technological innovations can address the complex challenges posed by overtourism while promoting sustainable urban development.

The volume is structured in four interconnected parts, each addressing distinct yet interrelated aspects of the overtourism phenomenon. The first part, *Understanding Overtourism*, establishes the theoretical foundation by examining fundamental concepts and presenting case studies from various global contexts, Tourism, once celebrated primarily for its economic benefits, has evolved into a complex phenomenon that challenges the very fabric of our architectural and urban spaces. This part presents five distinct perspectives on how excessive tourism impacts our built environment and cultural heritage, from religious monuments in Florence to the intricate canals of Venice. Authors examine how overtourism manifests across diverse settings: historic religious sites in Florence grappling with preservation and accessibility; traditional villages in Turkey and Indonesia balancing cultural authenticity with tourist demands; Venice's unique urban fabric facing pressures that threaten its community identity; and the emerging role of smart cities in managing tourist flows.

Through these varied lenses, authors examine innovative solutions that integrate strategic planning, technological innovation, and community engagement.

This part underscores a crucial message: the architectural response to overtourism must extend beyond physical infrastructure to encompass the social, cultural, and technological dimensions of our evolving cities and heritage sites. The authors present frameworks and strategies that aim to transform overtourism's challenges into opportunities for more sustainable and resilient urban environments, while preserving the authentic character that makes these destinations worth visiting in the first place.

The second part, *Technology and Overtourism*, explores the dual nature of technology in tourism management. While acknowledging technology's potential contribution to overtourism, this part also investigates how digital solutions, artificial intelligence, and smart infrastructure can be harnessed to mitigate its negative impacts. The chapters in this part present we delve into the complex relationship between technology and overtourism, exploring both its promises and pitfalls in shaping our architectural and urban landscapes. The chapters in this part present a critical examination of how technological solutions, particularly artificial intelligence and smart infrastructure, are being deployed to address the challenges of excessive tourism.

From the dark side of technology's influence on tourism trends to the practical application of AI in managing tourist flows in Nainital's lake town, these contributions reveal the multifaceted nature of technological interventions. Through case studies spanning from Florence's Historic Centre to Slovakia's High Tatras, our authors examine how smart tourism infrastructure and digital innovations are reshaping destination management strategies. However, this part goes beyond mere technological optimism. It critically evaluates whether technology truly offers solutions to overtourism or potentially exacerbates the problem. The authors present both the opportunities and limitations of digital tools in preserving architectural heritage, managing tourist flows, and maintaining community balance. Their insights suggest that while technology can be a powerful ally in addressing overtourism, it must be implemented thoughtfully and in conjunction with sound urban planning principles.

The third part, *Architectural Responses to Overtourism*, focuses on concrete design interventions and urban planning solutions. Through case studies from diverse geographical contexts, the chapters explore innovative architectural and urban planning solutions that respond to the challenges of excessive tourism, with a particular focus on emerging economies and cultural heritage sites. Through diverse case studies spanning from the historic Shahjahanabad in Delhi to the sensitive ecosystems of the Himalayas, and from traditional Asian markets to the urban centers of Northeast India, this part examines how architectural interventions can help balance tourism demands with community needs. Authors present thoughtful analyses of how sustainable design, smart technologies, and community engagement can work together to address overtourism's impacts. The chapters explore various approaches: the integration of digital technologies in traditional markets, the use of smart urban design in mountain destinations, the role of community participation through digital platforms, and the importance of understanding tourist psychology in architectural solutions.

A common thread running through these contributions is the emphasis on combining technological innovation with sensitive architectural design to preserve cultural heritage while managing tourist flows. The authors demonstrate how architectural solutions, when properly conceived and implemented, can help distribute tourism pressure, enhance visitor experiences, and protect local communities and their cultural assets. This part emphasizes that successful architectural responses to overtourism must be holistic, considering not just physical infrastructure, but also the social, cultural, and environmental dimensions of tourism development.

The final part, *The Future of Tourism*, looks ahead to emerging solutions and technologies, including the potential of the metaverse in promoting sustainable travel practices and the role of technology in cultural regeneration. Through two distinct but complementary perspectives, this part explores how emerging technologies might transform tourism experiences while addressing the persistent challenges of overtourism. The first contribution examines the intersection of technological innovation and cultural regeneration in Indian heritage precincts, from the bustling streets of Ahmedabad and Mumbai to the historic landscapes of Hampi and Mysore. The second chapter ventures into the virtual realm, investigating how Metaverse marketing could revolutionize sustainable travel practices and offer alternative ways to experience destinations without physical presence. Together, these chapters paint a picture of tourism's potential future—one where physical and virtual experiences coexist, where technology serves as a tool for both preservation and access, and where innovative solutions help balance the competing demands of tourism growth and heritage conservation. This part offers valuable insights for architects, urban planners, and policymakers working to shape the future of tourist destinations in an increasingly digital world.

This book serves as a resource for urban planners and architects seeking strategies to mitigate the negative impacts of excessive tourism. Tourism management professionals and policymakers responsible for developing sustainable tourism strategies will find guidance on best practices. Academics and researchers in urban studies, tourism, and sustainable development will benefit from the book's theoretical framework and empirical evidence. Technology professionals developing smart city solutions and digital tourism management tools will find relevant explorations of innovative technologies and their applications. Finally, the book offers a comprehensive overview of sustainable urban tourism for graduate students in architecture, urban planning, tourism management, and related disciplines. By bringing together perspectives from scholars and practitioners across multiple disciplines, this volume contributes to the growing body of knowledge on sustainable tourism development while offering practical solutions to one of the most significant challenges facing urban centers today. The book aligns with the United Nations Sustainable Development Goal 11, which aims to make cities inclusive, safe, resilient, and sustainable, while addressing the specific challenges posed by mass tourism.

As tourism continues to grow globally, the need for innovative solutions to manage its impact becomes increasingly urgent. This book provides a timely contribution to this critical discourse, offering both theoretical frameworks and practical approaches to creating more resilient and sustainable urban environments in the face of growing tourism pressures.

Mohali, India Vikas Chand Sharma
Amritsar, India Sandeep Dua

Contents

Editors and Contributors

About the Editors

Vikas Chand Sharma is an Associate Professor at Chandigarh University and a researcher at Guru Nanak Dev University, Amritsar, brings 13 years of experience in architectural design, urban planning, and research. He specializes in sustainable urban design strategies for historic cities, combining phenomenological approaches with digital mapping. His foundational research on religious spatial patterns and new urbanism was published in 2016. He serves on the Scientific Committee for Contesti. Città, Territori, Progetti and is a member of the Editorial Board for several international journals. His work appears in the International Journal of Construction Education and Research, Journal of Architecture & Urbanism, and Landscape Research.

Sandeep Dua is a Professor in the Department of Architecture at Guru Nanak Dev University, Amritsar. He holds a PG Diploma in Urban Design from CEPT University, Ahmedabad, and a PhD in Architecture from IIT Roorkee. His research interests encompass mathematics in architecture and urban form, history and theory of architecture and urban design. With a 30-year experience, Dr. Dua has published extensively on topics such as medieval domestic architecture, colonial and post-colonial architectural styles, and urban form and architecture. He has also undertaken research project funded by AICTE, New Delhi.

Pankaj Kumar Tyagi is an accomplished academician, researcher, and consultant. He is presently working as a professor at Chandigarh University, Mohali, India. He is an alumnus of Hemwati Nandan Bahuguna Garhwal University (Central University) and Kurukshetra University, India. He has more than 22 years of experience in the industry, teaching, training, consultancy, and expertise in destination management, travel process and disruptive technologies. He has published over 40 research papers in international journals indexed in the quality database and authored/edited

20 books. He is a reviewer for many international journals in tourism, management, and technology.

Elena Tarsi is an architect and has done Ph.D. in urban territorial and environmental planning. She is currently a researcher at the Department of Architecture of University of Florence where she is the coordinator of LabPSM—Laboratory City and territory in the Global South. She is also "collaborator researcher" at the Centre for Social Studies of University of Coimbra. Her research interests focus on informal settlements dynamics both in the Global South and the Global North. Recently, she has dealt with the impacts of climate change in cities and the role of cities and territories in the European Green Deal policies.

Contributors

Srinidhi D. Acharya Manipal School of Architecture and Planning, Manipal Academy of Higher Education, Manipal, Karnataka, India

Sobia Ahsan Department of Architecture, Jamia Millia Islamia, New Delhi, India

Bassam Samir Al-Romeedy Tourism Studies Department, Faculty of Tourism and Hotels, University of Sadat City, Sadat City, Egypt

Rashmi Ashtt Indira Gandhi Delhi Technical University for Women, Delhi, India

MinJoo Baek Department of Architecture, Xi'an Jiaotong-Liverpool University, Suzhou, China

Vanshika Bajaj Department of Architecture, Jamia Millia Islamia, New Delhi, India

Manju Bala University of Liverpool, Liverpool, UK

Sriya Banerjee Department of Architecture and Regional Planning, Indian Institute of Technology, Kharagpur, West Midnapore, India

Gianluca Belli University of Florence, Florence, Italy

Giada Cerri Dipartimento di Architettura, Università degli Studi di Firenze, Florence, Italy

Ana Rita Conde University Lusófona, Porto, Portugal

Marta Conte University of Florence, Florence, Italy

Giuseppe De Luca University of Florence, Florence, Italy

Carlo Francini Municipality of Florence, Florence, Italy

Martina Franco University of Florence, Florence, Italy

Tomáš Gajdošík Faculty of Economics, Matej Bel University in Basnká Bystrica, Banská Bystrica, Slovakia

Abraham George Department of Architecture and Regional Planning, Indian Institute of Technology, Kharagpur, West Midnapore, India

Shivani Goel Indira Gandhi Delhi Technical University for Women, Delhi, India

Ricardo Jorge da Costa Guerra Polytechnic University of Guarda, Seia, Portugal

Jit Kumar Gupta College of Architecture, IET Bhaddal Technical Campus, Ropar, Punjab, India

Prasoon Gupta Indira Gandhi Delhi Technical University for Women, Delhi, India

Zeeshan Ibrar Department of Architecture and Planning, Netaji Subhas University of Technology (NSUT), New Delhi, India

Nisar Khan Department of Architecture, Jamia Millia Islamia, New Delhi, India

Amit Kinjawadekar Manipal School of Architecture and Planning, Manipal Academy of Higher Education, Manipal, Karnataka, India

Maximiliano E. Korstanje University of Palermo, Buenos Aires, Argentina

Zvjezdana Krstić Independent Researcher (Previously affiliated with Faculty of Business Economics Bijeljina, University of East Sarajevo), Bijeljina, Bosnia and Herzegovina

Ahmet Kurnaz Istanbul Sabahattin Zaim University, Istanbul, Turkey

Jose Antonio Lara-Hernandez Auckland University of Technology, Auckland CBD, Auckland, New Zealand

Luísa Lopes Instituto Politécnico de Bragança, Mirandela, Portugal

Shem Wambugu Maingi Kenyatta University, Nairobi, Kenya

Mirjana Maksimović Faculty of Electrical Engineering, University of East Sarajevo, East Sarajevo, Bosnia and Herzegovina

Marco Martins CiTUR, Coimbra, Portugal

Iasef Md. Rian Department of Architecture, Xi'an Jiaotong-Liverpool University, Suzhou, China

Lara Santos Polytechnic University of Cávado and Ave, Barcelos, Portugal

Vikas Chand Sharma University Institute of Architecture, Chandigarh University, Mohali, 140413 Punjab, India

A. S. Sochipem Zimik SRM University, Gangtok, Sikkim, India

Pankaj Kumar Tyagi University Institute of Tourism and Hospitality Management, Chandigarh University, Mohali, 140413 Punjab, India

Monali Wankar Indira Gandhi Delhi Technical University for Women, Delhi, India

Architectural Responses to Overtourism

Part-III Architectural Responses to Overtourism, focuses on concrete design interventions and urban planning solutions. Through case studies from diverse geographical contexts, the chapters explore innovative architectural and urban planning solutions that respond to the challenges of excessive tourism, with a particular focus on emerging economies and cultural heritage sites. Through diverse case studies spanning from the historic Shahjahanabad in Delhi to the sensitive ecosystems of the Himalayas, and from traditional Asian markets to the urban centers of Northeast India, this part examines how architectural interventions can help balance tourism demands with community needs. Authors present thoughtful analyses of how sustainable design, smart technologies, and community engagement can work together to address overtourism's impacts. The chapters explore various approaches: the integration of digital technologies in traditional markets, the use of smart urban design in mountain destinations, the role of community participation through digital platforms, and the importance of understanding tourist psychology in architectural solutions. A common thread running through these contributions is the emphasis on combining technological innovation with sensitive architectural design to preserve cultural heritage while managing tourist flows. The authors demonstrate how architectural solutions, when properly conceived and implemented, can help distribute tourism pressure, enhance visitor experiences, and protect local communities and their cultural assets. This part emphasizes that successful architectural responses to overtourism must be holistic, considering not just physical infrastructure, but also the social, cultural, and environmental dimensions of tourism development.

Rethinking Tourism Development: Sustainable Architectural Interventions in Shahjahanabad and Beyond

Shivani Goel, Rashmi Ashtt, Monali Wankar, and Prasoon Gupta

Abstract Overtourism, characterized by the detrimental effects of excessive tourism on local environments and communities, has become a significant challenge in popular tourist destinations globally. This chapter examines how sustainable architectural design and interventions can ease these pressures, with a particular focus on Shahjahanabad, the historic walled city in Old Delhi, India. By reviewing urban and architectural approaches aimed at balancing visitor demands with the need to preserve cultural heritage and support community welfare, this research outlines a framework for evaluating tourism development through technological solutions. The study reviews notable examples from both Shahjahanabad and various global locations, assessing how creative architectural solutions have successfully mitigated the negative effects of overtourism while maintaining the historical integrity of cultural landmarks. Approaches such as smart technologies, effective crowd management, and the repurposing of heritage structures are explored as potential pathways to harmonize tourism with sustainability. Key questions guide the research: How can technology promote sustainable architectural approaches in historical cities? What are the broader impacts of overtourism on local populations, and how can design solutions be formulated to address these? The research methods include extensive literature analysis, case study comparisons, and fieldwork conducted in Shahjahanabad. The study's findings underscore the importance of integrating modern technology with sustainable design to curb the detrimental impacts of overtourism. The chapter ends by considering how lessons from Shahjahanabad can serve as a guide for future tourism planning in heritage-rich cities, advocating for models that prioritize environmental, cultural, and societal sustainability.

Keywords Overtourism · Sustainable design · Shahjahanabad · Urban development · Architectural interventions · Tourism management · Sustainable tourism

---

S. Goel (✉) · R. Ashtt · M. Wankar · P. Gupta
Indira Gandhi Delhi Technical University for Women, Delhi, India
e-mail: shivani004phd22@igdtuw.ac.in

V. C. Sharma et al. (eds.), *Architectural Vision of Overtourism*,
SpringerBriefs in Architectural Design and Technology,
https://doi.org/10.1007/978-981-96-5794-0_11

1 Introduction

Shahjahanabad, established in 1638 by Emperor Shah Jahan, represents the historic core of Old Delhi and serves as a living testament to India's rich cultural and architectural heritage. Designed as a walled city, it once flourished as a center of trade, culture, and power, with its intricate urban layout, grand monuments, and bustling bazaars [1]. Today, Shahjahanabad remains a vibrant cultural destination, attracting millions of visitors annually. However, the pressures of modern tourism have profoundly impacted its historic fabric and infrastructure.

The phenomenon of overtourism, defined as the excessive influx of tourists that overwhelms a destination's infrastructure, environment, and cultural authenticity, has become a global concern in heritage areas [2]. Cities like Venice, Dubrovnik, and Kyoto have long battled the adverse effects of overtourism, such as overcrowding, environmental degradation, and the loss of local character [3]. Similarly, Shahjahanabad grapples with these challenges as the growing number of visitors strains its already fragile infrastructure and disrupts the balance between tourism and daily life.

Sustainable tourism has emerged as a critical need for managing these pressures, particularly in heritage cities. By promoting responsible visitation, preserving cultural heritage, and addressing environmental concerns, sustainable tourism offers a pathway to harmonize the needs of visitors with those of local communities [4]. This chapter explores these issues through the lens of Shahjahanabad and addresses two key questions:

- How can sustainable architectural interventions mitigate the negative impacts of overtourism in historic urban areas?
- What specific strategies can be employed to balance the needs of tourists with the preservation of cultural heritage and local communities?

To address these questions, the chapter delves into the impacts of overtourism on Shahjahanabad, evaluates sustainable architectural and urban design practices, and draws lessons from international case studies. The aim is to propose actionable strategies for aligning tourism with urban sustainability, preserving Shahjahanabad's cultural identity while fostering its role as a thriving urban and tourist hub. The chapter also aligns with key Sustainable Development Goals (SDGs) to underscore the broader relevance of sustainable tourism. SDG 11 emphasizes the importance of making cities inclusive, safe, resilient, and sustainable, while SDG 12 promotes responsible consumption and production. SDG 8 highlights the potential of sustainable tourism to support decent work and economic growth, and SDG 13 addresses the environmental implications of overtourism by encouraging climate-conscious planning [5]. By addressing these critical questions and exploring innovative solutions, this chapter aspires to contribute to the broader discourse on heritage preservation and urban sustainability.

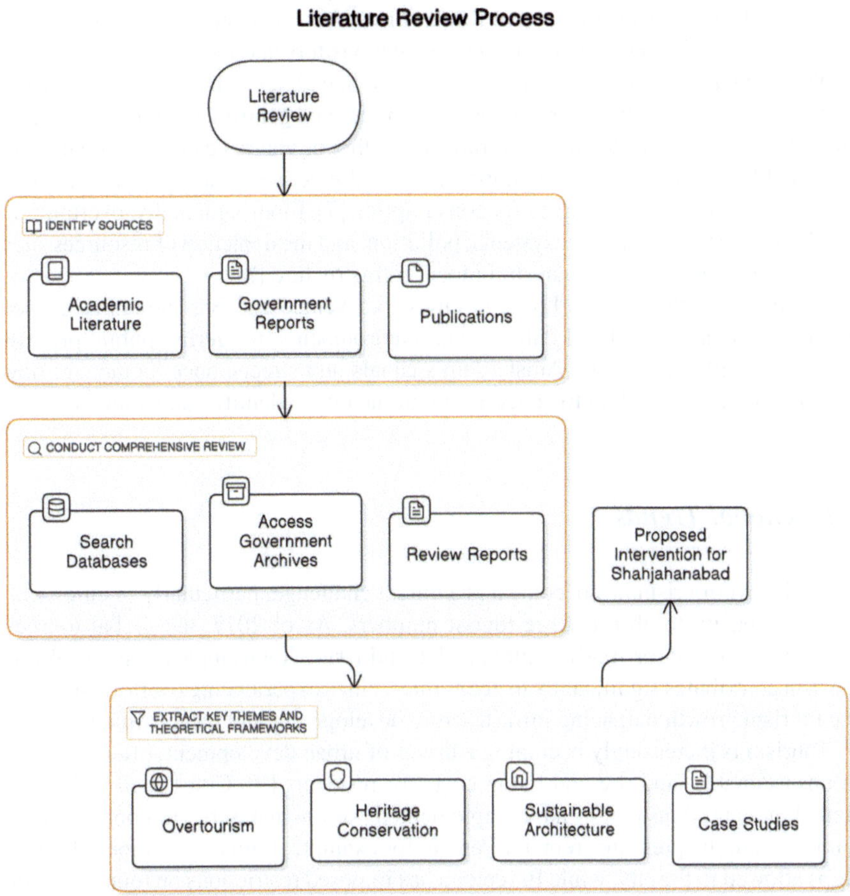

Fig. 1 Methodology chart. *Source* By Authors

2 Methodology

See Fig. 1.

3 Understanding Overtourism

Overtourism refers to a situation in which a destination experiences an unsustainable number of visitors, leading to negative social, environmental, and economic consequences [6]. While tourism can bring significant economic benefits, overtourism undermines the very resources and cultural heritage that make these destinations attractive. In cities worldwide, overtourism manifests as overcrowding, pollution, resource depletion, and social tensions between visitors and local residents.

The impacts of overtourism are multifaceted. Socially, it can lead to the displacement of local communities as rising property prices and gentrification force residents out of historic neighborhoods. Culturally, the influx of visitors can erode local traditions and lead to the commodification of cultural practices, often turning authentic experiences into mere attractions for consumption [7]. Environmentally, overtourism accelerates degradation of ecosystems, pollution, and the depletion of resources such as water and energy, which can strain local infrastructure [8].

For instance, the tourism boom in cities like Venice and Barcelona has caused irreversible changes to local cultures and environments, triggering public protests and policy shifts. Similarly, Amsterdam's canals and streets, once iconic, are now heavily congested, leading to concerns about the city's identity and livability.

3.1 Global Trends

Globally, overtourism has become a prominent challenge, particularly in cities with limited capacity to absorb large tourist numbers. As of 2017, the global tourism industry accounted for 10.4% of global GDP and 10% of total employment [5]. While tourism contributes significantly to economies, cities experiencing overtourism often see tourism growth outpacing infrastructure development and local resilience.

Tourism is increasingly becoming a driver of urban development, often resulting in unsustainable land use and rising costs for residents [9]. Cities such as Venice, Barcelona, and Amsterdam have implemented various measures to control tourist numbers, but the pressures remain. Venice, for example, limits the number of cruise ships allowed in the city, while Barcelona has imposed restrictions on tourist accommodations like Airbnb to protect neighborhoods from overdevelopment [10]. One of the key trends is the rise in long-term stays, which has contributed to the over-commercialization of once-authentic districts. The increasing dependence on tourism as an economic driver also exacerbates social inequities, as wealth generated by tourism tends to disproportionately benefit large-scale operators rather than local communities [11].

3.2 Overtourism in Shahjahanabad

Contextualizing Shahjahanabad: An Architectural and Cultural Treasure

Shahjahanabad, the historic walled city situated at the core of New Delhi, India, stands as a remarkable example of the Mughal period's architectural brilliance and cultural wealth. Yet, with the pressures of rapid urbanization and increasing tourism, the city's struggle to preserve its historical integrity while meeting contemporary needs has grown more challenging. Overtourism has become a significant issue, endangered the delicate structure of the city, and posed risks to both its heritage and the livelihoods of its inhabitants [12]. Shahjahanabad, the historic walled city of Old Delhi, is a significant cultural and architectural treasure that embodies the grandeur of the Mughal era. Established in 1638 by Emperor Shah Jahan, this vibrant city showcases a rich tapestry of history, art, and architecture. As a popular tourist destination, Shahjahanabad attracts millions of visitors each year, drawn to its iconic landmarks and bustling markets. Shahjahanabad is home to 783 heritage structures, including 229 historic buildings and 325 havelis that reflect the rich and diverse architectural legacy of the Mughal era. The city also features 10 major roads and over 60 main streets that connect various parts of the urban fabric. Known for its 15 prominent markets, Shahjahanabad receives heavy footfall daily, approximately 4–6 lakh visitors, offering a variety of goods and services to both locals and tourists alike [13].

These markets are famous for their wholesale and retail trade, handicrafts, food stalls, and cultural landmarks. Among these bustling markets, Chandni Chowk stands out as the most famous, while Dariba Kalan is renowned for its silver and gold jewelry. Kinari Bazaar is well-known for wedding attire and accessories, while Nai Sarak specializes in books and stationery. Additionally, Katra Neel is famous for textiles and embroidery, while Khari Baoli boasts Asia's largest spice market. Ballimaran is recognized for its optical shops and historical significance, and the Daryaganj Book Market is known for its second-hand and pirated books. With an annual footfall exceeding 10 million visitors, Shahjahanabad plays a crucial role in Delhi's tourism landscape [14]. Tourists flock to experience its vibrant street life, traditional crafts, and culinary delights. However, this influx poses significant challenges, including overcrowding and environmental degradation. The delicate balance between preserving cultural heritage and accommodating modern tourism demands is increasingly precarious. As global tourism trends evolve, cities like Shahjahanabad must navigate the complexities of overtourism while striving to maintain their unique identities. The historical significance and architectural beauty of Shahjahanabad make it imperative to implement sustainable tourism practices that protect its cultural integrity [1]. By examining the challenges and opportunities presented by overtourism in this context, stakeholders can work toward solutions that benefit both visitors and local communities alike. This chapter will explore how sustainable architectural interventions can address these challenges while ensuring that Shahjahanabad remains a vibrant cultural hub for future generations. Through innovative design strategies and community engagement, it is possible to create a responsible tourism model that honors the city's rich heritage while promoting economic viability (Figs. 2 and 3).

Connectivity in Shahjahanabad

Shahjahanabad boasts a well-developed connectivity network that enhances its accessibility for both locals and tourists. The city is strategically located in the heart of Delhi, making it easily reachable via various modes of transportation. The extensive road network includes 10 major roads and over 60 main streets, facilitating smooth

Fig. 2 Rooftop view of Khari Baoli, bustling Indian wholesale spice market, Old Delhi, India. *Source* Alamy

Fig. 3 Book shops at Daryaganj. *Source* so.city

movement within the city and connecting it to other parts of the capital. Public transportation options are abundant, with several bus routes and the Delhi Metro providing convenient access to Shahjahanabad.

The city is served by five metro stations inside the walled city: Kashmere Gate, Chandni Chowk, and Chawri Bazar on the Yellow Line, as well as Lal Qila and Jama Masjid on the Violet Line and two metro stations outside city walls, i.e., Delhi Gate on the violet line and New Delhi on the Yellow line. These stations are just a short walk from key attractions like the Red Fort and Jama Masjid, allowing visitors to navigate the bustling city with ease. Additionally, cycle rickshaws and auto-rickshaws are readily available, offering flexible transportation options for tourists exploring the narrow lanes and vibrant markets (Fig. 4).

The city's connectivity is further enhanced by its proximity to major railway stations, such as Delhi Junction, Sadar Bazaar, and New Delhi Junction, which is only a few kilometers away. This makes Shahjahanabad an attractive destination

Fig. 4 Connectivity map. *Source* IGDTUW unpublished work

for travelers arriving from other cities. The well-connected infrastructure not only facilitates tourism but also supports local businesses by ensuring a steady flow of customers to the area's famous markets, including Chandni Chowk, Dariba Kalan, and Kinari Bazaar.

In summary, Shahjahanabad's robust connectivity framework plays a vital role in its appeal as a tourist destination. By providing easy access to its rich cultural heritage and vibrant markets, the city can effectively manage visitor footfall while preserving its unique identity amid the challenges posed by overtourism [15].

Shahjahanabad, with its rich historical and cultural significance, has faced growing difficulties due to overtourism in recent years. As a popular tourist destination, the influx of visitors has exceeded the capacity of its historic infrastructure, leading to overcrowding, infrastructure strain, and a loss of authenticity in the city's heritage.

The city's narrow streets and alleys, designed for pedestrians and horse-drawn traffic, now struggle to accommodate vehicles and large tour groups. This results in congestion, air pollution, and the deterioration of historic structures, many of which are vulnerable to the wear and tear caused by increased foot traffic and vehicular activity. Reports show that over 30% of heritage structures are at risk due to neglect exacerbated by tourism.

The socio-economic fabric of Shahjahanabad is also disrupted by overtourism. Local artisans and small businesses have been forced to cater to the demands of tourists, leading to the commercialization of cultural offerings and the displacement of long-standing residents. A survey indicated that 65% of locals felt tourism negatively impacted their quality of life, with rising property prices and noise pollution being major concerns (Figs. 5 and 6).

The city's infrastructure, built to support a much smaller population, is overwhelmed by the surge in visitors. Overcrowding in areas like Chandni Chowk has become a major issue, and outdated water supply, waste management, and transportation systems are further strained. Environmental degradation is also evident, with increased foot traffic, vehicle emissions, and waste harming the city's heritage sites, including the Jama Masjid.

Fig. 5 Overcrowded streets. *Source* Earthtrekkers.com (left)

Fig. 6 Busy streets with shops in Old Delhi. *Source* India Stock (iStock photo) (right)

The commercialization of Shahjahanabad is reshaping its identity. Traditional shops and residences are being transformed into tourist accommodations or retail outlets, eroding the social fabric of the community that was once centered around craftsmanship and commerce.

The heritage of Shahjahanabad is at risk, with historic buildings facing irreversible damage due to overtourism. The combination of overcrowding, environmental damage, and a loss of cultural identity underscores the urgent need for sustainable management and conservation strategies to balance the benefits of tourism with the preservation of the city's heritage (Table 1).

In conclusion, the challenges of overtourism in Shahjahanabad demand a comprehensive and sustainable approach to protect the city's cultural and architectural heritage while improving the quality of life for its resident.

Table 1 Congestion patterns and visitor types in Shahjahanabad

Time Period	Congestion Level	Areas Affected	Types of Visitors/ Tourists	Specific Activities
2:30 am to 5:00 am	Low	Khari Baoli, Naya Bazaar, Chawri Bazar	Local traders, workers	Loading/Unloading starts
5:00 am to 9:00 am	Moderate	Matia Mahal, Chitli Qabar	Local residents, worshippers, tourists	Morning prayers, breakfast gatherings, early-morning sightseeing
9:00 am to 9:30 pm	High	Commercial streets, Hauz Qazi, Bharshbullah Chowk	Shoppers, traders, local residents, tourists	Selling specialized goods (hardware, spices, food grains, paper, electronics), on-street parking of commercial vehicles, visiting historical and cultural sites
9:30 pm to 2:30 am	Moderate	Food outlets, Ring Road	Late-night visitors, truck drivers, tourists	Late-night food outlets, heavy goods vehicles on Ring Road
Fridays, 12:30 pm to 1:30 pm	Extremely high	Jama Masjid, Lal Kuan, Charkhey Walan, Ballimaran, Chitli Qabar	Worshippers, local residents, tourists	Afternoon prayers, particularly high on Fridays due to Islamic religious structures
Sundays	Moderate	Red Fort–Nai Sarak–Jama Masjid loop, Daryaganj	Tourists, visitors, local residents	Visiting landmarks, Mandi, Weekly Book Market

Source DUAC Report, 2016

4 Sustainable Tourism Practices

Sustainable tourism refers to a form of tourism that considers its economic, social, and environmental impacts, both in the present and the future, while satisfying the needs of visitors. The goal of sustainable tourism is to reduce harmful effects on the environment and local cultures while enhancing economic benefits for host communities. Key elements include the following:

- Environmental Sustainability: Reducing resource use, limiting waste, and protecting natural ecosystems.
- Social Sustainability: Ensuring tourism benefits local populations while respecting their traditions.
- Economic Sustainability: Strengthening local economies through responsible tourism practices.

Sharpley and Telfer [12], in their work "Rethinking Tourism and Development," stress the importance of embedding sustainability within tourism development models. They argue that sustainable tourism should not only focus on environmental preservation but also ensure social fairness and economic viability. A comprehensive approach is critical to ensure tourism positively impacts local communities and safeguards cultural heritage.

Existing sustainable tourism frameworks emphasize collaboration among stakeholders and integrated planning. UNESCO's World Heritage and Sustainable Tourism Programme promotes a collective approach in which tourism management aligns with heritage conservation goals. This initiative works to foster sustainable tourism at World Heritage sites by increasing awareness, building capacity, and ensuring stakeholder involvement [16].

Additionally, numerous successful sustainable tourism efforts have emerged worldwide. Ecotourism initiatives in Costa Rica, for example, focus on conserving biodiversity while generating economic opportunities for local communities. In rural areas, community-based tourism invites visitors to engage directly with local cultures, encouraging mutual respect and deeper understanding [17].

Architectural Interventions

Architectural interventions aimed at mitigating the impacts of overtourism are essential for creating sustainable tourist destinations. Literature on this topic emphasizes innovative design solutions that integrate sustainability principles into historical contexts. For instance, adaptive reuse of heritage buildings allows for modern functions while preserving historical significance. This approach not only revitalizes underutilized spaces but also reduces the environmental footprint associated with new constructions [7]. Successful case studies demonstrate how architectural interventions can effectively address overtourism challenges. The Eden Project in Cornwall, UK, exemplifies a sustainable design that combines education with conservation by showcasing diverse ecosystems within geodesic domes. Similarly, the Soneva

Fushi Resort in the Maldives integrates luxury with sustainability by utilizing renewable energy sources and supporting local conservation initiatives. These examples highlight how thoughtful architectural design can enhance visitor experiences while protecting cultural heritage.

Another notable case is the Qasr al Sarab Desert Resort in Abu Dhabi, which employs traditional architectural styles while incorporating modern sustainability practices such as passive cooling and water conservation techniques. This project illustrates how blending contemporary design with local traditions can create environmentally friendly tourist accommodations that respect cultural heritage. Several innovative projects have emerged in Shahjahanabad that exemplify how adaptive reuse and modern design principles can harmonize with the city's rich cultural heritage. These interventions are crucial for creating sustainable tourist sites and showcase how carefully designed architecture can improve tourist experiences while safeguarding cultural heritage.

In Shahjahanabad, several forward-thinking projects demonstrate how adaptive reuse and modern design can coexist with the city's rich cultural heritage.

Adaptive Reuse of Havelis: One notable example is the adaptive reuse of traditional havelis (mansions) in Shahjahanabad. Many of these historic structures, which once served as residences for wealthy merchants and nobility, have been transformed into boutique hotels and cultural centers. For instance, the Haveli Dharampura has been renovated to offer a blend of heritage and modern hospitality [18].

- Design Approach: The renovation preserved original architectural features such as intricate jali work (lattice screens) and decorative frescoes while integrating modern amenities. This approach not only revitalizes underutilized spaces but also attracts tourists seeking authentic experiences without compromising the building's historical significance.
- Community Impact: By converting havelis into hospitality venues, local artisans are often employed for restoration work, thereby supporting traditional crafts and providing economic opportunities for residents (Figs. 7 and 8).

Fig. 7 Before restoration. *Source* Havelidharampura.com (left)

Fig. 8 After restoration. *Source* Havelidharampura.com (right)

Smart Infrastructure Initiatives: Shahjahanabad has also begun to implement smart infrastructure solutions aimed at managing tourist flows and enhancing visitor experiences. For example, the installation of smart signage throughout key tourist areas helps direct visitors to less crowded sites, thereby alleviating pressure on popular attractions like Jama Masjid and Chandni Chowk.

- Technology Integration: These signs can provide real-time information about crowd levels, nearby attractions, and cultural events, encouraging tourists to explore various facets of the city rather than congregating in high-traffic areas.
- Environmental Benefits: By dispersing tourist traffic, these smart solutions help reduce congestion and pollution in historically sensitive areas, contributing to better preservation of both the environment and cultural heritage.

Public Space Revitalization: Efforts to revitalize public spaces in Shahjahanabad have also been undertaken to enhance community engagement and improve the overall visitor experience. The transformation of Chandni Chowk [19] into a pedestrian-friendly zone is a prime example.

- Design Features: The redesign includes wider walkways, seating areas, and green spaces that encourage social interaction among residents and visitors alike. This initiative not only beautifies the area but also fosters a sense of community ownership over public spaces.
- Cultural Programming: Regular cultural events and markets are organized in these revitalized spaces, allowing local artisans to showcase their crafts while providing tourists with an immersive cultural experience (Fig. 9).

These architectural interventions in Shahjahanabad illustrate the potential for sustainable design practices to address the challenges posed by overtourism while preserving cultural integrity. By focusing on adaptive reuse, smart infrastructure, public space revitalization, and heritage conservation, Shahjahanabad can create a tourism model that benefits both visitors and local communities alike. Such initiatives not only enhance the visitor experience but also ensure that the city's rich history

Fig. 9 The revamped Chandni Chowk encapsulates a 1.3 km stretch connecting Red Fort and Fatehpuri Masjid (Photo I Shekhar Yadav, EPS)

continues to thrive amid modern demands. This section provides specific examples of architectural interventions in Shahjahanabad that align with sustainable design principles while addressing overtourism challenges. By highlighting these initiatives, it emphasizes practical applications of theory within the context of this historic city.

4.1 Case Studies

The Cinque Terre Sustainable Tourism Project

The Cinque Terre, located on the northeastern coast of Italy, is a prime example of a successful sustainable tourism initiative aimed at preserving cultural heritage and the environment while managing the impacts of overtourism. This region is famous for its five picturesque villages—Monterosso, Vernazza, Corniglia, Manarola, and Riomaggiore—set against dramatic cliffs overlooking the Ligurian Sea. Recognized as a National Park in 1999 and designated as a UNESCO World Heritage Site in 1997, Cinque Terre faces significant challenges due to its popularity among tourists [20].

Tourism Impact

The escalating influx of tourists in destinations like the Cinque Terre presents significant challenges, primarily through overcrowding that burdens infrastructure and diminishes residents' quality of life, particularly in its five villages. This tourism surge also causes severe environmental degradation, manifesting as erosion, pollution, and ecosystem damage from heavy foot traffic and boat activity. Furthermore, it poses a risk of cultural impact, potentially diluting local traditions. To counter these impacts, several sustainable tourism interventions have been implemented. The "Environmental Quality Brand" certifies accommodations for sustainable practices, while the "Cinque Terre Card" manages visitor numbers, regulates foot traffic, and funds conservation efforts. Clear guidelines for tourists promote responsible behavior and waste minimization, complemented by public awareness campaigns emphasizing conservation. Additionally, design and infrastructure solutions play a crucial role. This includes hiking path management to prevent overuse, encouraging eco-friendly accommodation through green technologies, and fostering public engagement programs to ensure active participation in conservation and tourism management. These comprehensive measures aim to mitigate negative impacts and secure the region's long-term sustainability.

Outcomes

- Balanced Tourism and Preservation: Successful integration of tourism management with environmental and cultural preservation, enhancing both tourist experience and local resident quality of life.
- Community and Environmental Benefits: The initiative has mitigated the negative effects of overtourism, contributing to the region's sustainable growth.

The Casa De La Memoria, Colombia

The Casa de la Memoria (House of Memory) in Medellín, Colombia, serves as an exemplary case of adaptive reuse that honors cultural heritage while addressing contemporary social issues. Originally constructed in the early twentieth century, this historic building was repurposed to serve as a cultural center dedicated to preserving the memory of the victims of violence in Colombia [21].

Tourism Impact

Uncontrolled tourism leads to overcrowding and cultural erosion, displacing locals and straining services. Governments are intervening by limiting short-term rentals and promoting responsible tourism to reduce consumption and encourage respectful behavior. Additionally, design and infrastructure solutions, like preserving local identity and implementing tourism zoning, aim to minimize disruption and balance development with community well-being. These efforts strive for sustainable tourism, benefiting both visitors and residents.

Outcomes

Strategic tourism interventions have improved resident quality of life by reducing overcrowding and preserving local culture. This shift fosters sustainable tourism growth, where managing visitor volume ensures mutual benefits for both tourists and residents, mitigating the negative consequences of overtourism.

The transformation of the Casa de la Memoria exemplifies how adaptive reuse can effectively honor cultural heritage while addressing contemporary social challenges. This case study highlights the potential for historic buildings to serve as spaces for community engagement, education, and healing. Through thoughtful design and community involvement, the Casa de la Memoria stands as an inspiring model for similar adaptive reuse projects worldwide, demonstrating that preserving architectural history can coexist with innovative solutions for social impact. The Casa de la Memoria contributes to SDG 11 (Sustainable Cities and Communities) by revitalizing urban spaces and creating a sustainable cultural center. It supports SDG 4 (Quality Education) by providing educational opportunities and promoting awareness of Colombia's history of violence, and plays a key role in peacebuilding efforts, aligning with SDG 16 (Peace, Justice, and Strong Institutions) through its focus on reconciliation and fostering inclusive societies.

Barcelona, Spain

Barcelona is another city grappling with overtourism, particularly in its popular neighborhoods like La Rambla and the Gothic Quarter [22].

Tourism Impact

Overtourism significantly impacts cities like Barcelona, causing overcrowding in hotspots like La Rambla and the Gothic Quarter, straining public services, and encroaching on residential areas. This influx also erodes local culture by displacing residents and businesses. In response, governments are limiting short-term rentals and launching responsible tourism campaigns to encourage respectful behavior and reduce resource consumption. Design and infrastructure solutions, such as preserving local identity and implementing tourism zoning, are also crucial for balancing development with minimizing disruption to local life, fostering a more sustainable future for both residents and visitors.

Outcomes

- Enhanced Resident Quality of Life: The reduction of overcrowding and the preservation of local culture through strategic interventions have helped improve the living conditions of residents.
- Sustainable Tourism Growth: Measures to manage tourism volume and ensure that it benefits both visitors and residents, reducing the negative impacts of overtourism.

Barcelona's efforts are closely aligned with SDG 11 (Sustainable Cities and Communities) by promoting sustainable urban tourism and preserving cultural heritage. The city's regulations on tourism also contribute to SDG 8 (Decent Work and Economic Growth) by ensuring that the economic benefits of tourism are more evenly distributed, reducing displacement and promoting local livelihoods.

These examples underscore the urgent need for sustainable tourism practices that prioritize both visitor satisfaction and community well-being. Each case study exemplifies how tourism-related initiatives can support multiple SDGs, contributing to sustainable, inclusive, and resilient communities, while addressing environmental and social challenges in heritage conservation and urban management.

5 Proposed Interventions for Shahjahanabad

5.1 Architectural Strategies

Adaptive Reuse of Historic Buildings:

While some historic buildings in Shahjahanabad have been repurposed, a more comprehensive approach should be adopted. Additional underutilized historic buildings should be repurposed for cultural centers, museums, or community spaces to preserve the city's rich history and address modern social issues more effectively.

Creation of Green Spaces and Public Plazas:

While some green spaces exist but are very less in number, more should be developed to provide additional recreational areas for both residents and tourists. This will enhance the overall quality of life and further promote sustainable tourism.

Implementation of Smart City Technologies:

Smart city technologies should be more widely integrated to improve traffic management, waste disposal, and energy efficiency. This will create a more sustainable and efficient urban environment.

5.2 Policy and Governance

Strong Regulatory Framework for Tourism Development:

A more comprehensive regulatory framework should be developed to guide sustainable tourism development. This will ensure that tourism activities are managed in a way that respects both local communities and cultural integrity.

Public–Private Partnerships to Fund Sustainable Projects:

Public–private partnerships should be established more robustly to fund sustainable tourism projects (e.g., Red Fort adopted by Dalmia) [23]. Leveraging both government and private sector resources will support conservation and sustainable development more effectively.

Capacity-Building for Local Communities and Stakeholders:

Capacity-building programs for local communities and stakeholders should be implemented more widely. Providing them with the skills and knowledge needed to participate effectively in sustainable tourism initiatives will be crucial for long-term success.

Establish Policies to Cap Visitor Numbers:

Policies should be established to cap visitor numbers during peak seasons. This will help manage overcrowding and ensure a better experience for both tourists and local residents.

Regulate Short-Term Rentals:

Policies should be enacted to regulate short-term rentals, similar to Barcelona's tourism zoning policies. This will help curb the rapid growth of short-term rental properties that displace locals and ensure more stable housing for residents.

Enforce Responsible Tourism Practices:

Responsible tourism practices should be enforced, encouraging tourists to behave respectfully, reduce excessive consumption of local resources, and promote sustainable activities. This will help minimize the negative impacts of tourism on the local environment and community.

These proposed interventions should be implemented to build upon existing efforts and address the specific challenges and opportunities in Shahjahanabad. By doing so, a more sustainable and inclusive tourism model can be achieved, benefiting both visitors and local residents.

6 Conclusion

This chapter has explored the critical need for sustainable strategies to address overtourism in Shahjahanabad, drawing parallels with international case studies to propose actionable interventions. Shahjahanabad, as a historic urban area, faces unique challenges stemming from its dense population, fragile heritage structures, and high tourist footfall. Overtourism has exacerbated issues such as overcrowding, environmental degradation, and the erosion of cultural identity, threatening the delicate balance between preserving the city's heritage and supporting economic growth through tourism. Case studies from Cinque Terre, Barcelona, and Medellín provide valuable insights into effective strategies for balancing tourism with cultural preservation and community well-being. These examples demonstrate the potential of adaptive reuse, sustainable infrastructure, and innovative governance models to mitigate overtourism's adverse impacts. Architectural strategies such as adaptive reuse and eco-sensitive designs can enhance the utility of historic structures while preserving their cultural and historical significance. Urban planning interventions, including traffic management, zoning regulations, and community-based initiatives, can create a more sustainable tourism ecosystem. Finally, strong policy frameworks and stakeholder engagement are pivotal for long-term success, ensuring that both residents and visitors benefit from tourism development.

For Shahjahanabad, a comprehensive approach incorporating these strategies can foster a resilient model of heritage conservation and tourism management. Implementing adaptive reuse, creating green spaces, and employing smart city technologies can significantly improve the urban environment. Zoning regulations and sustainable transportation systems can alleviate congestion while protecting cultural assets. Community involvement and public–private partnerships will further ensure equitable benefits and sustained engagement in preserving the city's unique character. By adopting these interventions, Shahjahanabad can transition toward a sustainable future where tourism supports, rather than undermines, its heritage and residents' quality of life.

References

1. Blake SP (2002) Shahjahanabad: the sovereign city in Mughal India 1639–1739 (No 49). Cambridge University Press
2. Amore A (2019) Overtourism: issues, realities and solutions. In: Dodds R, Butler RW (eds) Berlin, de Gruyter, 2019, vi + 288 pp, £ 98.00 (hbk), £ 39.00 (pbk), £ 98.00 (ebk), ISBN 978–3–11–060408–5 (hbk), ISBN 978–3–11–062045–0 (pbk), ISBN 978–3–11–060570–9 (ebk). J Herit Tour 15(5):592–593. https://doi.org/10.1080/1743873X.2019.1701818
3. Seraphin H (2019) Overtourism: excesses, discontents and measures in travel and tourism. J Tour Futur 5(3):295–296. https://doi.org/10.1108/JTF-09-2019-088
4. Becken S, Hay JE (2007) Tourism and climate change: risks and opportunities (Vol 1). Multilingual Matters
5. UNWTO Tourism Highlights: 2017 Edition (2017). https://doi.org/10.18111/9789284419029
6. Duignan M (2019) 'Overtourism'? Understanding and managing urban tourism growth beyond perceptions: Cambridge case study: Strategies and tactics to tackle overtourism. In: 'Overtourism'? Understanding and managing urban tourism growth beyond perceptions: case studies, pp 34–39. United Nations World Tourism Organisation (UNWTO)
7. Timothy DJ (2014) Contemporary cultural heritage and tourism: development issues and emerging trends. Public Archaeol 13(1–3):30–47
8. Mowforth M, Munt I (2015) Tourism and sustainability: development, globalisation and new tourism in the third world. Routledge
9. Binggeli U, Chen Z, Kopke S, Yu J (2023) The Future of Tourism: Bridging the labor gap, enhancing customer experience. McKinsey & Company. https://www.mckinsey.com/industries/travel-logistics-and-infrastructure/our-insights/future-of-tourism-bridging-the-labor-gap-enhancing-customer-experience
10. Dredge D, Airey D, Gross MJ (eds) (2015) The Routledge handbook of tourism and hospitality education. Routledge, London
11. Bramwell B, Lane B (eds) (2000) Tourism collaboration and partnerships: Politics, practice and sustainability (Vol 2). Channel View Publications
12. Sharpley R, Telfer DJ (eds) (2014) Tourism and development: Concepts and issues. Multilingual Matters
13. Goel S, Ashtt R, Wankar M, Gupta P (2024) Impact of tourism on residents' quality of life in Shahjahanabad. In: International conference on trends in architecture and construction, pp 1017–1030. Singapore: Springer Nature Singapore
14. Revitalization of shahjahanabad (Walled City of Delhi) (n.d.). https://srdc.delhi.gov.in/sites/default/files/generic_multiple_files/revitalization_of_shahjahanabad_project_concept_proposal.pdf
15. McKercher B, Du Cros H (2002) Cultural tourism: the partnership between tourism and cultural heritage management. Routledge
16. UNESCO (2020) World heritage and sustainable tourism programme
17. Bermúdez-Elizondo M, Araúz-Beita I, Arias-Navarro A, Borbon-Sandí R (2020) Ecotourism development in Costa Rica. Tourism planning and development in Latin America. CABI, Wallingford UK, pp 66–85
18. Haveli Dharampura—Restoration Story (n.d.). https://havelidharampura.com/restoration-story.html
19. Redevelopment of chandni chowk | shahjahanabad redevelopment corporation (n.d.-a) https://srdc.delhi.gov.in/en/srdc/redevelopment-chandni-chowk
20. UNESCO (1997) Cinque Terre world heritage site. https://whc.unesco.org/en/list/826/
21. Casa de la Memoria (n.d.) About casa de la memoria. http://www.museocasadelamemoria.org/
22. Barcelona City Council (n.d.) Strategies for managing overtourism in popular neighborhoods. https://ajuntament.barcelona.cat/turisme/en/strategic-plan.

23. Dalmia Bharat Group (n.d.) Adoption of red fort. https://www.dalmiabharat.com/press_release/dalmia-bharat-limited-ministry-of-tourism-and-archaeological-survey-of-india-sign-mou-to-take-up-the-red-fort-delhi-and-gandikota-fort-kadapa/
24. United Nations (n.d.) The 17 goals | sustainable development. United Nations. https://sdgs.un.org/goals

Urban Planning Solutions to Overtourism in Himalayan Tourist Destinations Through Smart Technologies

Sriya Banerjee and Abraham George

Abstract Overtourism in the Himalayas presents intricate issues for urban planners, legislators, and local populations. Prominent locales like Shimla, Gangtok, Darjeeling, and Manali endure considerable pressure from the swift surge of tourists, affecting urban infrastructure, local ecosystems, and cultural heritage sites. Primary issues encompass overcrowding, environmental degradation, water scarcity, and inadequate transit infrastructure. These mountainous locales are especially sensitive due to their delicate ecosystems, inadequate infrastructure, and economies reliant on seasonal variations, rendering them more prone to challenges such as soil erosion, deforestation, and loss of biodiversity. The seasonal characteristics of tourism exacerbate demand during peak times, diminishing the quality of guest experiences and depleting local resources. Although tourism generates significant economic advantages, many locations are at risk of enduring unsustainability without adequate management. This study examines how intelligent urban design, enhanced by technologies like the Internet of Things (IoT) and Geographic Information Systems (GIS), might alleviate the adverse impacts of overtourism while promoting sustainable growth. Case studies from Himalayan locales illustrate how strategic urban design may manage tourist flow, enhance resource efficiency, and safeguard delicate ecosystems, thus harmonising the interests of visitors with those of local inhabitants and fostering enduring sustainability.

Keywords Sustainable tourism · Smart urban design · Mountain ecosystems

S. Banerjee (✉) · A. George
Department of Architecture and Regional Planning, Indian Institute of Technology, Kharagpur, West Midnapore, India
e-mail: sriyab@iitkgp.ac.in

V. C. Sharma et al. (eds.), *Architectural Vision of Overtourism*,
SpringerBriefs in Architectural Design and Technology,
https://doi.org/10.1007/978-981-96-5794-0_12

1 Introduction

Mountain destinations encounter specific issues related to overtourism, differing from those in metropolitan or coastal regions. These locations are frequently defined by delicate ecosystems, inadequate infrastructure, and economies reliant on seasonal fluctuations, rendering them more susceptible to the impacts of mass tourism. A significant issue is the environmental degradation resulting from the inflow of visitors, which causes soil erosion, deforestation, and harm to biodiversity (Bourdeau et al. 2015). The fragile alpine ecosystems are particularly vulnerable to human activity, and heightened foot traffic can adversely affect the flora and fauna essential to the natural equilibrium of these regions [11, 26, 45, 48, 54].

Another concern is the pressure on infrastructure. Mountain areas frequently possess restricted capacity regarding lodging, trash management, and transportation, intensifying the adverse impacts of overtourism [50]. The cyclical characteristics of tourism in these regions, marked by significant surges during winter sports or summer hiking seasons, exacerbate the strain during brief intervals [20]. This results in overpopulation, diminished visitor experience, and increased deterioration of facilities.

In contrast to metropolitan locales, where overtourism frequently results in cultural erosion and housing deficits, mountainous regions are more susceptible to ecological harm and the depletion of natural resources [31]. Moreover, local people in mountainous areas may rely significantly on tourism for economic viability, complicating the equilibrium between visitor influx and sustainable practices [49]. In certain regions, global measures such as tourist limitations, seasonal constraints, and encouragement of off-peak travel have been enacted to alleviate these impacts [34, 40].

Overtourism in the Himalayas presents intricate challenges for urban planners and legislators. The rapid expansion of tourism in this area has imposed considerable pressure on urban infrastructure, local ecosystems, and cultural heritage sites. Locations such as Shimla, Gangtok, Darjeeling, and Manali are facing issues including overpopulation, ecological degradation, water shortages, and inadequate transportation infrastructure. While tourism provides significant economic benefits to these communities, it also endangers their long-term sustainability if inadequately managed [29, 42, 58, 60, 66, 68, 75, 76].

This chapter explores how intelligent urban design, enhanced by emerging technologies such as the Internet of Things (IoT), Geographic Information Systems (GIS), and Artificial Intelligence (AI), can provide innovative solutions to the issues associated with overtourism. This chapter employs case studies from Shimla, Gangtok, Darjeeling, and Manali to illustrate how technology-driven urban planning methods can harmonise the needs of tourists with those of local inhabitants and ecosystems. This illustrates the incorporation of intelligent technologies into urban planning to manage tourist flow, improve resource efficiency, and foster sustainable development in sensitive ecosystems.

The discussion will focus on the central question: How can urban planning, enhanced by smart technologies, alleviate the negative impacts of overtourism while promoting sustainable and equitable development in Himalayan tourist destinations?

2 Context

Overtourism implies the excessive arrival of tourists to a location, leading to detrimental effects on local communities, infrastructure, and the environment. The phenomenon has attracted global attention as famous places endeavour to reconcile economic advantages with the adverse impacts of mass tourism. Overpopulated urban areas, such as Venice, Barcelona, and Dubrovnik, together with natural landmarks like Machu Picchu, have experienced significant pressure from overtourism. The problem is evident in the deterioration of natural resources, erosion of cultural authenticity, heightened pollution, and reduced quality of life for inhabitants. As overtourism intensifies, international countermeasures have diversified, including the establishment of visitation limits, encouragement of off-peak travel, and demarketing initiatives to reduce tourist influx [13, 40, 46].

The Tourism Area Life Cycle (TALC) model, established by Butler in 1980, is one of the most commonly analysed frameworks for comprehending the lifecycle of tourism sites. The TALC model posits that tourism destinations progress through six phases: discovery, involvement, development, consolidation, stagnation, and post-stagnation (which may result in either rejuvenation or decline). Overtourism typically arises during the consolidation or stagnation phases, when tourist numbers meet or surpass the destination's carrying capacity. At this juncture, the destination gets saturated, and the adverse effects start to surpass the economic advantages. Venice has reached a critical state of stagnation, wherein the excessive influx of visitors strains local infrastructure and jeopardises the city's cultural heritage [32, 65]. Employing the TALC model enables stakeholders to more effectively predict these phases and execute prompt measures to avert overtourism and steer the destination towards revitalisation.

Diverse international reactions to overtourism demonstrate an awareness of this lifecycle. Venice has instituted entry fees and limited access to its most frequented areas during high tourist seasons to regulate visitor influx [20]. Iceland has implemented a scheme to distribute tourists among lesser visited areas to alleviate pressure on its most frequented natural sites [63]. These measures, according to the TALC concept, aim to transition destinations from stagnation to rejuvenation. Alongside the TALC model, Social Exchange Theory (SET) offers an alternative perspective for comprehending resident attitudes towards overtourism. Social Exchange Theory asserts that individuals and societies assess their relationships by weighing the perceived costs and benefits of interactions [24]. In the realm of tourism, locals evaluate the economic advantages against the social, environmental, and cultural detriments. Effective management of tourism typically outweighs the drawbacks, fostering positive sentiments among residents. In cases of excessive tourism, the negative impacts such as traffic congestion, environmental degradation, and loss of cultural identity may surpass the benefits, leading to hostility and opposition from the local community [51]. This phenomenon has been noted in numerous over-touristed locales, where inhabitants have objected to the adverse effects of mass tourism [44]. SET elucidates the reasons communities that initially endorse tourism may become antagonistic when overtourism intensifies.

Policymakers and stakeholders must incorporate both the TALC model and Social Exchange Theory into their plans to alleviate the impacts of overtourism. The TALC concept facilitates proactive destination management, enabling stakeholders to identify when a location is susceptible to stagnation and overtourism. SET emphasises the necessity of acknowledging local communities' opinions and ensuring equitable distribution of tourism development advantages [59]. Strategies that neglect local issues are likely to encounter opposition and may jeopardise the destination's long-term sustainability. However, despite the clear advantages of these models in guiding policy, there remain gaps in their application, particularly in terms of long-term planning and the integration of smart technologies. Existing literature stresses the need for adaptive, community-inclusive, and data-driven tourism management practices. To address overtourism in mountain destinations, it is crucial to implement short-term solutions like carrying capacity assessments and sustainable tourism models, while medium-term and long-term goals should focus on real-time resource management systems and the establishment of monitoring authorities. By incorporating these strategies, mountain destinations can manage overtourism more effectively and ensure sustainable development.

2.1 Overtourism Mitigation Strategies

Overtourism mitigation strategies offer distinct approaches to address the negative impacts of excessive tourism. Tourism Degrowth advocates for reducing tourist activities to alleviate environmental degradation like pollution and resource depletion, and to foster resilient local economies prioritising well-being over constant growth [8, 9, 23, 25, 36, 43, 46, 52, 73]. However, it presents economic challenges for tourism-dependent regions, risking job losses and reduced revenue, and could exacerbate inequality by limiting access for less affluent populations. Demarketing tourism aims to diminish demand to prevent overexploitation of resources. Its advantages include alleviating pressure on local infrastructure, mitigating environmental degradation, and enhancing resident quality of life by regulating visitor numbers [35, 56, 78]. Yet, it can lead to economic detriment for local businesses, diminished employment, and potentially intensify disparities by confining travel to wealthier tourists. Successful demarketing requires meticulous planning to avoid alienating demographics or impeding recovery during economic downturns.

Regenerative tourism goes beyond sustainability, aiming to restore and rejuvenate natural and cultural environments impacted by tourism. This proactive approach involves tourists and locals in restoration activities, promoting deeper community involvement and long-term sustainability [10, 21, 22, 27, 55, 62]. Challenges include the risk of "greenwashing" by businesses, significant investment requirements, and the potential commodification of genuine regenerative efforts. Implementation of IoT offers a transformative technological solution. IoT devices like smart sensors and interconnected systems track tourist behaviour, optimise energy use, and streamline services, enhancing operational efficiency and sustainability [6, 15, 30, 33, 70, 72]. Smart cities and hotels use IoT for monitoring energy consumption, automating lighting, and managing waste. It also improves visitor experiences through personalised services and real-time transportation data. However, IoT implementation faces challenges with data privacy, cybersecurity, high installation costs, and the need for robust infrastructure, particularly in less developed regions.

3 Research Methodology

Figure 1 shows the span of the Indian Himalayan Regions and the selected regions for study, e.g., Himachal Pradesh, Sikkim, and the Darjeeling District of West Bengal. Manali and Shimla are two prominent hilly tourist towns of Himachal Pradesh, while Darjeeling and Gangtok are situated in the Darjeeling District of West Bengal and Sikkim, respectively. The rationale for choosing these specific sites was to cover their diverse qualities in terms of geographical location, elevation, and climate.

Data was collected for the study from government statistics, supported by literature review and the authors' primary observation from field surveys. The secondary and primary data is supported by the observation of satellite imagery (see Fig. 2).

4 Results and Discussion

The following Figure (Fig. 3) shows the tourism intensity in the study regions; the tourism intensity is calculated as the total number of tourists and local population per sq. km. of area. It shows the rising trend in anthropogenic pressure in the study area pre- and post-Covid-19 pandemic, which warrants immediate attention.

Figure 4 shows the tourist-to-population ratio, which is quite high for Manali, Shimla, and Gangtok, which causes concern for the local community; although Darjeeling shows a lower tourist-to-local ratio, the high volume of anthropogenic pressure as seen in indicates the possible pressure on the resources. As evidenced in Fig. 4, the built-up area in the Darjeeling municipal area is increasing at a high rate, which causes environmental concern. All of these hill towns show high water demand and high volume of municipal solid waste production (see Fig. 5), which further reinforces the need for planning intervention.

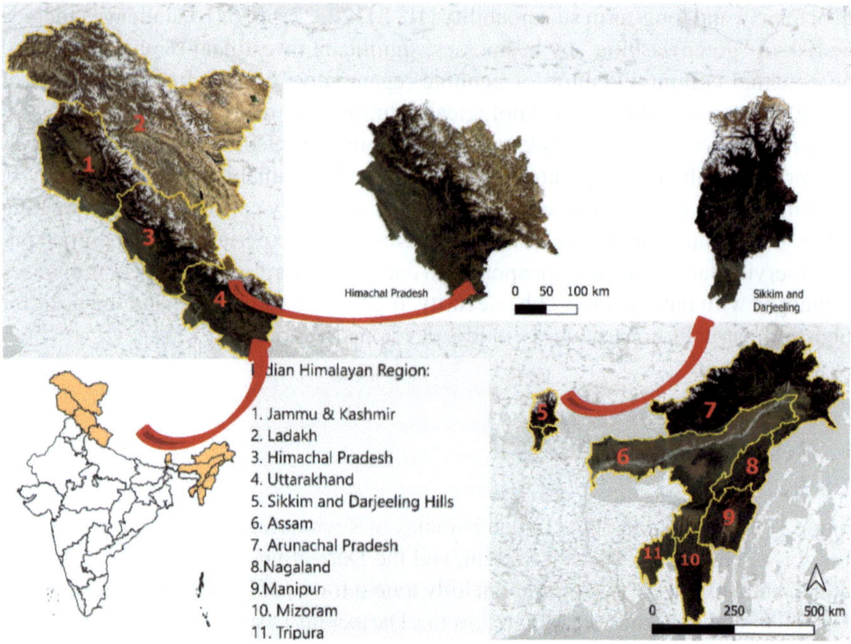

Fig. 1 Geographical location of the study area (*Source* Author)

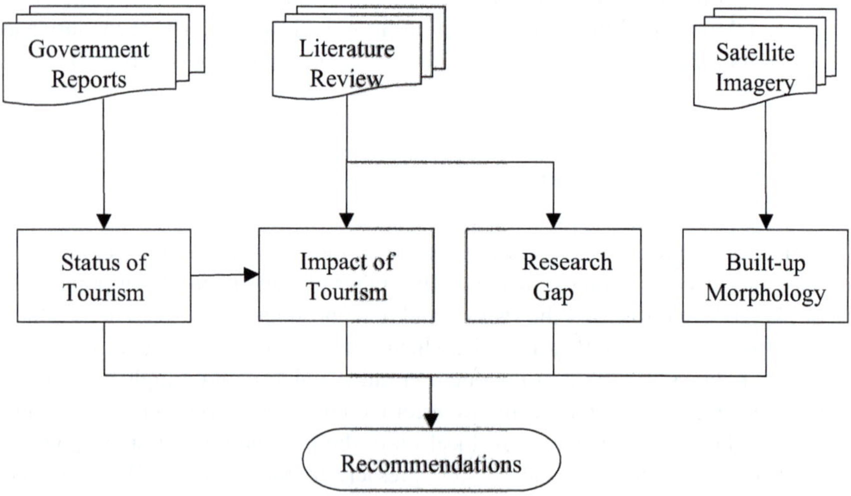

Fig. 2 Research methodology adopted by authors

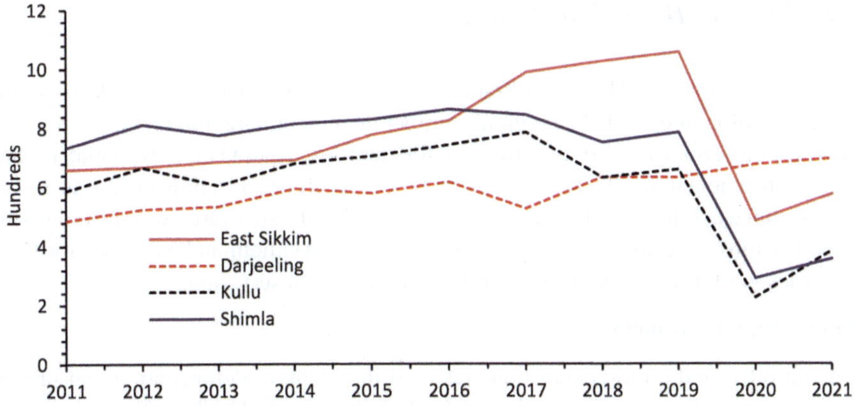

Fig. 3 Tourism intensity in the study regions (*Source* Author's Analysis)

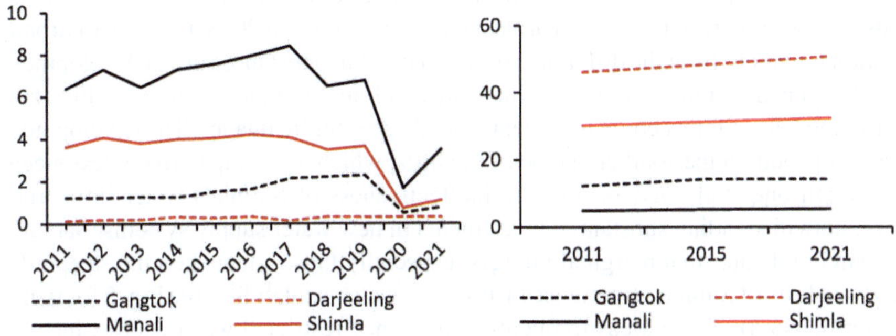

Fig. 4 Tourist-to-population ratio (left); change in built-up area (right) (*Source* various government databases)

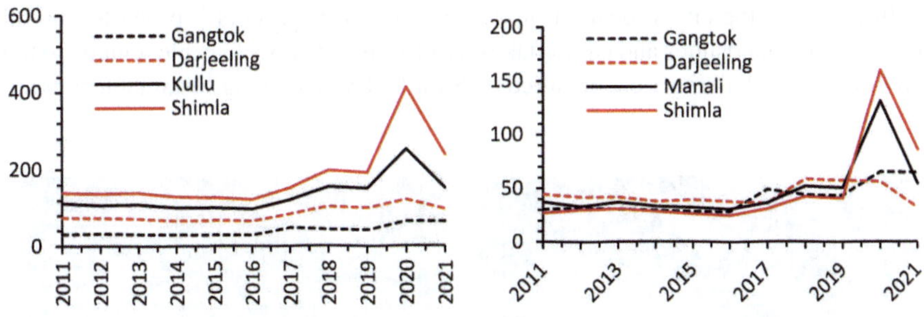

Fig. 5 Water demand (left) and municipal solid waste generation (right) in the study area (*Source* various government databases)

4.1 *Shimla, Himachal Pradesh*

Shimla, the capital of Himachal Pradesh, has been a prominent tourist destination since the colonial period. The region's temperate temperature, old architecture, and picturesque terrain have drawn tourists for decades, leading to substantial urban and environmental transformations. Nonetheless, swift urbanisation and the surge of tourists have exerted significant strain on the city's infrastructure. Shimla's tourism infrastructure underwent fast expansion in the late twentieth and early twenty-first centuries, with less consideration for environmental conservation.

Environmental Concern

Uncontrolled and erratic expansions resulted in extensive deforestation and urban sprawl, especially in regions adjacent to the city centre. The first expansion was largely spurred by increasing demand for hotels, resorts, and vacation residences, frequently infringing upon green spaces and arable land [17]. Figure 6 show the false-colour image of the Shimla urban area; from 2011 and 2023, the level of urbanisation is evident from the difference between these images; more urban development has happened, centred around the roadways and near the city centre (see Fig. 10). Also, the loss of forest cover is evident from these satellite images. The development has happened on the southern slope of the hills, which are comparatively less steep. Although one study underscores the ineffectiveness of Shimla's water infrastructure, notwithstanding substantial investments in new water supply systems. Sarkar's research indicates that marginal villages, especially those situated on the steep hills and outskirts of Shimla, experience insufficient water availability (see Fig. 5 for rising water demand). These areas frequently experience restricted water access, intensifying inequality within the metropolitan area. The water shortage crisis in Shimla is a critical issue for the city, primarily attributable to its population growth and tourism demand [64].

In addition, the city's location on vulnerable slopes renders it prone to natural disasters, like landslides and earthquakes. Ecotourism offers a feasible remedy to the environmental challenges encountered by Shimla. By promoting sustainable tourism

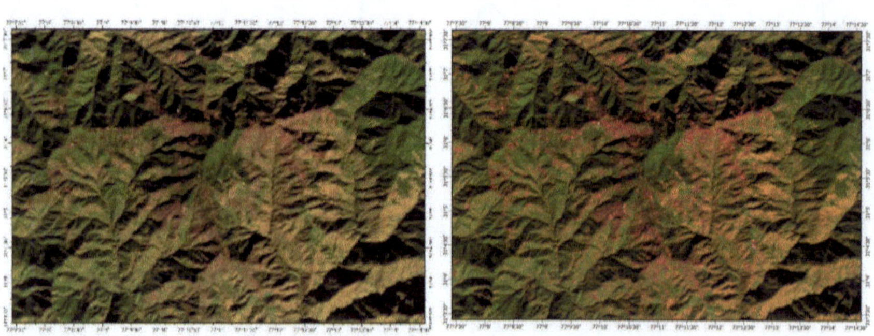

Fig. 6 False-colour Landsat imagery of Shimla: 2011 (left) and 2023 (right) (*Source* Author)

habits, such as walking tours and cycling, Shimla could reduce its carbon footprint while enhancing the tourist experience. The lack of pedestrian-friendly infrastructure and pollution-free zones impedes the effectiveness of ecotourism initiatives [4].

A significant issue in Shimla is the concentration of tourist activity in designated areas, especially the Ridge, which becomes congested during high tourist seasons, resulting in a decline in visitor experience, as the city's infrastructure fails to accommodate the elevated tourist influx. The researchers promote the regulation of visitor flow and the enhancement of tourist attractions in underserved regions to distribute the effects of tourism more equitably [1].

Social and Economic Impacts

Although tourism has generated economic advantages for Shimla, it has further increased the disparity between tourist zones and residential regions. Tourism-focused development frequently disregards the requirements of local inhabitants, leading to disputes over resources, especially water and housing. The swift urbanisation of Shimla has resulted in the erosion of traditional architecture, with heritage buildings being supplanted by contemporary high-rise constructions to cater to the increasing population and tourist influx [4].

Research Gap

A primary research gap in the current literature is the insufficient emphasis on the long-term effects of policies. Although research has proposed short-term remedies for the water problem and environmental deterioration, there is a paucity of studies examining the long-term efficacy of these measures. Additionally, the socio-economic disparities exacerbated by tourism development, particularly for access to resources for marginalised communities, warrant further investigation.

4.2 Manali, Himachal Pradesh

Manali has experienced a significant rise in tourism in the recent past, leading to considerable economic and environmental shifts in the region (see Fig. 3). Research shows the adverse impact of tourism on the region's ecosystem, noting the significant decrease in arable and pastureland resulting from the construction of hotels and roads. The establishment of tourism infrastructure in formerly productive agricultural areas has increased dependence on foreign food and energy resources, exacerbating environmental deterioration. Tourism activities, such as trekking and excursions to Rohtang Pass, have been shown to impose environmental stress [14]. Also, community concerns over the long-term sustainability of tourism in the region, namely its impact on natural resources and the environment, are addressed in studies as early as 2002. Also, community concerns over the long-term sustainability of tourism in the region, namely its impact on natural resources and the environment, are addressed in studies as early as 2002 [28]. Despite these concerns, the economic benefits of tourism have impeded local stakeholders from adopting stringent environmental regulations.

Environmental Concerns

The initial phase of Manali's tourism can be characterised as "circumstantial alternative tourism" (CAT), typified by an unregulated, small-scale tourism sector that superficially resembles alternative tourism [67]. Land-use alteration in Manali has been a notable concern, as evidenced by a study using remote sensing data from 1989 to 2012, which indicates a consistent expansion of built-up regions in Manali, especially in formerly undeveloped steep slope zones (see Fig. 7). The expansion has increased the risk of landslides and other natural disasters, indicating that more stringent zoning laws are necessary to avert additional development in ecologically sensitive regions [74]. Water scarcity has increasingly become a significant issue. Figure 5 shows the water demand in Manali. Detrimental effects of tourism and urbanisation on water resources in Manali result in the demand for potable water from tourists and inhabitants often surpassing supply, resulting in water shortages during high tourist seasons. The heightened utilisation of non-biodegradable materials by tourists has exacerbated the pollution of the Beas River, hence intensifying the region's water crisis [39]. Manali has identified traffic congestion as a significant problem. The town's primary node experiences substantial traffic loads, especially during peak tourist seasons. The establishment of designated parking facilities at prominent tourist sites could mitigate congestion and enhance the overall visitor experience Manali has identified traffic congestion as a significant problem. The town's primary node experiences substantial traffic loads, especially during peak tourist seasons. The establishment of designated parking facilities at prominent tourist sites could mitigate congestion and enhance the overall visitor experience [5]. The satellite imagery reveals encroachment on the Beas River basin (Fig. 7).

Social and Economic Impacts

Considering the environmental challenges, the local people in Manali largely favour the expansion of the tourism sector. Community members perceive tourism as a catalyst for employment creation and enhancement of the region's overall image. Concerns have emerged over the strain on public utilities and the degradation of cultural treasures resulting from the growth of the tourism sector [61]. Ineffective

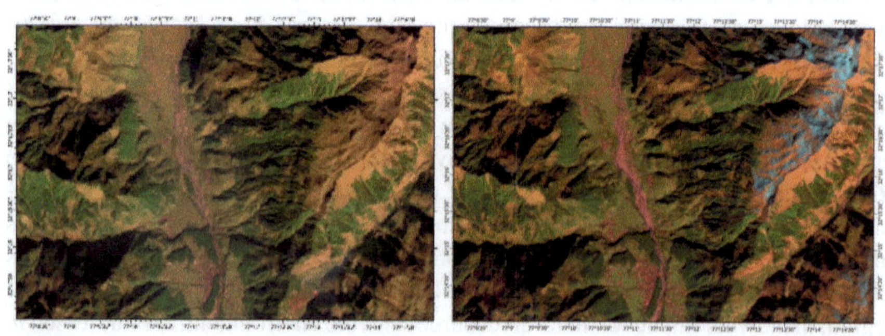

Fig. 7 False-colour Landsat imagery of Manali: 2011 (left) and 2023 (right) (*Source* Author)

administration has allowed mass tourism to grow unchecked. The community-driven governance model in Manali has faced challenges in mitigating the environmental and social repercussions of tourism since local stakeholders have favoured economic benefits over sustainable development [69].

Research Gap

Although extensive research has concentrated on the immediate environmental effects of tourism in Manali, further investigation is required to understand how climate change may intensify these challenges. Due to the rising incidence of landslides and other natural catastrophes, further research is essential to evaluate the integration of climate resilience into tourism planning and development. Furthermore, the long-term viability of Manali's tourism-dependent economy requires examination, especially concerning the environmental repercussions of tourism.

4.3 *Darjeeling, West Bengal*

Darjeeling, an iconic hill station in India, has undergone swift urbanisation in recent decades, propelled by tourists and rural-to-urban migration. This extensive migration put considerable strain on Darjeeling's infrastructure, especially its water supply and housing systems. The growing population has led to overpopulation, intensifying the city's environmental and infrastructure issues [7]. The unrestrained expansion of Darjeeling has been associated with zoning deficiencies. The lack of appropriate zoning regulations has led to disorganised urban development, where residences, utility lines, and water pipes are intertwined without any planning. The unregulated expansion has led to the deterioration of Darjeeling's natural environment and infrastructure, resulting in an "urban jungle" in the mountains [77].

Environmental and Infrastructure Issues

Water scarcity remains a critical concern in Darjeeling, despite the region's substantial yearly rainfall. The city's haphazard urbanisation has compromised natural water catchment zones, leading to recurrent water shortages, especially during the peak tourist season. The increasing disparity between water demand and supply has exacerbated the city's infrastructure, resulting in recurrent shortages of water that affect both inhabitants and tourists [41, 47]. The demands of urbanisation have impacted Darjeeling's heritage sites and cultural landscape. Contemporary constructions have replaced numerous colonial-era edifices and heritage residences in the city, as the high maintenance costs make these buildings unaffordable for the majority of middle-class residents. The Darjeeling Municipality is deficient in money required to conserve these heritage assets, resulting in the progressive deterioration of the city's historical identity [18]. The satellite imageries reveal the expansion of the urban built-up area in the span of a decade (Fig. 8).

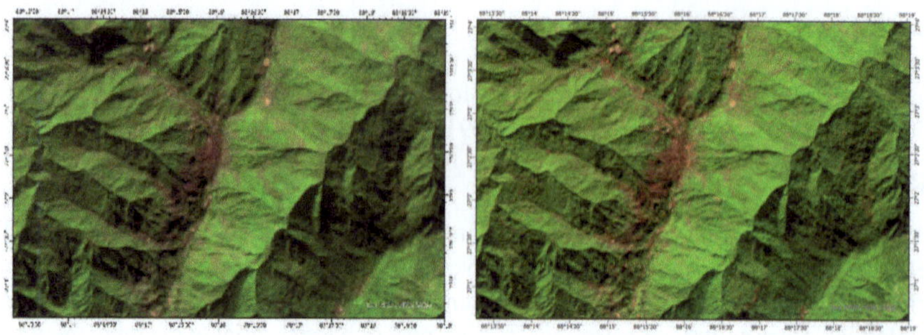

Fig. 8 False-colour Landsat imagery of Darjeeling: 2011 (left) and 2023 (right) (*Source* Author)

Social and Economic Impacts

The swift growth of tourism in Darjeeling has yielded both beneficial and detrimental impacts on the local economy. The emergence of rural tourism in the adjacent Himalayan region has generated economic prospects for rural communities, facilitating their integration into the tourism industry. The commercialisation of local cultures has elicited worries regarding the erosion of authenticity in rural tourism experiences, as traditional lifestyles are further commodified for tourist consumption [3, 38]. The pressure on Darjeeling's infrastructure has considerably affected the quality of life for local inhabitants.

Research Gap

Despite extensive documentation of the environmental and physical issues facing Darjeeling, more research is required to ensure the economic sustainability of the region's tourism sector. While tourism has generated considerable economic advantages, it has also fostered reliance on external resources, especially regarding water and food. Additional research is required to investigate how Darjeeling might attain a more sustainable and self-sufficient tourism industry, especially considering its increasing water scarcity challenges.

4.4 Gangtok, Sikkim

Researchers have linked the rapid expansion of tourism in Gangtok to various environmental and infrastructural challenges. The adverse impacts of overtourism, including the deterioration of urban ecosystems resulting from trekking, grazing, and improper waste management, are addressed by researchers [12]. Mass tourism has exacerbated the burden on local infrastructure, overwhelming waste disposal systems and contributing to pollution and additional environmental degradation. Other researchers have expressed additional worries, cautioning that Gangtok might become as congested as Darjeeling without appropriate interventions. It was advised that the construction of multistoreyed buildings on hill slopes be suspended, especially given that Gangtok has been designated a seismic zone [53]. The potential for urban expansion to intensify natural catastrophe risks in the region was emphasised, and Paul's study recommended more stringent urban planning laws to avert overcrowding and environmental degradation. Proactive zoning measures and the establishment of new tourist circuits to redirect tourist traffic from ecologically sensitive locations are recommended [71]. The study proposed a destination development strategy that integrates architectural design appropriate for the local environment, safeguarding against further degradation of the natural landscape by tourists. These solutions were suggested as essential measures to regulate increasing tourist volumes while safeguarding the environment.

Environmental and Social Impacts

There are inefficiencies in Gangtok's waste management system due to the prevalence of open waste disposal, particularly near natural drains. The city's steep inclines and restricted access in specific regions have impeded efficient rubbish collection. Additionally, the Gangtok Municipal Corporation (GMC) has encountered financial limitations that have hindered the use of superior waste management methods. The deficiency of resources and infrastructure has resulted in considerable environmental degradation in the city [57]. Urban sprawl has emerged as a significant concern in Gangtok, where the influx of individuals from rural regions and the growth of the tourism industry have exerted significant strain on Gangtok's land resources (see Fig. 9). Settlements have been constructed in susceptible areas, including ridge tops, shoulder slopes, and foot slopes. These regions, although topographically more stable, have become congested, leading to encroachment into more perilous locations, hence heightening the risk of landslides and exacerbating environmental degradation [16].

Research Gap

A noteworthy gap in the literature persists concerning the function of community-driven governance in alleviating environmental challenges in Gangtok. Despite the

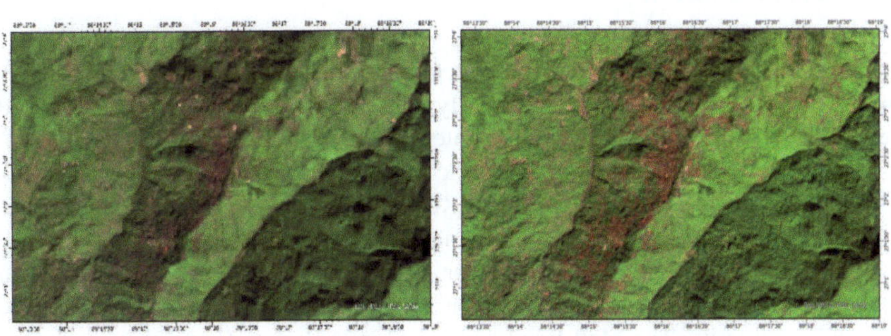

Fig. 9 False-colour Landsat imagery of Gangtok: 2011 (left) and 2023 (right) (*Source* Author)

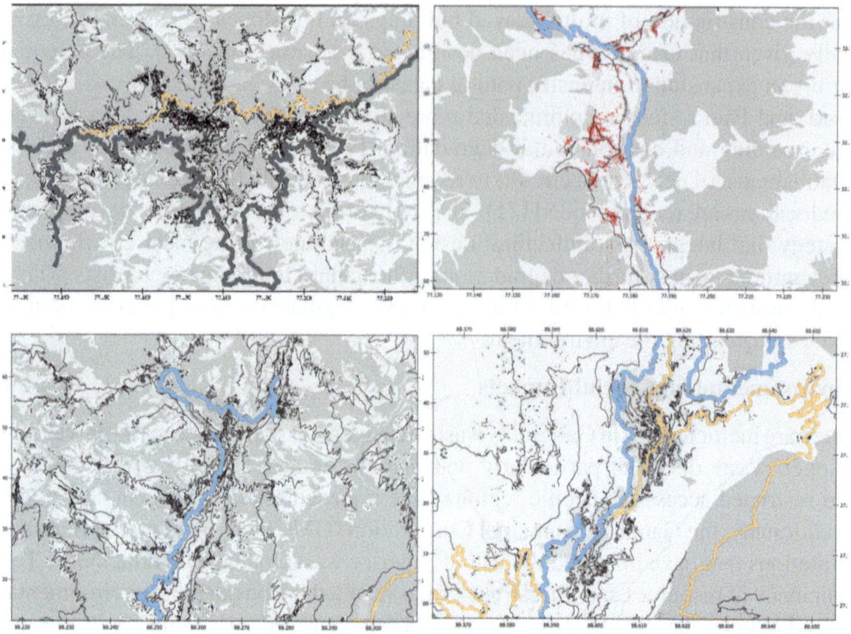

Fig. 10 Road network of the study regions: **a** Shimla (top left), **b** Manali (top right), **c** Darjeeling (bottom left), and d. Gangtok (bottom right) (*Source* openstreetmap.org)

suggestion of technical solutions for waste management and urban planning, there is a lack of research on the involvement of local people in these initiatives. Further research is necessary to evaluate the long-term effectiveness of the tourism zoning and destination development strategies proposed by [71], especially in mitigating environmental stresses and enhancing urban sustainability.

4.5 Urban Pattern

A comparative analysis has been conducted between the hill stations to identify morphological differences, and the underlying causes of development pattern. Road network data from OpenStreetMap and satellite imagery has been used for this analysis. Figure 10 illustrates the road networks of the four hill stations. For Shimla, the road network acts as part of the Shimla–Kullu–Manali circuit, which gained popularity due to the unrest in the neighbouring state of Kashmir, and the Lahaul–Spiti circuit. Shimla also serves as a node for the neighbouring districts of Himachal Pradesh. Additionally, the British colonial heritage significantly contributes to Shimla's popularity. The Mall Road and the Shimla Bypass Highway serve as the focal points for urban development, primarily to accommodate the tourism industry. Although the urban development had started on the southern side of the ridge, in recent years, sprawl of the urban area has also been observed on the northern side of the Mall Road, which gives cause for concern, as it results in deforestation, and development on the steeper slopes increases land vulnerability. Due to overcrowding in Shimla, over the years, other regions such as the Spiti Valley, Manali, and Kalpa–Kinnaur Valley are experiencing more footfall and built-up area development [37]. The tourism intensity in Shimla, on the other hand, is experiencing a decline due to the overcrowding discussed in the literature, which results in loss of revenue and eventually economic degrowth.

Manali is also showing similar patterns in terms of tourism intensity (see Fig. 3), but it is caused by recent natural calamities, such as the flooding caused by cloudbursts in Himachal Pradesh in the years 2022, 2023, and 2024. Manali is developed in the Beas Valley, connected by the Kullu–Manali Highway, which follows the river. The geographical constraints render the destination inaccessible in the events of landslides and flooding (Fig. 10). On the other hand, construction of new underground tunnels and routes to improve accessibility increases the risk of land vulnerability. Additionally, guiding the development away from Beas Valley poses the risk of encroachment into the forested pristine areas, which needs to be tackled mindfully.

Darjeeling, a hill station and former summer capital of British rulers before Shimla, has a similar urban form; it is developed on a ridge with the Mall Road at the very top and littered with the British colonial heritage. However, unlike Shimla, Darjeeling only has access to Hill Cart Road and Lebong Cart Road, which serve as connectors to the Siliguri Highway, due to its geographical constraints of altitude and slope. The development in Darjeeling is centred around the urban centre of Mall Road, with most of the development observed on the western slope of the ridge facing the Mt. Kanchenjunga range. The eastern slope has tea estates due to its climatological needs. Even within the tea estate areas, we've seen development in recent years (see Fig. 11), and in the neighbouring rural areas, as a spillover of the tourism intensity in the region [2], the authors identified this in their previous research. The development of Gangtok also follows a pattern similar to Darjeeling, centred around M.G. Marg (mall road), but the development is mostly limited to the western slopes due to the views of the snowy caps of the Mt. Kanchenjunga range. Additional development leading away from Gangtok is observed on the Tsomgo Lake–Nathu La route.

5 Future Recommendations and Conclusion

The literature assessment on tourism in mountain areas has identified several significant research gaps that this project intends to solve. Despite acknowledging environmental deterioration, urban sprawl, waste management challenges, and insufficient governance frameworks, the execution of flexible, community-inclusive, and data-driven sustainable tourism management methods remains stagnant. Current research highlights the necessity for proactive zoning and infrastructure management [71]. However, there hasn't been a thorough, long-term assessment of these systems'

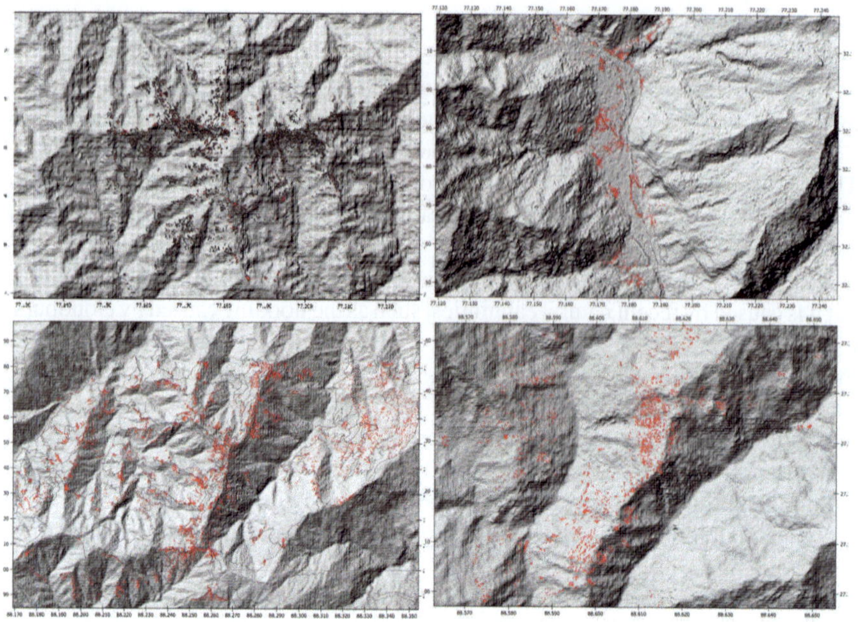

Fig. 11 Distribution of built-up area in the study regions: **a** Shimla (top left), **b** Manali (top right), **c** Darjeeling (bottom left), and **d** Gangtok (bottom right) (*Source* openstreetmap.org)

effectiveness. The study underscores the isolation of communities from tourist governance, wherein local inhabitants are frequently marginalised by tourism-focused development initiatives [4, 57]. Furthermore, real-time data solutions and intelligent resource management have been recognised as future goals, yet, a definitive framework for incorporating these technologies into local tourist practices is lacking. Addressing these deficiencies necessitates an innovative, scalable strategy that integrates real-time data surveillance, community-driven governance, and intelligent infrastructure management.

The Himachal Pradesh government has delineated a series of long-term and short-term objectives within its sustainable tourism development framework, focusing on carrying capacity [19]. However, the governments of Sikkim and Darjeeling have yet to take any action. Informed by the literature study and the instruction from the Himachal Pradesh government, a series of proposals and a framework have been developed to tackle the issue of overtourism in Indian Himalayan destinations. The framework is categorised into three temporal phases: short term (to be implemented immediately), medium term (to be adopted following the attainment of short-term objectives), and long term (to be embraced for future advancement).

Short-Term Objectives

- Evaluation of the present carrying capacity of destinations via a localised sustainable tourism index to benchmark carrying capacity levels and find opportunities for improvement.
- Identification of alternative tourist venues for crowd dispersal.
- Creation of a tailored sustainable tourism model for locations, addressing the specific needs and unique problems of each location.

Medium-Term Objectives

- Development of a Real-Time Tourism Recommendation System to efficiently manage visitor traffic and optimise time utilisation while preserving the destinations' carrying capacity.
- The administration and oversight of designated alternative tourist attractions is crucial. Without efficient administration, these alternative destinations may face the same challenges as the primary tourist sites.
- Deployment of intelligent water, solid waste, and traffic management systems to facilitate a real-time suggestion system and optimal resource utilisation.

Long-Term Objectives

- Establishment of a Development Authority for benchmarking and monitoring tourism in the mountain areas.
- Capacity building in terms of natural, cultural, and human resources.

In conclusion, the proposed framework (see Fig. 12) for addressing overtourism in Indian Himalayan destinations represents a comprehensive, multi-tiered approach to sustainable tourism management. By focusing on short-term objectives like assessing current carrying capacity, identifying alternative tourist destinations, and developing localised sustainable tourism models, the framework offers immediate solutions to alleviate overcrowding and environmental degradation. The medium-term objectives, such as implementing a real-time tourism recommendation system and integrating smart resource management, aim to enhance operational efficiency while maintaining ecological balance. Over the long term, establishing a dedicated development authority and investing in capacity building will ensure that tourism growth is monitored and aligned with sustainability goals. This approach not only addresses the immediate concerns of overtourism but also lays the groundwork for future resilience by integrating smart technologies and community-led governance. Ultimately, this framework strives to balance economic benefits with environmental conservation and cultural preservation, ensuring that mountain destinations can thrive sustainably while maintaining their unique natural and cultural assets for future generations. The successful implementation of these strategies will require strong collaboration between governments, local communities, and industry stakeholders to ensure lasting and meaningful change. The proposed framework for addressing the negative impacts of overtourism in Indian Himalayan destinations aligns strongly with the United Nations' Sustainable Development Goals (SDGs), ensuring that solutions are globally relevant and impactful.

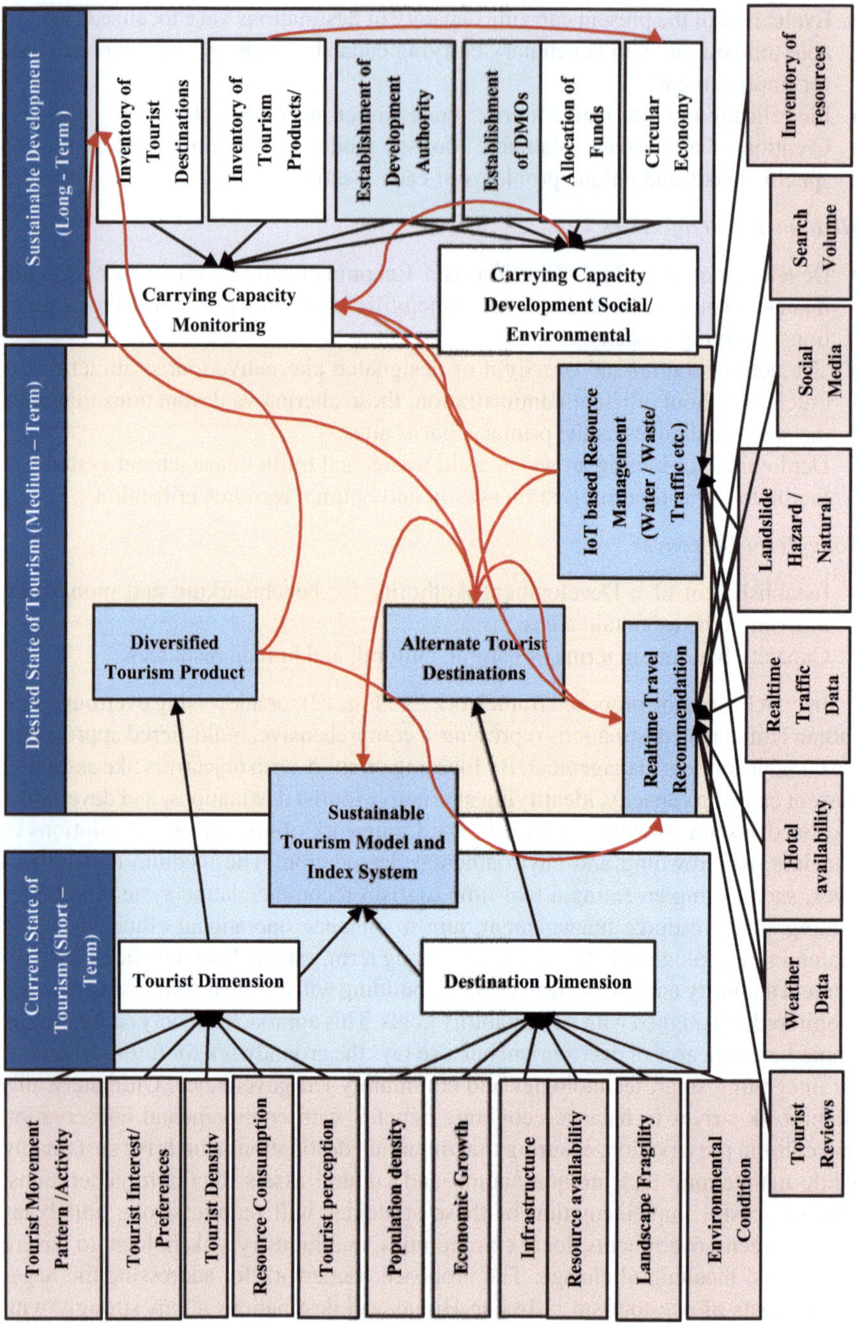

Fig. 12 Proposed framework (*Source* Author)

By addressing overtourism through a multi-tiered approach that harmonises economic growth with ecological preservation and community well-being, this research offers a holistic solution to one of the most pressing challenges in sustainable tourism. The proposed framework not only mitigates immediate impacts but also lays the groundwork for long-term resilience through the integration of smart technologies, real-time resource management, and community-driven governance. The alignment with the SDGs underscores the broader relevance of this study, highlighting its potential for replicability and scalability in other vulnerable mountain ecosystems worldwide. Ultimately, this research demonstrates that with collaborative efforts, thoughtful planning, and the adoption of innovative technologies, it is possible to balance tourism development with the preservation of natural and cultural heritage, fostering a sustainable future for Himalayan destinations and beyond.

References

1. Badar RN, Bahadure SP (2020). Assessing tourism sustainability in hill towns: case study of Shimla, India. J Mountain Sci 17(9). https://doi.org/10.1007/s11629-019-5683-5
2. Banerjee S, George A (2024) Identifying overtourism & spill-over tourism using ST-DBSCAN analysis for sustainable management of tourism. Current Issues in Tourism, 1–21.https://doi.org/10.1080/13683500.2024.2382378
3. Basak D, Bose A, Roy S, Chowdhury IR, Sarkar BC (2021) Understanding sustainable homestay tourism as a driving factor of tourist's satisfaction through structural equation modelling: A case of Darjeeling Himalayan region, India. Current Res Environ Sustain, 3. https://doi.org/10.1016/j.crsust.2021.100098
4. Bhardwaj A, Gupta AK (2023) Tourism sustainability in hilly regions—a review for Shimla. Lect Notes Civil Eng, 277. https://doi.org/10.1007/978-981-19-4040-8_70
5. Bharmoria R, Sharma V (2023) Urban sustainable intervention to address the physical factor for degradation of visual place quality in the hilly urban region: a case of Manali town, Himachal Pradesh. IOP Conference series: earth and environmental science, 1279(1). https://doi.org/10.1088/1755-1315/1279/1/012015
6. Bhatia A (2023) Measuring ICT proficiency and evaluating hindrance among rural inhabitants of selected rural tourism (pong dam wetland) sites of Himachal Pradesh. Towards Excellence. https://doi.org/10.37867/te150113
7. Brown T, Ganguly-Scrase R, Scrase TJ (2016) Urbanization, rural mobility, and new class relations in Darjeeling, India. Critical Asian Studies 48(2). https://doi.org/10.1080/14672715.2016.1164392
8. Butcher J (2023) Covid-19, tourism and the advocacy of degrowth. Tourism Recreation Res 48(5). https://doi.org/10.1080/02508281.2021.1953306
9. Çakar K, Uzut İ (2020) Exploring the stakeholder's role in sustainable degrowth within the context of tourist destination governance: the case of Istanbul, Turkey. J Travel Tourism Marke 37(8–9). https://doi.org/10.1080/10548408.2020.1782307
10. Cave J, Dredge D (2020) Regenerative tourism needs diverse economic practices. Tourism Geogr 22(3). https://doi.org/10.1080/14616688.2020.1768434
11. Chambru M, Claeys C, Lewis N (2022) The development of scientific tourism in mountain regions: challenges and issues for territories in transition. Revue de Géographie Alpine, 110–1.https://doi.org/10.4000/rga.10310
12. Chandran A (2008) Dynamics of tourism growth and it's impact on an emerging destination: a study on Gangtok, Sikkim. Atna J Tourism Studies 3(1):138–148. https://doi.org/10.12727/ajts.3.12

13. Cheer JM, Milano C, Novelli M (2019) Tourism and community resilience in the Anthropocene: accentuating temporal overtourism. J Sustain Tourism 27(4). https://doi.org/10.1080/09669582.2019.1578363
14. Cole V, Sinclair AJ (2002) Measuring the ecological footprint of a Himalayan tourist center. Mountain Res Dev 22(2). https://doi.org/10.1659/0276-4741(2002)022[0132:MTEFOA]2.0.CO;2
15. Cuomo MT, Colosimo I, Celsi LR, Ferulano R, Festa G, La Rocca M (2022) Enhancing traveller experience in integrated mobility services via big social data analytics. Technol Forecasting Soc Change 176:121460. https://doi.org/10.1016/j.techfore.2021.121460
16. Das N, Chatterjee S, Roy U (2013) Habitat-based ecological analysis of urban space in mountain environment: an appraisal for Gangtok town, Sikkim, India. Ecol Environ Conserv 19(3)
17. Das P Kr (2014) Mass tourism & environ-infrastructural crises of Shimla city: a case study. Int J Sci Res 3(11)
18. Dasgupta S, Garg P (2021) The urban development and heritage conundrum: the challenges of heritage conservation in the hill-station of Darjeeling. J Urban Design 26(2). https://doi.org/10.1080/13574809.2020.1770584
19. Department of Tourism & Civil Aviation, G. of H. P. (2019) The Himachal Pradesh tourism policy (2019). https://himachaltourism.gov.in/wp-content/uploads/2019/09/Himachal-Pradesh-Tourism-Policy-2019.pdf
20. Dodds R, Butler R (2019) Overtourism. In: Dodds R, Butler R (Eds), Overtourism: Issues, realities and solutions (Vol 1). De Gruyter. https://doi.org/10.1515/9783110607369
21. Dredge D (2022) Regenerative tourism: transforming mindsets, systems and practices. J Tourism Futures 8(3). https://doi.org/10.1108/JTF-01-2022-0015
22. Duxbury N, Bakas FE, Vinagre de Castro T, Silva S (2020) Creative tourism development models towards sustainable and regenerative tourism. Sustainability 13(1). https://doi.org/10.3390/su13010002
23. Dwyer L (2023) Tourism Degrowth: painful but necessary. Sustainability (Switzerland) 15(20). https://doi.org/10.3390/su152014676
24. Emerson RM (1976) Social exchange theory. Ann Rev Sociol 2(1):335–362. https://doi.org/10.1146/annurev.so.02.080176.002003
25. Fletcher R, Murray Mas I, Blanco-Romero A, Blázquez-Salom M (2019) Tourism and degrowth: an emerging agenda for research and praxis. J Sustain Tourism 27(12). https://doi.org/10.1080/09669582.2019.1679822
26. Fuchs M, Peters M, Weiermair K (2002) Tourism sustainability through destination benchmarking indicator systems: the case of alpine tourism. Tourism Recreation Res 27(3). https://doi.org/10.1080/02508281.2002.11081371
27. Fusté-Forné F, Hussain A (2022) Regenerative tourism futures: a case study of Aotearoa New Zealand. J Tourism Futures 8(3). https://doi.org/10.1108/JTF-01-2022-0027
28. Gardner J, Sinclair J, Berkes F, Singh RB (2002) Accelerated tourism development and its impacts in Kullu-Manali, H.P., India. Tourism Recreation Res 27(3). https://doi.org/10.1080/02508281.2002.11081370
29. Geneletti D, Dawa D (2009) Environmental impact assessment of mountain tourism in developing regions: a study in Ladakh, Indian Himalaya. Environ Impact Assess Rev 29(4). https://doi.org/10.1016/j.eiar.2009.01.003
30. Gössling S (2021) Technology, ICT and tourism: from big data to the big picture. J Sustain Tour 29(5):849–858. https://doi.org/10.1080/09669582.2020.1865387
31. Gössling S, Peeters P (2015) Assessing tourism's global environmental impact 1900–2050. J Sustain Tourism 23(5). https://doi.org/10.1080/09669582.2015.1008500
32. Gowreesunkar V, Seraphin H (2019) Conclusion: local communities' quality of life: what strategy to address overtourism? Worldwide Hosp Tourism Themes 11(5). https://doi.org/10.1108/WHATT-06-2019-0042
33. Gretzel U (2019) The role of social media in creating and addressing overtourism. In: De Gruyter Studies in Tourism (Vol 1). https://doi.org/10.1515/9783110607369-005

34. Hall CM, Scott D, Gössling S (2020) Pandemics, transformations and tourism: be careful what you wish for. Tour Geogr 22(3):577–598. https://doi.org/10.1080/14616688.2020.1759131
35. Hall CM, Wood KJ (2021) Demarketing tourism for sustainability: Degrowing tourism or moving the deckchairs on the titanic? Sustainability (Switzerland) 13(3). https://doi.org/10.3390/su13031585
36. Higgins-Desbiolles F, Carnicelli S, Krolikowski C, Wijesinghe G, Boluk K (2019) Degrowing tourism: rethinking tourism. J Sustain Tour 27(12):1926–1944. https://doi.org/10.1080/09669582.2019.1601732
37. Jangra R, Kaushik SP (2018) Analysis of trends and seasonality in the tourism industry: the case of a cold desert Destination-Kinnaur, Himachal Pradesh. African J Hosp Tourism Leisure 7(1)
38. Kannegieser I (2015) A Home in the hills : examining the socioeconomic benefits of homestay tourism on rural women and their communities in the Darjeeling district. School for International Training India: Sustainable Development and Social Change
39. Kirch A (2002) Impact of tourism and urbanization on water supply and water quality in Manali, northern India. Canadian Water Resour J 27(4). https://doi.org/10.4296/cwrj2704383
40. Koens K, Postma A, Papp B (2018) Is overtourism overused? Understanding the impact of tourism in a city context. Sustainability (Switzerland) 10(12). https://doi.org/10.3390/su10124384
41. Koner K, Samanta G (2021) 'Where does water go'? A critical analysis of nature of water crisis in Darjeeling city, India. Appl Water Sci 11(12). https://doi.org/10.1007/s13201-021-01523-0
42. Kr Singh A, Goswami M (2018) Life in the Himalaya: an Ecosystem at Risk. Delhi Bus Rev 19(1). https://doi.org/10.51768/dbr.v19i1.191201818
43. Langer M., Schmude J (2024) COVID-19 as a paradigm shift? Insights from the degrowth debate in tourism. Europ J Tourism Res, 36. https://doi.org/10.54055/ejtr.v36i.3137
44. Martín JMM, Martínez JMG, Fernández JAS (2018) An analysis of the factors behind the citizen's attitude of rejection towards tourism in a context of overtourism and economic dependence on this activity. Sustainability (Switzerland) 10(8). https://doi.org/10.3390/su10082851
45. Martini U, Buffa F (2016) Creating tourist experiences in European alpine areas: beyond mass tourism. In: Mountain tourism: experiences, communities, environments and sustainable futures. https://doi.org/10.1079/9781780644608.0036
46. Milano C, Novelli M, Cheer JM (2019) Overtourism and degrowth: a social movements perspective. J Sustain Tourism 27(12). https://doi.org/10.1080/09669582.2019.1650054
47. Mondal TK, Roychowdhury P (2019) Water scarcity in Himalayan Hill Town: a study of Darjeeling municipality, India. https://doi.org/10.1007/978-981-10-8947-3_21
48. Morellato M (2008) Alpine destination management: case studies of successful strategies for mountain destinations and venues under impact of mass tourism flows. In: XVII International tourism and leisure symposium ….
49. Nepal SK (2013) Mountain tourism and climate change: implications for the Nepal Himalaya. Nepal Tourism Dev Rev 1(1). https://doi.org/10.3126/ntdr.v1i1.7367
50. Nepal SK, Chipeniuk R (2005) Mountain tourism: toward a conceptual framework. Tourism Geogr 7(3). https://doi.org/10.1080/14616680500164849
51. Nunkoo R, Ramkissoon H (2011) Developing a community support model for tourism. Ann Tourism Res 38(3). https://doi.org/10.1016/j.annals.2011.01.017
52. Panzer-Krause S (2022) Rural Tourism in and after the COVID-19 Era: "Revenge Travel" or Chance for a Degrowth-oriented restart? cases from Ireland and Germany. Tourism Hosp 3(2). https://doi.org/10.3390/tourhosp3020026
53. Paul S (2013) Analysis of tourism attractiveness using probabilistic travel model: a study on Gangtok and its surroundings. Europ J Geogr 4(2)
54. Perry EE, Xiao X, Iretskaia TA, Li P, Manning RE, Valliere WA, Reigner NP (2022) A review of digitalization and sustainability in parks and recreation indicators and thresholds research. J Outdoor Recreation Tourism, 39. https://doi.org/10.1016/j.jort.2022.100550

55. Pung JM, Houge Mackenzie S, Lovelock B (2024) Regenerative tourism: Perceptions and insights from tourism destination planners in Aotearoa New Zealand. J Destination Market Manag, 32.https://doi.org/10.1016/j.jdmm.2024.100874
56. Raab K, Wagner R, Ertz M, Salem M (2023) When marketing discourages consumption: demarketing of single-use plastics for city tourism in Ottawa, Canada. J Ecotourism 22(3). https://doi.org/10.1080/14724049.2022.2028794
57. Rai B, Goswami S (2019) Urbanization and tourism induced challenges in waste management in hill towns: Case of Gangtok. In: Waste management and resource efficiency. https://doi.org/10.1007/978-981-10-7290-1_34
58. Rashid I, Romshoo SA, Amin M, Khanday SA, Chauhan P (2017) Linking human-biophysical interactions with the trophic status of Dal Lake, Kashmir Himalaya, India. Limnologica, 62. https://doi.org/10.1016/j.limno.2016.11.008
59. Rasoolimanesh SM, Jaafar M, Kock N, Ramayah T (2015) A revised framework of social exchange theory to investigate the factors influencing residents' perceptions. Tourism Manag Perspect 16. https://doi.org/10.1016/j.tmp.2015.10.001
60. Rather RA, Bano H, Padder SA, Baba TR, Ara S, Lone FA, Nazir S (2022) Impact of anthropogenic pressure on physico-chemical characteristics of forest soils of Kashmir Himalaya. Bull Environ Contam Toxicol 108(6):1088–1097. https://doi.org/10.1007/s00128-022-03458-x
61. Ravi Bhushan K, Manish K (2020) Socio-cultural impacts of tourism in Himalayan Hill stations: a case study of Kullu-Manali (Himachal Pardesh), India. Res Rev Int J Multidiscip 5(7). https://doi.org/10.31305/rrijm.2020.v05.i07.018
62. Ravichandran SS (2023) Branding regenerative tourism for sustainable rural destinations : a critical reflection perspective. Indian J Market 53(6). https://doi.org/10.17010/ijom/2023/v53/i6/172768
63. Saeporsdottir AD, Hall CM, Wendt M (2020) Overtourism in iceland: fantasy or reality? Sustainability 12(18):25. https://doi.org/10.3390/su12187375
64. Sarkar S (2023) Urban water crisis and the promise of infrastructure: a case study of Shimla. India. Frontiers in Water 5. https://doi.org/10.3389/frwa.2023.1051336
65. Seraphin H, Gowreesunkar V, Zaman M, Lorey T (2019) Limitations of Trexit (tourism exit) as a solution to overtourism. Worldwide Hosp Tourism Themes 11(5). https://doi.org/10.1108/WHATT-06-2019-0037
66. Sharma R, Kumar A (2023) Determinants of quantification of tourism waste in the hilly terrain of Himalayas: a spatial approach. Waste Manage 166. https://doi.org/10.1016/j.wasman.2023.04.039
67. Singh S (2008) Destination development dilemma—Case of Manali in Himachal Himalaya. Tourism Manag 29(6):1152–1156. https://doi.org/10.1016/j.tourman.2008.02.018
68. Singh T, Kaur J (1986) Paradox of mountain tourism: case references from the himalaya. Tourism Recreation Res 11(1). https://doi.org/10.1080/02508281.1986.11014413
69. Singh TV, Naqvi MA, Gowreesunkar G (Vanessa) (2018) What tourism can do: the fall of pastoral Manali resort in the Kulu Valley of the Indian Himalayas. In: Perspectives on Asian Tourism: Vol Part F182. https://doi.org/10.1007/978-981-10-8426-3_9
70. Stylos N, Zwiegelaar J, Buhalis D (2021) Big data empowered agility for dynamic, volatile, and time-sensitive service industries: the case of tourism sector. Int J Contemp Hosp Manag 33(3):1015–1036. https://doi.org/10.1108/ijchm-07-2020-0644
71. Swamy GA, Batabyal D, Gantait A (2021) Measuring the guest-host interaction an empirical analysis for sustainable tourism development in Sikkim. SSRN Electron J. https://doi.org/10.2139/ssrn.3835609
72. UNWTO (2021) Big data for better tourism policy, management, and sustainable recovery from COVID-19. https://doi.org/10.18111/9789284423095
73. Valdivielso J, Moranta J (2019). The social construction of the tourism degrowth discourse in the Balearic Islands. J Sustain Tourism 27(12). https://doi.org/10.1080/09669582.2019.1660670
74. Vijay R, Kushwaha VK, Chaudhury AS, Naik K, Gupta I, Kumar R, Wate SR (2016) Assessment of tourism impact on land use/land cover and natural slope in Manali, India: a geospatial analysis. Environ Earth Sci 75(1). https://doi.org/10.1007/s12665-015-4858-9

75. Wang B (2019) Sacred trash and personhood: Living in daily waste-management infrastructures in the Eastern Himalayas. Cross-Currents 8(1). https://doi.org/10.1353/ach.2019.0009
76. Wani MS, Bhat MS, Alam A, Mir SA (2023) Assessing indigenous community's perspectives and attitudes toward tourism development impacts in the northwestern Himalayas, India. Socio-Ecol Pract Res 5(1). https://doi.org/10.1007/s42532-022-00134-6
77. Yu R, Wang Q, Cheong KC (2022) More than tea—environmental decay, administrative isolation and the struggle for identity in Darjeeling. Malaysian J Econ Studies 59(1). https://doi.org/10.22452/MJES.VOL59NO1.6
78. Zeidabadi G, Mahmoudzadeh SM, Hemati M (2021) Demarketing strategies for tourism demand management: case study of Kashan, Iran. Acta Turistica 33(2). https://doi.org/10.22598/at/2021.33.2.167

Sriya Banerjee is a Doctoral Research Scholar in the Department of Architecture and Regional Planning at the Indian Institute of Technology Kharagpur. Her doctoral research focuses on carrying capacity-based sustainable tourism development in mountain destinations of the Indian Himalayan Region. She has published in renowned journals and presented at international conferences. Her research interests include overtourism, geospatial data science, regional planning, policy development, and ecotourism for community development in mountainous regions.

Abraham George is a distinguished academic and Professor of Architecture at the Indian Institute of Technology Kharagpur. As a former faculty member at NIT Calicut (1988–2009), he contributed to institutional development, architectural projects, research, and pedagogy. A recipient of the John Crawford Scholarship, Fulbright Research Fellowship, and Nehru Trust research grant, he actively serves on academic boards and editorial panels of esteemed architectural journals. His Research interests include, Cultural Landscape of Coastal Zones, Sustainable Community Planning, Architectural Design, Age-friendly design.

Revitalising Traditional Markets with Digital Technologies to Enhance Architectural and Cultural Tourism: Selected Asian Cities

MinJoo Baek and Iasef Md. Rian

Abstract Traditional markets in Asian cities have historically been pivotal to local communities, serving as vibrant centres for commerce, social interaction, and cultural exchange. However, these markets are increasingly threatened by rapid urban development and shifting modern lifestyles, leading to their decline. This research investigates innovative strategies to preserve and rejuvenate these traditional markets by integrating advanced digital technologies. By harnessing the potential of mobile applications, e-commerce platforms, and social media, these markets can be reimagined as dynamic destinations for architectural and cultural tourism. This transformation aims to provide unique and immersive experiences that appeal to contemporary tourists while safeguarding the cultural and economic significance of these markets. Through an analysis of successful case studies and speculative applications, this study offers a comprehensive framework for revitalising traditional markets, ensuring their continued relevance within the cultural and economic fabric of Asian cities. This approach not only enhances the visitor experience but also bolsters the economic sustainability of market vendors, thereby maintaining the integral role of these markets in urban landscapes. The findings underscore the importance of digital innovation in preserving cultural heritage and promoting sustainable economic development in rapidly urbanising regions.

Keywords Traditional markets · Urban development · Digital technologies · Cultural tourism

M. Baek (✉) · I. Md. Rian
Department of Architecture, Xi'an Jiaotong-Liverpool University, Suzhou, China
e-mail: minjoo.baek@xjtlu.edu.cn

I. Md. Rian
e-mail: iasef.rian@xjtlu.edu.cn

V. C. Sharma et al. (eds.), *Architectural Vision of Overtourism*,
SpringerBriefs in Architectural Design and Technology,
https://doi.org/10.1007/978-981-96-5794-0_13

1 Introduction

Traditional markets in Asian cities have historically played a central role in community life, acting as hubs for commerce, social interaction, and cultural exchange. These markets are much more than economic centres; they embody the architectural, historical, and cultural heritage of their regions, making them integral to the social fabric of urban environments [1]. However, in recent decades, the rapid pace of urbanisation, the proliferation of modern shopping malls, and the shifting preferences of consumers towards convenience-driven retail environments have placed immense pressure on these markets. Several studies highlight this ongoing decline, pointing out how traditional markets are losing their relevance in contemporary urban life [2–4]. Moreover, the rapid pace of urban expansion has intensified spatial and infrastructural challenges, diminishing the prominence of these once-vibrant marketplaces [5].

Despite the increasing number of literature on urban development and the decline of traditional markets, there remains a substantial gap in research exploring how digital technologies can be leveraged to revitalise these markets and, more importantly, how they can enhance their potential as tourism destinations. The role of digital tools in reshaping traditional market spaces has been largely overlooked, with few comprehensive studies addressing how digital and new media could offer innovative solutions to the challenges faced by these markets.

This chapter seeks to address this gap by investigating how digital technologies can catalyse revitalising traditional markets in Asia, turning them into vibrant architectural and cultural tourism attractions. It explores the transformative potential of mobile applications, e-commerce platforms, and other digital interventions in preserving and promoting these markets. Through a brief theoretical analysis and case studies, this chapter highlights the importance of integrating digital tools to safeguard the cultural, economic, and architectural heritage of these markets. In doing so, it aims to propose strategies that ensure the survival of traditional markets and extend their role in promoting tourism in urban environments.

2 The Decline of Traditional Markets

2.1 Urban Development and Lifestyle Changes

The decline of traditional markets in many Asian cities is a pressing issue, primarily driven by rapid urbanisation and the emergence of modern retail environments such as shopping malls and supermarkets. These contemporary retail spaces offer unparalleled convenience, providing a wide array of services and products under one roof, which aligns with the fast-paced lifestyle of today's consumers [6, 7]. In contrast, with their smaller scale and specific product offerings, traditional markets are often perceived as less convenient and lacking modern facilities.

For instance, the mega-malls in cities like Bangkok and Kuala Lumpur have significantly impacted local markets, drawing consumers away with promises of convenience and variety [8]. Similarly, in Seoul, the proliferation of supermarkets has shifted consumer preferences towards one-stop shopping experiences, further contributing to the decline of traditional market spaces [9].

In India, this trend is evident as well. Traditional markets are struggling to compete with rapid urbanisation and the rise of malls and supermarkets [10]. Markets like Delhi's Chandni Chowk face declining foot traffic as modern retail chains attract consumers [11, 12]. Yet, these markets remain culturally and economically vital, preserving local craftsmanship and heritage. To stay relevant, digital tools and innovative strategies are essential [13–15].

Despite these challenges, traditional markets remain vital to their communities' cultural and economic fabric. They offer unique cultural experiences and preserve their regions' architectural and historical heritage. However, the literature reveals a gap in exploring how digital technologies can revitalise these markets and enhance their tourism potential. This chapter aims to address this gap by examining the impact of urbanisation on traditional markets, assessing the role of digital tools in market revitalisation, and presenting case studies and speculative applications for digital interventions. Doing so offers a strategic framework for preserving the cultural, economic, and architectural significance of traditional markets in Asian cities.

2.2 *Cultural and Economic Implications*

The decline of traditional markets signifies more than just a reduction in commerce; it represents a profound loss of community spaces where people have historically gathered to interact and exchange cultural practices. These markets are custodians of local craftsmanship and unique architectural heritage, embodying the cultural identity of their regions [16, 17].

For instance, the mega-malls in cities like Bangkok and Kuala Lumpur have significantly impacted local markets, drawing consumers away with promises of convenience and variety [8]. Similarly, in Seoul, the proliferation of supermarkets has shifted consumer preferences towards one-stop shopping experiences, further contributing to the decline of traditional market spaces [9]. In India, a similar trend has been noticed; the decline of traditional markets threatens cultural heritage and community identity [14]. Small vendors and artisans face economic hardship due to reduced foot traffic and competition from modern retail [18].

The economic implications are equally significant. Small vendors and artisans face mounting economic struggles as foot traffic dwindles and consumer preferences shift towards modern retail options [19]. This shift affects their livelihoods and erodes community identity and local interaction, as these markets have historically served as vital hubs for social engagement and cultural exchange.

Without concerted revitalisation efforts, there is a real risk of losing these invaluable spaces, leading to a diminished cultural identity and reduced economic diversity in urban settings. This underscores the urgent need for strategies to preserve and enhance traditional markets' role in contemporary urban life.

3 Role of Traditional Markets in Tourism

3.1 Architectural Significance

Traditional markets in many Asian cities are housed in structures of significant architectural importance, serving as iconic representations of local history and styles. These markets function as commercial hubs and embody their regions' historical and cultural values, making them vital to architectural tourism [20, 21].

For example, the Ben Thanh Market in Ho Chi Minh City is a prime example of French colonial architecture, offering visitors a unique insight into the city's colonial past and architectural evolution [22]. Similarly, the Dongdaemun Market in Seoul is renowned for blending traditional Korean architectural elements with modern influences, reflecting the dynamic interplay between Korea's rich heritage and contemporary design [23]. Traditional markets in India, like Delhi's Chandni Chowk and Jaipur's Johari Bazaar, are cultural and architectural landmarks, preserving Mughal and Rajput heritage while attracting tourists [14].

These markets connect to the region's architectural heritage, attracting tourists eager to explore and appreciate the unique architectural styles that define these vibrant urban spaces. By preserving and promoting these architectural treasures, traditional markets can continue to play a crucial role in enhancing the cultural tourism appeal of their cities.

3.2 Cultural Tourism

Beyond their architectural charm, traditional markets serve as dynamic cultural hubs that offer tourists an authentic taste of local life. These markets are pivotal to cultural tourism, providing immersive experiences that allow visitors to connect with the region's culture and traditions [24, 25].

The Chatuchak Weekend Market in Bangkok exemplifies this role, attracting global tourists with its diverse products, from handmade crafts to local delicacies. This market not only highlights the rich cultural tapestry of Thailand but also offers an interactive environment where tourists can engage with local vendors and artisans, gaining insights into the cultural nuances of the region [26]. The market's vibrant atmosphere and wide-ranging offerings make it a must-visit destination for those seeking genuine cultural experiences. Traditional markets in India, like Agra's Sadar Bazaar, offer rich cultural experiences through local crafts, foods, and festivals, contributing significantly to tourism [14].

By facilitating these cultural exchanges, traditional markets like Chatuchak significantly enhance their cities' cultural tourism appeal, drawing visitors eager to explore and appreciate the authentic aspects of local life.

To further enhance the tourism potential of traditional markets, integrating contemporary facilities like digital technologies and new media platforms can help attract a broader audience, improve accessibility, and offer virtual experiences that showcase their architectural and cultural significance to a global audience.

4 Digital Technologies to Revitalise Traditional Markets

Asian traditional markets are rich in cultural heritage, serving as representations of the unique identity of their respective countries, cities, cultures, and customs. These markets have gained international attention through various media platforms. For example, Netflix has featured them in several popular Asian series, movies, and documentaries, which has sparked increased interest in tourism. Additionally, the portrayal of these markets on Instagram has captivated a global audience. Consequently, new media and social media platforms have significantly contributed to the growth of tourism in Asian cities. This underscores the critical role of digital technologies in revitalising traditional markets and enhancing their appeal to tourists.

4.1 Role of Digital Technologies

Digital technologies hold significant potential to revitalise the traditional market experience, offering innovative ways to engage both locals and tourists. In today's digital age, tools such as augmented reality (AR), virtual reality (VR), mobile applications, and social media platforms can create immersive experiences that appeal to a new generation of visitors [27, 28].

Digital innovations offer the potential to create immersive and engaging experiences, transforming traditional markets into dynamic destinations that attract tech-savvy tourists. By integrating digital tools, markets can enhance their cultural and economic appeal, ensuring relevance in a rapidly evolving urban landscape.

4.2 Mobile Applications and Digital Guides

Mobile applications transform the traditional market experience as digital companions for tourists, enhancing convenience and personalisation. These apps are comprehensive navigation tools, providing detailed market maps that guide visitors through bustling stalls and hidden gems. By offering personalised recommendations based on user preferences and past interactions, these digital guides help tourists discover local vendors and unique products that align with their interests.

For instance, the Borough Market in London has utilised an audio guide app for mobile phones to enhance visitor engagement by providing interactive maps, vendor information, and real-time updates on events and promotions. This digital approach improves the visitor experience and helps vendors connect with a broader audience [29]. This allows tourists to explore the market's history and cultural significance through interactive digital overlays, enriching their understanding and appreciation of the space.

Moreover, mobile apps facilitate seamless cashless transactions, a feature particularly beneficial for international tourists who might otherwise face challenges with currency exchange. This functionality not only simplifies the shopping experience but also encourages more spontaneous purchases, as visitors can easily pay with a tap of their smartphone.

By integrating these features, mobile applications enrich the market visit, making it more accessible and enjoyable for tourists. They bridge the gap between traditional market charm and modern technological convenience, ensuring visitors fully immerse themselves in the local culture without logistical concerns (Fig. 1).

4.3 Social Media, Digital Display, and E-Commerce

Social media platforms have become indispensable tools for promoting traditional markets, enabling vendors to reach and engage with a global audience. Through vibrant posts and engaging stories, market vendors can showcase their unique products and share the rich narratives behind them, effectively drawing in potential customers from around the world. For instance, markets like the Central Market in Kuala Lumpur leverage social media to captivate tourists by highlighting special events, workshops, and cultural festivals, enhancing their appeal and visibility.

Digital displays and kiosks are interactive tools that provide tourists with easy access to information about traditional markets. These digital interfaces can offer maps, event schedules, and product highlights, enhancing the visitor experience by making navigation and exploration more intuitive. By integrating digital displays within market environments, tourists can effortlessly discover points of interest, learn about the cultural significance of various market offerings, and plan their visits more efficiently. This digital wayfinding approach enriches the tourist experience and encourages deeper engagement with the market's cultural and historical context.

In addition to social media and digital displays, e-commerce platforms offer traditional market vendors the opportunity to expand their reach beyond physical boundaries. By selling their products online, vendors can tap into a global customer base, allowing international audiences to experience the charm and diversity of traditional markets from afar. This digital expansion boosts sales and preserves and promotes cultural heritage worldwide, ensuring that traditional markets remain vibrant and relevant in the digital age.

Fig. 1 Borough market in London audio guide app[1]

5 Digital Technologies to Revitalise Traditional Markets

5.1 Chatuchak Weekend Market, Bangkok, Thailand

As one of Asia's largest and most renowned markets, the Chatuchak Weekend Market in Bangkok has adeptly embraced digital integration to enhance the visitor experience. Recognising the challenges posed by its vast and bustling environment, the market developed a comprehensive mobile app, "Chatuchak Guide," which serves as an essential tool for tourists. This app offers detailed maps that guide visitors through

[1] *Source* https://www.pelago.com/en/activity/p0gxy-borough-market-audio-guide-app-for-your-smartphone-london/.

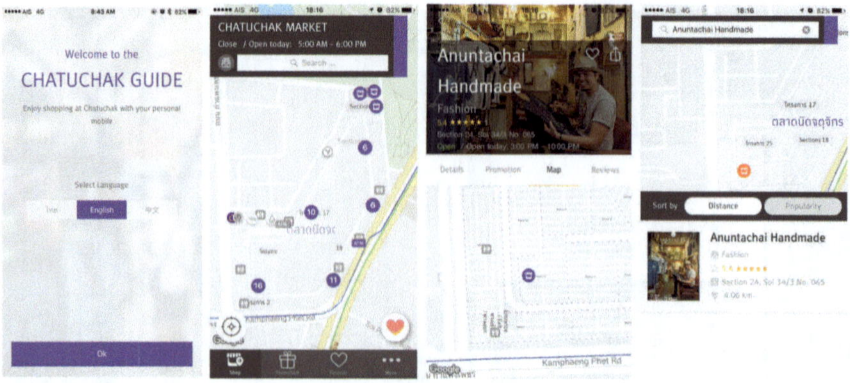

Fig. 2 Chatuchak weekend market mobile app "Chatuchak Guide" interface, downloaded from http://chatuchak-guide.appstor.io

the market's extensive network of stalls, ensuring they can easily locate specific vendors and areas of interest. Additionally, the app provides in-depth vendor information, including product offerings and special deals, alongside personalised recommendations based on user preferences and browsing history. This tailored approach enhances the shopping experience, making it more efficient and enjoyable for visitors (Fig. 2).

Beyond navigation, Chatuchak has harnessed the power of social media to bolster its online presence. Platforms such as Instagram and Facebook are utilised to showcase the market's vibrant atmosphere, unique products, and cultural events. By promoting special events, workshops, and seasonal offerings, the market maintains a dynamic and engaging online profile that attracts both local and international tourists. This strategic use of social media drives foot traffic and fosters a sense of community and anticipation among potential visitors, effectively bridging the gap between the digital and physical realms of the market. Through these digital initiatives, Chatuchak Weekend Market continues to thrive as a must-visit destination, seamlessly blending traditional market charm with modern technological advancements.

5.2 *Dongdaemun Market, Seoul, South Korea*

The Dongdaemun Market in Seoul exemplifies the successful fusion of traditional market culture with cutting-edge technology, creating a unique and engaging visitor experience. Strategically placed digital kiosks throughout the market serve as interactive hubs, offering visitors access to detailed maps that simplify navigation through the market's extensive layout [30]. This integration of digital kiosks enhances the visitor experience by overlaying digital information onto the physical environment, making it both educational and immersive [31] (Fig. 3).

Fig. 3 Customers at a Juicy branch in Dongdaemun District, eastern Seoul, place orders on an electronic kiosk. The branch is one of roughly 50 stores in the juice bar chain that has adopted the payment kiosks[2]

In addition to on-site digital enhancements, Dongdaemun Market has cultivated a robust online presence. The market showcases its vibrant atmosphere, diverse product offerings, and cultural events to a global audience through active engagement on social media platforms like Instagram, Facebook, and YouTube. This digital strategy is complemented by virtual shopping events, which allow international tourists to explore and purchase products from the market remotely. These events often feature live-streamed tours and interactive sessions with vendors, providing a virtual yet authentic market experience [32].

The integration of these digital tools has modernised the market and expanded its reach, attracting a diverse array of visitors from around the world. By blending traditional charm with technological innovation, Dongdaemun Market has successfully positioned itself as a leading destination for both cultural tourism and digital commerce, demonstrating the effectiveness of this approach in enhancing market appeal and accessibility [33].

5.3 Nishiki Market, Kyoto, Japan

Nishiki Market, affectionately known as "Kyoto's Kitchen," has effectively integrated digital technologies to enhance its appeal as a cultural tourism destination. The market offers live streaming on a website accessible on mobile, allowing potential visitors to explore its layout and delve into its rich history from anywhere in the world.

[2] *Source* https://koreajoongangdaily.joins.com/2018/10/14/industry/The-Future-Is-Now-Its-shopping-but-not-as-we-know-it/3054253.html.

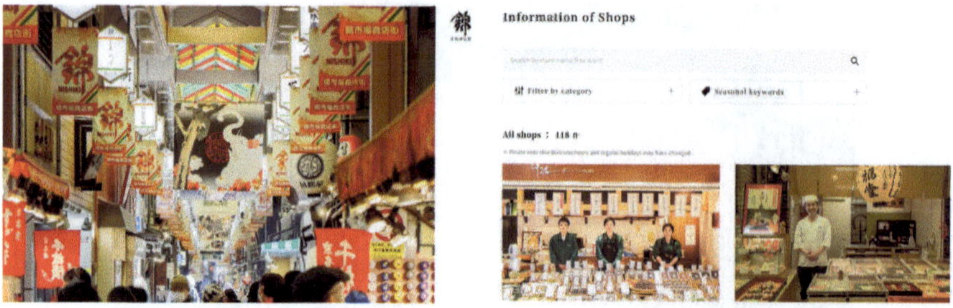

Fig. 4 Nishiki Market view and the online shop

This digital preview entices tourists by providing a glimpse of the market's vibrant atmosphere and diverse offerings (Fig. 4).

Its website and online shop are central to Nishiki Market's digital strategy, which provides comprehensive information about each vendor and its products. These sites offer real-time cultural insights and personalised recommendations, enhancing the visitor experience by tailoring suggestions to individual interests. A standout feature of the site is that it provides translations and detailed explanations of local customs and traditions. This functionality is particularly beneficial for international visitors, as it facilitates a deeper appreciation of the market's cultural significance by breaking down language barriers and offering context to the market's unique practices.

The success of this digital approach is evidenced by increased visitor engagement and positive feedback from tourists who appreciate the seamless blend of traditional market charm with modern technology. The user reviews and market studies highlight the site's value it adds to the visitor experience. By leveraging these online platforms, Nishiki Market preserves its cultural heritage and enhances its accessibility and appeal to a global audience, solidifying its status as a must-visit destination in Kyoto.

5.4 Central Market, Kuala Lumpur, Malaysia

Central Market in Kuala Lumpur has successfully evolved into a vibrant cultural and tourism hub by integrating an online shop. Central Market leverages social media platforms to promote its diverse array of cultural events and workshops. These platforms are used to highlight upcoming activities, such as traditional craft workshops, cultural performances, and seasonal festivals, thereby attracting both local residents and international tourists. The strategic use of social media helps to maintain a dynamic online presence, fostering a sense of community and anticipation among potential visitors.

Through these digital initiatives, Central Market has successfully enhanced its appeal as a cultural destination, drawing in a diverse audience and ensuring that its rich heritage is celebrated and preserved for future generations (Fig. 5 and Table 1).

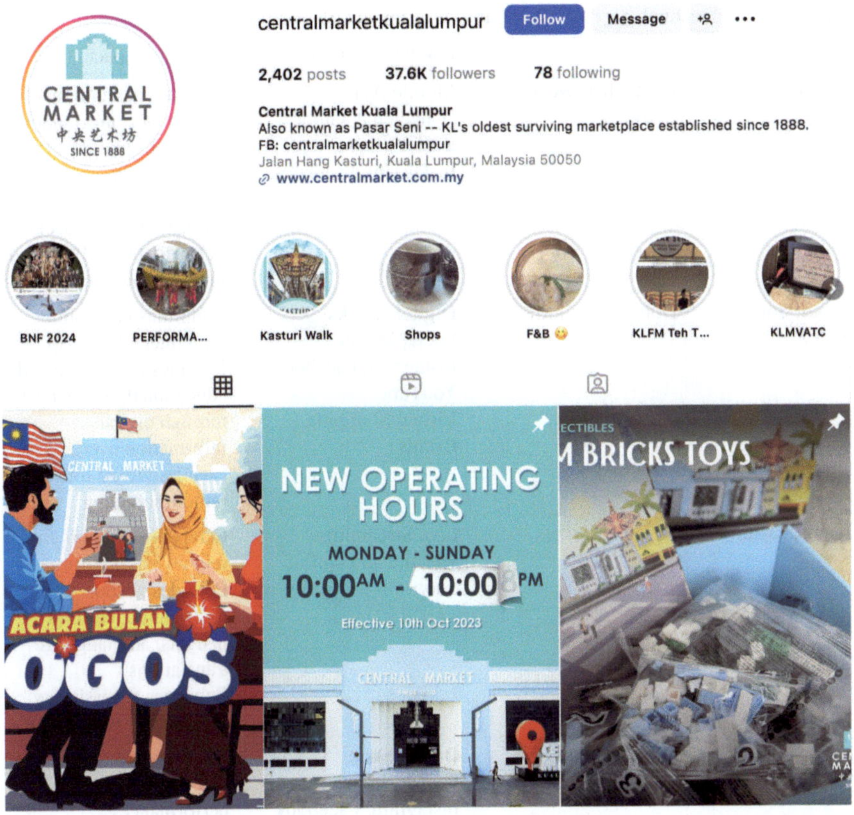

Fig. 5 Central Market account on the Instagram link

6 Benefits of Promoting Traditional Markets as Cultural Tourism

6.1 Economic Benefits

Integrating digital technologies into traditional markets offers significant economic benefits, such as increased tourist footfall and higher sales for vendors. This boost in tourism helps preserve traditional crafts and creates job opportunities by providing artisans a platform to showcase their work. Digital marketing and social media extend the market's reach to a global audience, attracting international tourists and diversifying revenue streams. Studies show that digital adoption enhances business innovation and growth, particularly in emerging markets, leading to improved performance and profitability. Digital transformation sustains and revitalises traditional markets, ensuring financial sustainability and cultural preservation.

Table 1 Digital technologies and their purpose or impact on traditional markets

City	Market	Digital technologies	Purpose/Impact
Bangkok, Thailand	Chatuchak Weekend Market	Mobile App: "Chatuchak Guide" Social Media: Instagram, Facebook	Provides detailed maps and vendor information Personalised recommendations Promotes events, workshops, and special offers to attract local and international tourists
Seoul, South Korea	Dongdaemun Market	Digital Kiosks Social Media: Instagram, Facebook, YouTube Virtual Shopping Events	Simplifies navigation with interactive maps Provides immersive and educational experiences through digital information overlays Enables virtual tours and remote shopping
Kyoto, Japan	Nishiki Market	Live Streaming Website and Online Shop	Offers virtual tours for remote exploration Provides vendor information, cultural insights, and language translations Enhances accessibility for a global audience
Kuala Lumpur, Malaysia	Central Market	Online Shop Social Media: Instagram, Facebook	Promotes cultural events, workshops, and performances Maintains online presence to attract both local and international visitors

Research indicates that integrating digital technologies into cultural tourism can significantly enhance economic outcomes. For instance, Thanh (2018) highlights that traditional markets in the Mekong Delta attract considerable tourist interest when promoted as cultural tourism destinations, thereby boosting local economies [34]. Furthermore, adopting digital tools in tourism has been shown to improve business innovation and growth, particularly in emerging markets, leading to enhanced performance and profitability [35].

Moreover, the socio-cultural benefits of tourism development are well-documented. A study on tourism in a mountain area in Romania found that residents perceived tourism as a factor that enhances cultural identity and improves the quality of recreational services, indirectly boosting the standard of living [36]. This underscores the potential of digital transformation in sustaining and revitalising traditional markets, ensuring their financial sustainability and cultural preservation.

6.2 Cultural Preservation

Promoting traditional markets through digital technologies plays a vital role in cultural preservation by attracting tourists eager to experience local traditions and customs. This interaction strengthens local cultural practices and generates revenue that can be reinvested into maintaining market structures, thus preserving both architectural heritage and cultural practices. For example, the Grand Bazaar in Istanbul has successfully implemented digital strategies to increase visitor numbers, providing crucial financial support for the upkeep of its historic architecture [37]. Similarly, Borough Market in London has leveraged digital platforms to enhance tourist experiences, increasing funds for market maintenance and cultural preservation [38]. These cases illustrate how digital transformation can economically sustain traditional markets while protecting cultural heritage.

Moreover, preserving cultural heritage, including traditional markets, can greatly benefit from digital technologies that lower costs and enhance access to information. This advancement can result in improved management and promotion of cultural sites, attracting more visitors and generating revenue for maintenance and preservation [39].

6.3 Social Benefits

Revitalising traditional markets with digital technologies transforms them into lively community centres encouraging social interaction and cultural exchange. In Asian societies, these markets play a crucial role in preserving social bonds and cultural heritage. By promoting them as cultural tourism destinations, cities can enhance community pride and motivate younger generations to engage with traditional crafts.

For instance, the Nishiki Market in Kyoto has effectively attracted a diverse audience by using digital platforms, reinforcing its status as a cultural and social hub. This market, celebrated for its historical importance and local delicacies, employs online tools to broaden its reach, amplifying its cultural influence.

Similarly, the Chatuchak Weekend Market in Bangkok has evolved into a vibrant social centre, where digital promotion has boosted tourist visits and inspired younger generations to reconnect with their cultural roots. The market's strategic use of social media and online platforms has made it a popular spot for locals and tourists, enhancing cultural exchange and community involvement.

These examples, along with the revitalisation of Queen Victoria Market in Melbourne [40] and Mercado de San Miguel in Madrid [41], demonstrate how digital transformation can rejuvenate traditional markets, fostering social benefits and cultural continuity across various cultural settings. This approach is particularly significant in Asian societies, where community and tradition are deeply valued.

7 Challenges and Considerations

Balancing the preservation of cultural character with the integration of modern technologies is a key challenge in revitalising traditional markets. While digital tools can enhance the visitor experience, it's crucial they don't overshadow the market's essence. Engaging vendors and local communities in the modernisation process is essential to preserving the unique cultural identity of these markets.

The influx of tourists offers opportunities and challenges, boosting sales but potentially leading to overcrowding and infrastructure strain. Effective crowd management and educating tourists on respectful behaviour are vital to addressing these issues. For instance, the Tsukiji Fish Market in Tokyo and the Chatuchak Weekend Market in Bangkok have successfully implemented strategies to manage tourist impact while maintaining cultural significance [42, 43]. These examples underscore the importance of balancing tourist engagement with market culture and infrastructure preservation.

Ensuring inclusivity involves maintaining accessibility and affordability for local residents while integrating digital advancements. Keeping market prices affordable prevents displacement, allowing residents to continue benefiting from these cultural hubs. As digital tools become more prevalent, making them accessible to less tech-savvy vendors and visitors is important. Training programmes and user-friendly digital interfaces empower all stakeholders to participate in the market's modernisation, ensuring everyone benefits from the enhanced market experience without losing the essence of traditional commerce.

8 Conclusion

This study underscores the transformative potential of digital technologies in revitalising traditional markets, positioning them as pivotal cultural and architectural tourism destinations. The integration of these technologies can significantly enhance the visitor experience while safeguarding the cultural and architectural heritage intrinsic to these markets. By leveraging digital tools, cities can stimulate economic growth, sustain local crafts, and foster deeper community engagement, reinforcing urban environments' socio-economic fabric.

The future of traditional market revitalisation is contingent upon meticulous planning and a nuanced balance between tradition and modernisation. It is imperative to ensure that digital innovations are harmoniously integrated, preserving these markets' authenticity and cultural essence. Strategic partnerships among governments, local communities, and technology providers are essential to facilitating this integration, ensuring that technological advancements enhance rather than erode cultural heritage.

The revitalisation of traditional markets necessitates a concerted effort from all stakeholders to secure their survival and relevance in the contemporary urban landscape. While digital technologies present novel opportunities for preservation and economic sustainability, it is crucial to recognise and address their inherent limitations and potential drawbacks. Over-reliance on digital solutions may weaken these markets' authenticity and cultural significance. To counteract these challenges, it is vital to implement solutions such as inclusive community participation in digital planning, comprehensive education programmes on respectful tourist behaviour, and targeted infrastructure enhancements. By addressing these considerations, we can effectively harness digital innovations to preserve and invigorate the rich cultural heritage of traditional markets, ensuring their enduring vibrancy and relevance in the modern era.

References

1. Ng WK, Hsu FT, Chao CF, Chen CL (2023) Sustainable competitive advantage of cultural heritage sites: three destinations in East Asia. Sustainability 15(11):8593
2. Barata-Salgueiro T, Erkip F (2014) Retail planning and urban resilience—an introduction to the special issue. Cities 36:107–111
3. Rao F, Dovey K, Pafka E (2018) Towards a genealogy of urban shopping: types, adaptations and resilience. J Urban Des 23:544–557
4. Ariefiani RN, Burhan MU (2013) The analysis on consumer behavior towards modern retail shop and traditional shop (Case Study of Young Consumers' Shopping Preference in Lowokwaru District, Malang). Jurnal Ilmiah Mahasiswa FEB2(2)
5. Lee S (2017) A Study on traditional market decline and revitalization in Korea—improving the Iksan Jungang traditional market. J Asian Archit Build Eng 16:455–462
6. Cheng S (2016) Urbanisation and the decline of traditional markets in Asia. J Urban Studies 53(4):567–589
7. Kim J, Park H (2017) The impact of modern retail on traditional markets in South Korea. Asian Econ J 31(2):123–145
8. Wong T (2018) Shopping malls and the transformation of retail in Southeast Asia. Southeast Asian J Soc Sci 46(3):234–250
9. Lee S (2019) Consumer preferences and the rise of supermarkets in Seoul. Int J Retail Distrib Manag 47(5):456–472
10. Basu D (2015) Malls vs. traditional markets: the transformation of retail in India. Retail Res Strategy J 12(2):201–220
11. Gupta P (2020) Consumer preferences in Indian retail: the shift from traditional to modern markets. Econ Pol Wkly 55(8):121–133
12. Kumar R, Sharma P, Verma A (2017) The impact of supermarkets on traditional markets in Indian cities. J Retail Consumer Serv 34(2):95–103
13. Bhatnagar S (2021) Revitalising traditional markets in India: digital interventions and their role. Asian Urban Studies 45(3):78–92
14. Banerjee A, Dutta S (2016) The impact of urbanisation on traditional bazaars: a case study of Kolkata. J Urban Studies 23(4):512–530
15. Mishra A (2019) The future of traditional bazaars: policy perspectives for sustainable urban development. India Dev Rev 22(1):45–59
16. Zukin S (2010) Naked city: the death and life of authentic urban places. Oxford University Press, New York

17. Watson S, Studdert D (2006) Markets as sites for social interaction: spaces of diversity. Policy Press, Bristol
18. Singh R, Kaur M (2018) The changing landscape of Indian retail: traditional vs. modern marketplaces. Indian J Econ Dev 14(1):34–42
19. Porter ME (1995) The competitive advantage of the inner city. Harv Bus Rev 73(3):55–71
20. Logan WS (2002) The disappearing 'Asian' city: Protecting Asia's urban heritage in a globalizing world. Oxford University Press, Oxford
21. Henderson JC (2011) Traditional markets and tourism: the case of Dongdaemun. Seoul. Tourism Geographies 13(3):417–435
22. Gillen J (2016) Urban Vietnam: Negotiating the city. National University of Singapore Press, Singapore
23. Kim H (2013) The architectural evolution of Dongdaemun market. Korean J Archit 33(2):89–102
24. Richards G (2018) Cultural tourism: a review of recent research and trends. J Hosp Tour Manag 36:12–21
25. Timothy DJ, Ron AS (2013) Understanding heritage cuisines and tourism: identity, image, authenticity, and change. J Herit Tour 8(2–3):99–104
26. Cohen E (2015) Tourism and development in Southeast Asia. Routledge, Abingdon
27. Guttentag DA (2010) Virtual reality: applications and implications for tourism. Tour Manage 31(5):637–651
28. Tussyadiah IP, Wang D, Jia CH (2018) Exploring the persuasive power of virtual reality imagery for destination marketing. Tour Manage 65:140–154
29. Smith A (2019) Digital engagement in traditional markets: the case of Borough Market. J Urban Technol 26(2):3–18
30. Korean JoongAng Daily: Dongdaemun Market's digital kiosks enhance the visitor experience (2018). https://koreajoongangdaily.joins.com/. Accessed 20 Oct 2023
31. Aljohi AS, Alzaabi SS, Almahri RS, Tsaramirsis G, Hamid OH (2024) Navigating the divide: digital kiosks and mobile apps as complementary human-centered self-service technologies. Eng Proc 59(1):162
32. Barbosa B, Santos JD (2023) Bibliometric study on the social shopping concept. Administ Sci 13(10):213
33. Miah MR, Hossain A, Shikder R, Saha T, Neger M (2022) Evaluating the impact of social media on online shopping behavior during COVID-19 pandemic: a Bangladeshi consumers' perspectives. Heliyon 8(9)
34. Thanh T (2023) Analysis of social results of tourism performing arts industry in cultural tourism convergence using fuzzy comprehensive analysis methods. Econ Res 26(4):234–250
35. Varotsis N (2022) Digital entrepreneurship and creative industries in tourism: a research agenda. Economies 10(7):167
36. Muresan IC, Harun R, Arion FH, Fatah AO, Dumitras DE (2021) Exploring residents' perceptions of the socio-cultural benefits of tourism development in the mountain area. Societies 11(3):83
37. Çakmak E, Isaac RK (2012) What destination marketers can learn from their visitors' blogs: an image analysis of Bethlehem, Palestine. J Destin Mark Manag 1(1–2):124–133
38. Jones P, Evans J (2013) Urban regeneration in the UK: Boom, bust and recovery. SAGE Publications, London
39. Hung CS, Chen TL, Lee YC (2021) From cultural heritage preservation to art craft education: a study on Taiwan traditional lacquerware art preservation and training. Educ Sci 11(12):801
40. Davies A, Prentice R (2015) Conceptualising the social value of place: a framework for the strategic management of cultural heritage assets. J Cultural Heritage Manag Sustain Dev 5(2):130–147
41. González S, Waley P (2013) Traditional retail markets: the new gentrification frontier? Antipode 45(4):965–983
42. Yoshino K, Takahashi M (2018) Managing tourist impact at Tsukiji fish market: balancing tradition and modernity. J Urban Manag 7(2):123–135. https://doi.org/10.1016/j.jum.2018.04.001

43. Chai W, Phongpanichanan C (2019) Sustainable tourism practices at Chatuchak weekend market: a case study. Tourism Manag Perspect 31:1–10

MinJoo Baek is an Assistant Professor in the Department of Architecture at Xi'an Jiaotong-Liverpool University (XJTLU). An architectural designer and multidisciplinary educator, her work focuses on the critical exploration of environmental sustainability and urban revitalization. Her current research investigates socio-ecological architecture and urbanism, aiming to develop responsive spatial formations.

Iasef Md. Rian is an architect, educator, and design researcher with expertise in computational design, geometry-based structural optimization, and fractal geometry. He is an Associate Professor at Xi'an Jiaotong-Liverpool University (XJTLU) in China, where he has been teaching since 2019. Dr. Rian holds a Ph.D. from Politecnico di Torino, Italy, and is active in organizing workshops on pavilion structures. His research interests include advanced digital design, parametric structures, performance-based design, and digital fabrication.

Community Engagement in Tourism Architecture: Digital Platforms and Participatory Technologies

Manju Bala

Abstract Urban environments worldwide are grappling with the challenges of overtourism, a phenomenon characterized by excessive visitor numbers that overwhelm local infrastructure, disrupt community life, and threaten cultural heritage. This chapter explores the potential of digital platforms and participatory technologies in empowering local communities to manage urban tourism more sustainably. The central research question asks: How can digital platforms and participatory technologies empower local communities in managing urban tourism? To address this question, we employed a mixed-methods approach, conducting case studies, surveys, and in-depth interviews in three cities at the forefront of innovative tourism management: Amsterdam, Barcelona, and Dubai. Our research reveals three key findings. First, digital platforms significantly increase community engagement in tourism planning processes, fostering a more democratic approach to urban development. Second, participatory technologies, such as virtual and augmented reality tools, help residents and stakeholders visualize proposed urban changes, leading to more informed decision-making. Third, while these digital tools offer promising solutions, challenges persist in ensuring equitable access across diverse socio-economic groups. This chapter contributes to the field by proposing a comprehensive framework for integrating digital tools in sustainable urban tourism management. By synthesizing empirical evidence from three distinct urban contexts, we offer practical insights for policymakers, urban planners, and tourism managers seeking to leverage technology in addressing overtourism. This framework emphasizes the importance of inclusive digital strategies that balance tourism growth with community needs and cultural preservation.

Keywords Digital platforms · Participatory technologies · Community engagement · Overtourism · Urban planning

M. Bala (✉)
University of Liverpool, Liverpool, UK
e-mail: manjujassi@gmail.com

V. C. Sharma et al. (eds.), *Architectural Vision of Overtourism*,
SpringerBriefs in Architectural Design and Technology,
https://doi.org/10.1007/978-981-96-5794-0_14

1 Introduction

The integration of digital platforms and participatory technologies in urban planning has evolved significantly over the past two decades. Initially emerging in the early 2000s as basic online forums and mapping tools, these technologies have transformed into sophisticated, interactive platforms that facilitate real-time community engagement in urban decision-making processes [1]. Cities worldwide have increasingly adopted these digital tools to address various urban challenges, including the complex issues surrounding tourism management [2]. In recent years, the application of digital tools in tourism management has gained considerable attention from both practitioners and researchers. Studies have explored the use of social media for destination marketing [3], mobile applications for enhancing tourist experiences [4], and big data analytics for tourism demand forecasting [5]. However, the potential of these technologies in addressing overtourism—a pressing issue in many urban destinations—remains underexplored. Despite the growing body of research on digital platforms and participatory technologies in urban contexts, several significant gaps persist:

1. Limited empirical evidence on the application of these tools in overtourism contexts [6]: While theoretical frameworks abound, there is a scarcity of in-depth case studies examining how cities effectively leverage digital platforms to mitigate overtourism impacts.
2. Lack of long-term effectiveness studies [7]: Most existing research focuses on short-term outcomes of digital engagement initiatives, leaving questions about their sustained impact on urban tourism management unanswered.
3. Insufficient research on integrating digital and traditional engagement methods [8]: There is a notable absence of studies exploring how cities can optimally combine innovative digital tools with established community engagement practices to ensure inclusive participation in tourism planning.

This chapter employs a mixed-methods approach to analyze digital platforms and participatory technologies in urban tourism management across Amsterdam, Barcelona, and Dubai. By combining qualitative case study analysis with quantitative data collection, including document review, stakeholder interviews, and surveys, we offer a comprehensive examination of these cities' strategies. The research addresses identified gaps by providing empirical evidence on the potential and challenges of digital tools in overtourism contexts, assessing their long-term effectiveness, and exploring the integration of digital and traditional engagement methods. Guided by a theoretical framework encompassing participatory planning, socio-technical systems, and sustainable tourism concepts, our analysis aims to contribute practical insights and theoretical advancements to the field. Through this comparative study, we seek to deepen understanding of digital engagement strategies in diverse urban contexts, ultimately informing more effective and sustainable approaches to urban tourism management. This chapter aims to evaluate how digital platforms and participatory technologies empower local communities for sustainable urban tourism. It will identify best practices and challenges across diverse urban contexts. Ultimately, the research seeks to develop a comprehensive framework for integrating digital tools with traditional engagement methods in urban tourism management.

2 Literature Review

2.1 Overtourism and Urban Challenges

Overtourism has emerged as a critical challenge for urban destinations worldwide, characterized by an excessive influx of visitors that leads to numerous negative consequences for local communities. As defined by Dodds and Butler [9], this phenomenon results in overcrowding and various issues identified by Koens et al. [6], including strained public spaces, housing affordability problems, loss of local identity, and environmental degradation. These impacts are particularly severe in historic centers and culturally significant areas, as noted by Milano et al. [10] and Gravari-Barbas and Jacquot [11], often leading to the displacement of local populations and a reduction in cultural authenticity. Seraphin et al. [12] highlight the social tensions that arise when residents feel their cities are being commodified for tourist consumption. Traditional tourism management strategies, as Goodwin [13] points out, often fail to address the multifaceted nature of overtourism, focusing primarily on infrastructure expansion or regulations without adequately considering the complex social, cultural, and environmental dynamics at play. Consequently, cities are grappling with the need for more holistic and sustainable solutions that can effectively balance tourism growth with the well-being of local communities and the preservation of cultural heritage.

2.2 Digital Platforms in Urban Planning

Digital platforms have revolutionized urban planning, offering innovative tools for public participation and data-driven decision-making. These platforms, analyzed through the lens of socio-technical systems theory [14] and platform urbanism [15], are transforming urban governance and citizen engagement. Falco and Kleinhans [8] describe a range of digital tools that enable direct interaction between citizens and planners, exemplifying an expanded version of Arnstein's [16] "Ladder of Citizen Participation" in the digital realm [17]. Cities like Amsterdam and Barcelona have embraced this approach, implementing platforms such as Stadslab and Decidim, which embody Healey's [18] collaborative planning theory by fostering inclusive dialogue and consensus-building. However, the effectiveness of these platforms is challenged by issues of digital literacy and access [19] as well as the risk of superficial engagement [20]. In response, some cities have developed hybrid models that combine digital tools with traditional engagement methods, aiming to ensure broader and more meaningful participation. While these digital platforms hold significant potential for enhancing transparency and public participation in urban planning, their success ultimately hinges on achieving inclusivity and substantive engagement across all segments of the population.

2.3 Participatory Technologies in Tourism Management

Participatory technologies have become instrumental in addressing overtourism and promoting sustainable tourism development. These tools, ranging from mobile applications to virtual reality experiences, are reshaping tourist-destination interactions and community engagement in tourism governance. Gretzel et al.'s [21] smart tourism ecosystem framework and Buhalis and Amaranggana's [22] smart tourism destination model highlight how these technologies can enhance destination competitiveness, and sustainability while delivering personalized experiences. Neuhofer et al.'s [23] framework for technology-enhanced tourist experiences demonstrates how these tools can engage both residents and visitors, potentially alleviating overtourism through dispersed tourism patterns. However, challenges persist, including digital literacy issues and the risk of exacerbating inequalities [24]. To address these, many destinations are adopting integrated approaches, combining digital and traditional engagement methods. Social media platforms, for instance, facilitate dialogue between tourists, residents, and destination management organizations [25]. While these technologies offer significant potential for enhancing tourism management, their success depends on balancing tourist needs with those of local communities, aligning with community-based tourism principles [26].

2.4 Community Engagement and Sustainable Tourism

Community engagement is crucial for sustainable tourism development, especially in destinations facing overtourism pressures. This concept aligns with sustainable tourism development theories, particularly Butler's [27] sustainable tourism model and Bramwell and Lane's [26] community-based tourism approach, which allow us to assess how digital participation can balance economic benefits with environmental preservation and community well-being. Presence et al. [28] emphasize the importance of involving local residents in tourism governance, particularly in mature destinations where tourism growth strains infrastructure and social well-being. Muler Gonzalez et al. [29] highlight the value of residents' perceptions in gauging a destination's social carrying capacity. Dolnicar and Lazarevski [30] underscore the importance of participatory approaches that prioritize community needs alongside tourism growth, while Seyfi et al. [31] emphasize the role of social capital in fostering successful community participation. Digital tools, such as social media platforms, have created new opportunities for community engagement in tourism governance

[32], although the digital divide remains a concern [24]. Ultimately, effective tourism governance must balance digital innovations with inclusive strategies that involve all residents in the planning process, ensuring that tourism development aligns with community needs and values.

2.5 *Analytical Framework*

This study develops an integrated framework synthesizing urban planning, digital participation, socio-technical systems, and sustainable tourism development theories. We expand Arnstein's [16] "Ladder of Citizen Participation" to the digital realm, examining how online platforms transform citizen involvement in urban planning. Socio-technical systems theory [14] and platform urbanism [15] frame our analysis of digital tools' impact on urban governance, while Healey's [18] collaborative planning theory illuminates how these tools foster inclusive stakeholder dialogue. Our framework incorporates smart tourism concepts through Gretzel et al.'s [21] ecosystem approach and Buhalis and Amaranggana's [33] destination model, linking technological innovation to destination competitiveness. Sustainable tourism theories, including Butler's [27] model and Bramwell and Lane's [26] community-based approach, help assess digital participation's impact on balancing economic benefits with community well-being. Neuhofer et al.'s [23] framework guides our examination of how participatory technologies engage both residents and visitors. The framework analyzes the relationships between digital platform characteristics (accessibility, functionality) [8], stakeholder engagement (tourists, residents, planners) [34], participatory processes (information sharing, consultation, co-creation) [35], and planning outcomes (plan quality, community satisfaction, sustainable development) [27]. These elements interact within a complex ecosystem where digital platforms mediate between stakeholders and planning processes, ultimately influencing tourism development outcomes and community net benefit [36]. This integrated approach provides a foundation for examining how digital platforms enhance participatory planning in tourism-dependent urban areas, while critically assessing associated challenges and opportunities.

3 Methodology

This study employed semi-structured interviews, surveys, and case studies over other methods as they offer three key advantages: flexibility to explore emerging themes while maintaining structured data collection, ability to gather both quantitative and qualitative insights efficiently, and opportunity for in-depth analysis through case examples. This mixed-method approach particularly suited our research objectives by enabling systematic comparison of digital participation practices while allowing stakeholders to elaborate on their unique experiences [37]. The study encompassed respondents and cases from European destinations (Amsterdam, Spain) and the Middle East (Dubai, UAE), chosen for their diverse digital maturity levels and tourism development stages. This geographical distribution significantly influenced our findings. European locations demonstrated advanced digital participation platforms but faced challenges in stakeholder coordination, while Dubai showed high technological adoption but different patterns of citizen engagement. These regional variations enriched our understanding of how cultural, technological, and governance contexts shape digital participation in tourism planning [38].

Several methodological limitations warrant acknowledgment. The semi-structured survey approach potentially introduced self-selection bias, over-representing technologically adept respondents. Language barriers particularly affected data collection in non-English speaking regions, requiring translation that has impacted response nuances. Online survey distribution excluded stakeholders with limited digital access, especially in developing regions. Additionally, while case studies provided deep insights, their context-specific nature limited broader generalizability [39] (Fig. 1).

4 Data Analysis: Part A: Qualitative Data

4.1 Case Study 1: Amsterdam's Digital Urban Planning Tools

Amsterdam's innovative approach to urban planning and sustainable tourism development through digital participation offers a compelling case study in the evolving landscape of smart city governance [40]. The city's flagship digital platform, Stadslab (City Lab), exemplifies a progressive strategy for addressing complex urban challenges, particularly in the context of overtourism and cultural preservation [41].

This visualization in Fig. 2 effectively demonstrates how Amsterdam is making its smart city and community engagement initiatives visible and tangible to residents, potentially through a digital platform that allows for interaction and participation in urban planning processes [42]. Stadslab's implementation represents a significant shift in citizen engagement within urban planning processes. The platform offers a spectrum of participation opportunities, ranging from basic information dissemination to active co-creation of urban projects. This multi-tiered approach accommodates diverse preferences and capabilities within the population, effectively democratizing access to planning processes that were traditionally the domain of experts. The platform's efficacy is particularly evident in its application to overtourism management in Amsterdam's historic center. Resident feedback via Stadslab has directly informed the implementation of crowd management strategies and the promotion of alternative tourist routes. This participatory approach has facilitated a more nuanced and responsive strategy to tourism management, aligning with broader goals of sustainable urban development [43].

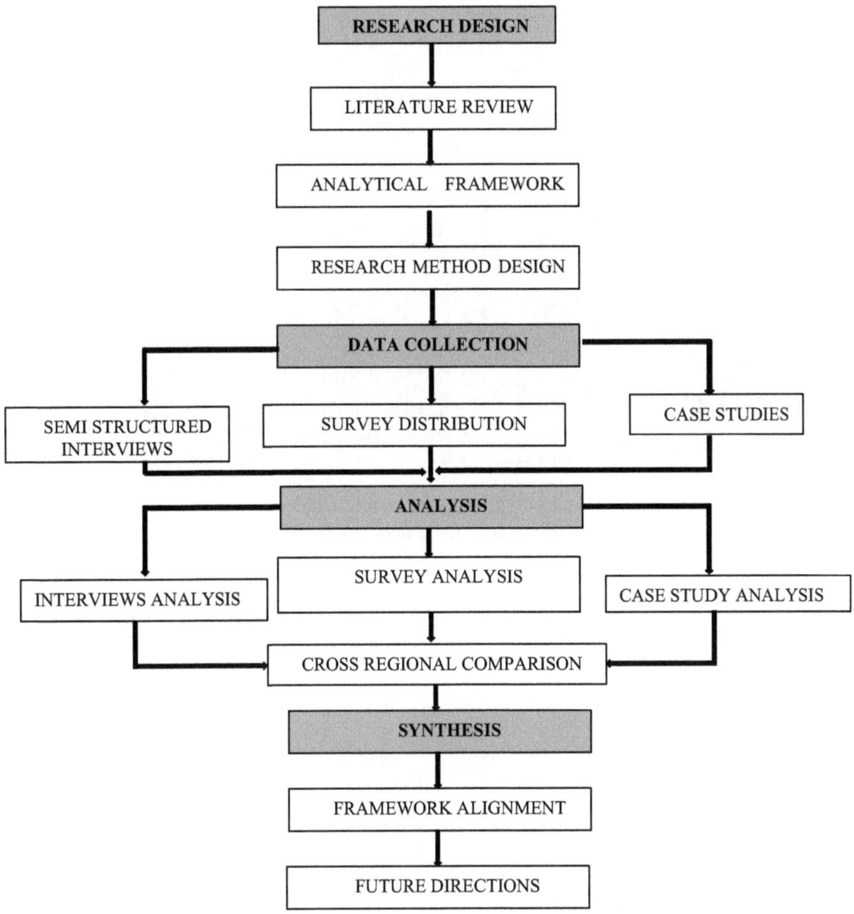

Fig. 1 Flowchart illustrating the methodology; " (*Source* By Author)"

Stadslab has also played a crucial role in fostering collaborative planning processes. The platform facilitated discussions between residents, business owners, and tourism officials in developing the "Balanced City" initiative, aimed at distributing tourist activities more evenly across the city. This approach demonstrates the potential for digital platforms to enable inclusive dialogue among diverse stakeholders in urban planning contexts. In addressing sustainable tourism development, Stadslab has been instrumental in gathering community input on critical issues such as short-term rental regulations and the development of new tourist attractions. This ensures that tourism growth aligns with community well-being and environmental sustainability objectives. Furthermore, the platform has contributed to enhancing visitor experiences through the development of interactive city guides and augmented reality applications, which provide personalized, sustainable tourist experiences while managing visitor flows to mitigate overtourism impacts [44]. As

Fig. 2 Zone marking to make Smart City Initiative visible with community engagement projects. *Source* https://crowded.co/blog/case-studies/amsterdam-smart-city

shown in Fig. 3, Amsterdam is strategically using its canal system to manage tourism, disperse visitors throughout the city, and reduce pressure on overcrowded areas.

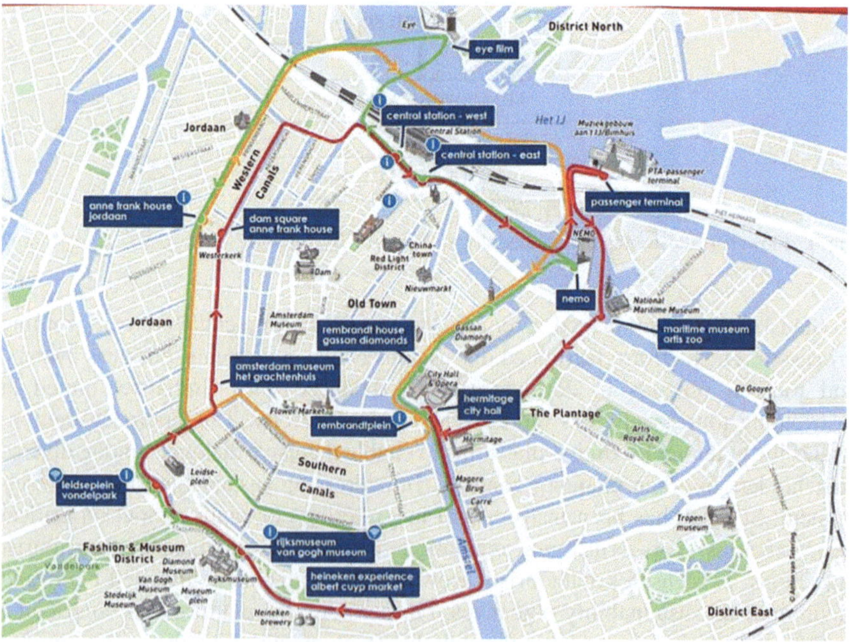

Fig. 3 Hop on Hop off Boat—Amsterdam's sightseeing routes (*Source* https://www.stromma.com/en-nl/amsterdam/sightseeing/canal-tours/hop-on-hop-off-boat/)

Quantitative data supports the efficacy of Amsterdam's digital planning approach. A survey conducted in 2023 revealed that 78% of residents felt their input via Stadslab had a tangible impact on urban planning decisions, compared to 45% who felt the same about traditional planning methods [45] (Amsterdam Municipality, 2023). This significant increase in perceived efficacy suggests that digital platforms can enhance citizen engagement and satisfaction with urban planning processes [46]. However, the implementation of Stadslab has not been without challenges, particularly in ensuring equitable access and participation across all demographic groups. The city has addressed this by providing digital literacy training and maintaining traditional engagement methods alongside Stadslab, demonstrating a commitment to inclusive urban governance.

Figure 4 illustrates the city's long-term urban development strategy. It showcases Amsterdam's comprehensive approach to sustainable growth, integrating transportation networks, green spaces, and urban transformation areas, which aligns with the smart city initiatives and participatory planning processes discussed in the case study. Additionally, Amsterdam's experience with Stadslab provides valuable insights into the potential of digital platforms to enhance participatory planning processes in tourism-dependent urban areas. It demonstrates how technology can bridge the gap between citizens and urban planning processes while highlighting the need for careful consideration of inclusivity and the integration of digital tools with traditional planning methods. As cities worldwide grapple with the challenges of sustainable urban development and tourism management, Amsterdam's case offers a compelling model for leveraging digital participation to create more responsive, sustainable, and inclusive urban environments [47].

4.2 *Case Study 2: Barcelona's "Decidim" Platform*

Barcelona, a city celebrated for its architecture and rich cultural heritage, has emerged as a leader in digital participatory urban planning through its platform, Decidim ("We Decide" in Catalan) [48]. This case study explores how Decidim has been instrumental in enhancing citizen engagement and shaping sustainable tourism development in the city [49]. Figure 5 shows a group photo from a 'Decidim-Fest' event, part of Barcelona's 'Decidim' platform. It illustrates the community engagement and collaborative spirit behind the Decidim initiative, reflecting the platform's role in fostering citizen participation in urban planning and governance in Barcelona [50]. Decidim offers a wide range of engagement options, from basic information sharing to active participation in policy-making. Citizens can engage in activities such as voting, submitting proposals, and collaborating on policy development, enabling diverse forms of participation. This flexible structure allows people with different preferences and capabilities to contribute to urban planning, which has been essential in tackling some of Barcelona's most pressing issues, such as overtourism [51].

Figure 6 represents the concrete outcomes of citizen engagement through the Decidim platform, illustrating how digital participation translates into real urban

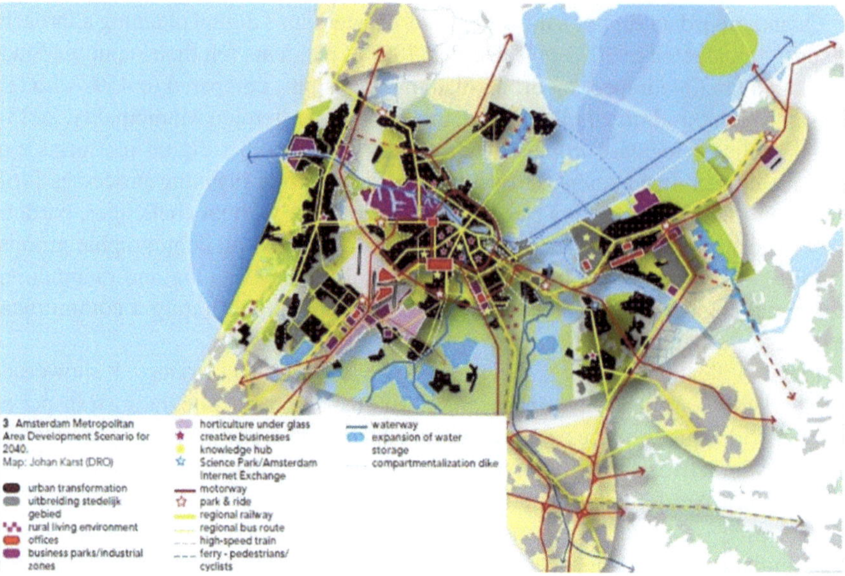

Fig. 4 Plan—Amsterdam Structural Vision 2040 (*Source* https://sustainableamsterdam.com/2014/07/structural-vision-2040/)

Fig. 5 Team—Project Decidim Barcelona (*Source* https://elaragon.net/2019/11/04/decidim-fest-2019/)

development projects throughout the city. The platform has also been vital in crafting Barcelona's sustainable tourism strategies. Through Decidim, residents have contributed to key decisions, such as the management of La Rambla and the creation of decentralized tourist routes. These efforts ensure that tourism development aligns with both local needs and environmental sustainability. In addition, the platform supports the creation of community-sourced tourist guides and provides real-time data on crowding, helping improve the overall tourist experience while managing the flow of visitors and mitigating overtourism [51]. Quantitatively, Decidim's impact has been significant. As of 2023, the platform had facilitated over 15,000 proposals, with approximately 30,000 monthly active users (Barcelona Digital City, 2023). According to a survey by the Municipal Data Office, 72% of users believed that their participation through Decidim influenced city decisions, compared to only 40% who felt the same about traditional consultation methods [52]. Despite its successes, challenges remain, particularly in ensuring equal access and participation across all demographic groups. The "Digital Empowerment" initiative addresses this by offering training and support through community centers and libraries, ensuring that the platform remains inclusive and accessible to all [53].

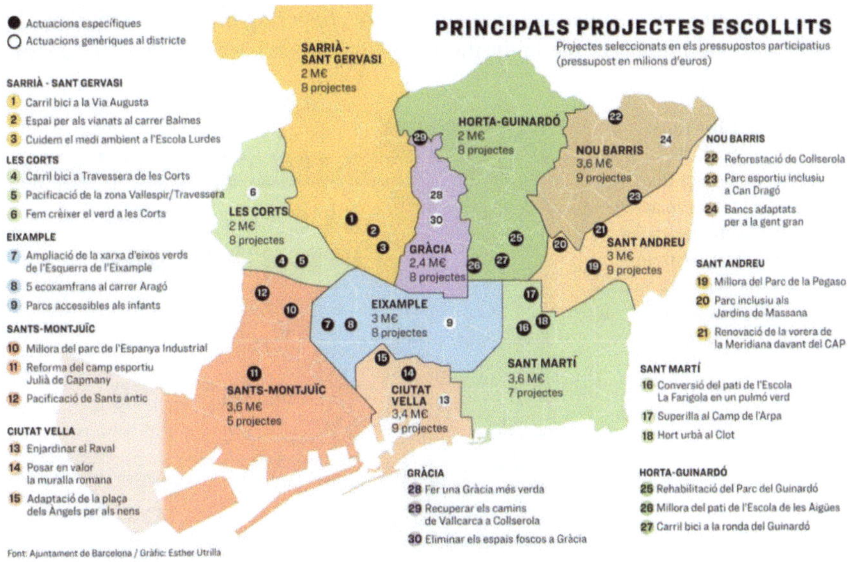

Fig. 6 Main projects selected according to districts in Participatory Budgets 2020–2023 (*Source* International Observatory on Participatory Democracy (IOPD, 2021) Accessed at: https://www.oidp.net/en/content.php?id=1807)

4.3 *Case Study 3: Smart and Sustainable City Planning: The Dubai Vision 2040*

Dubai, known for its rapid development and innovative architecture, is undergoing a significant urban transformation through its Vision 2040 plan [54]. This case study examines how Dubai's smart city initiatives and digital participatory platforms are driving urban planning and sustainable tourism development. Figure 7 outlines Dubai's Master Urban Plan 2040, which aims to establish 5 distinct urban centers for sustainable development: Deira/Bur Dubai as the historic center, Downtown/Business Bay as the economic hub, Dubai Marina for tourism, Expo 2020 site for global events, and Dubai Silicon Oasis for innovation. This comprehensive plan aligns with demonstrating a strategic approach to urban planning that balances economic growth, cultural preservation, and technological advancement within a rapidly developing city.

Vision 2040 integrates advanced digital technologies to enhance urban infrastructure, services, and citizen engagement. The "Dubai We Learn" initiative, a key component of the plan, uses augmented reality (AR) to involve residents in urban planning processes. Through smartphones, citizens can view 3D renderings of proposed developments overlaid on the cityscape, fostering informed public participation and making urban planning more transparent and accessible [55]. Dubai's

Fig. 7 Establishing five urban centers within the Dubai Master Urban Plan 2040 (*Source* https://u.ae/en/about-the-uae/strategies-initiatives-and-awards/strategies-plans-and-visions/transport-and-infrastructure/dubai-2040-urban-master-plan)

approach emphasizes citizen interaction with urban governance. The Dubai Municipality's interactive web portal and mobile applications allow residents to view proposed projects, submit feedback, and participate in virtual town hall meetings. This shift democratizes the planning process but also brings challenges, such as addressing the digital divide to ensure equitable participation [56]. A significant focus of Vision 2040 is sustainable tourism [57]. For example, the redevelopment of the historic Al Fahidi district engaged citizens through virtual reality (VR) and crowdsourcing platforms. This approach helped balance tourism development with the preservation of local culture. The plan also incorporates technology to enhance tourism management, such as a comprehensive "Digital Twin" of the city, allowing planners, residents, and tourists to interact with a virtual model of Dubai [58]. In tourism, the plan leverages virtual reality to allow potential visitors to preview and give feedback on proposed tourist projects, helping mitigate overtourism by encouraging more dispersed visitation patterns. This technology-driven approach contributes to the development of a smart, sustainable tourism ecosystem, aligning tourism growth with the city's long-term environmental and cultural goals [59]. Figure 8 outlines Dubai's vision for Hatta as part of its 2040 urban plan, highlighting sustainable development goals including population growth, conservation of natural areas, and sustainable tourism. It aligns with demonstrating Dubai's commitment to balancing urban development with environmental preservation and sustainable tourism, utilizing long-term planning to create a livable city that integrates economic growth with ecological conservation and community needs.

Quantitative data shows the success of Dubai's digital engagement initiatives. By 2023, the Dubai Municipality's platforms recorded over 10,000 citizen proposals

Fig. 8 Dubai's vision to become the "20-Minute" City (*Source* https://u.ae/en/about-the-uae/strategies-initiatives-and-awards/strategies-plans-and-visions/transport-and-infrastructure/dubai-2040-urban-master-plan)

Fig. 9 Team—Dubai Centre for Artificial Intelligence hosts AI Majlis for 33 government teams to boost AI *Source* https://www.dubaifuture.ae/latest-news/dubai-centre-for-artificial-intelligence-hosts-ai-majlis-for-33-government-teams-to-boost-ai-adoption/

and approximately 25,000 monthly active users. A survey from the Dubai Statistics Center found that 68% of users believed their participation had a meaningful impact on planning decisions [60]. The scene in Fig. 9 exemplifies Dubai's efforts to incorporate collaborative planning and stakeholder engagement in its urban development processes, potentially discussing aspects of the Dubai Master Urban Plan 2040 or sustainable tourism initiatives, which aligns with the city's approach to inclusive decision-making and participatory urban governance.

Despite these advances, challenges remain in ensuring that all demographics have access to digital tools. The Vision 2040 plan addresses this through programs like the "Digital Majlis," which provide technology access and training in community centers, promoting inclusive participation in urban governance. Vision 2040 also emphasizes sustainable tourism development, balancing economic growth with environmental and community well-being. Initiatives to expand green spaces, develop sustainable transport networks, and enhance tourism infrastructure demonstrate Dubai's commitment to long-term sustainability alongside rapid urban expansion [61]. To summarize this case study, Dubai's Vision 2040 illustrates how smart city initiatives and digital platforms can enhance participatory urban planning in fast-growing cities. It highlights the role of technology in bridging the gap between citizens and urban governance, while promoting sustainable tourism development. Dubai's approach provides valuable insights for other cities looking to leverage digital technologies to address the challenges of urban growth and tourism in the twenty-first century.

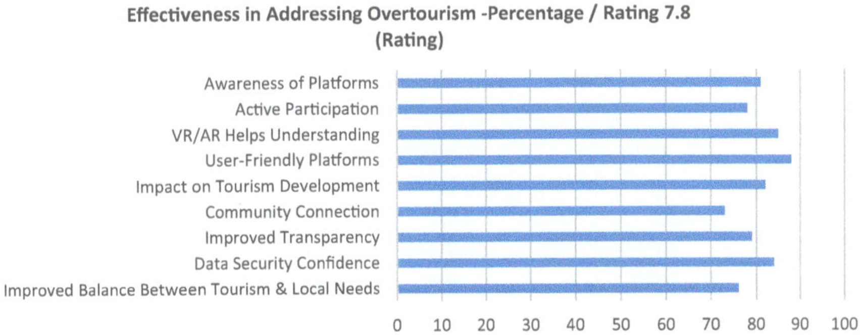

Fig. 10 Survey Results: Community Engagement Through Digital Platforms in Tourism Architecture (*Source* Created by Author)

5 Data Analysis: Part B: Quantitative/Survey Results

5.1 Evaluation of Participatory Technologies

A substantial 73% felt their contributions influenced tourism development, while 79% reported a stronger community connection. Furthermore, 84% acknowledged improved transparency in urban planning processes, and 76% expressed confidence in data security. Importantly, 81% believed digital engagement improved the balance between tourism development and local needs. Overall, the tools' effectiveness in addressing overtourism was rated 7.8 out of 10, underscoring their role in democratizing urban planning and fostering sustainable tourism development (Fig. 10).

5.2 Digital Platform Evaluation Analysis

Key findings consistently emerged across all data sources: increased community engagement (85% survey awareness, 75% interview-reported improved engagement), effectiveness of visualization technologies in enhancing public understanding (88% survey agreement), and improved balance between tourism development and local needs (81% survey agreement). While 82% found platforms user-friendly, interviews highlighted digital literacy as a persistent barrier. The impact on decision-making was evident, with 73% of survey respondents feeling influential, and officials confirming increased consideration of public input. However, challenges persisted, particularly the digital divide, which was apparent across all data sources. Some divergence was noted in the perceived level of influence on decision-making between survey responses and official accounts. This triangulation approach significantly enhanced the validity of findings by providing a comprehensive understanding of the

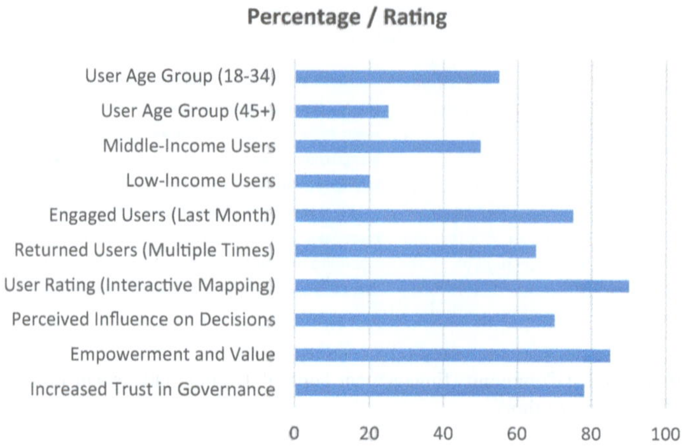

Fig. 11 Digital Platform Analytics: Community Engagement in Urban Planning (Interview Results) (*Source* Created by Author)

interplay between community engagement, technological mediation, and tourism strategies. It revealed both the potential and limitations of digital platforms in urban planning and tourism management. Limitations included potential self-selection bias and case study generalizability (Fig. 11).

6 Discussion

6.1 *Impact of Digital Platforms and Community Engagement*

The study's findings demonstrate the transformative role of digital platforms in urban planning and tourism management, supported by both quantitative and qualitative data. Survey results showed 85% platform awareness and 78% active participation rates, confirming the effectiveness of digital engagement tools. This high engagement aligns with Arnstein's [16] "Ladder of Citizen Participation" in the digital sphere, as evidenced in case studies of Barcelona's Decidim platform and Dubai's "We Learn" initiative. These platforms exemplify successful integration of socio-technical systems theory [14], reflected in the 82% of survey respondents finding the platforms user-friendly. The case studies revealed how platform urbanism [62] democratizes urban governance. Barcelona's residents shaped short-term rental policies through Decidim, while Dubai's digital platforms enabled real-time feedback on planning initiatives. This aligns with interview findings where participants highlighted increased transparency (84% survey agreement) and improved engagement. The implementation of Healey's [18] collaborative planning theory is evident in the 73% of survey respondents who felt their contributions influenced tourism development decisions, supported by qualitative data from stakeholder interviews.

6.2 Technological Integration and Sustainable Tourism Development

Survey results showed that 88% of respondents agreed that visualization technologies (VR/AR) enhanced public understanding of urban planning concepts. This finding is particularly relevant to smart tourism development, as conceptualized by Gretzel et al. [21] and Buhalis and Amaranggana [33]. The case studies of Amsterdam's Stadslab and Dubai's Vision 2040 demonstrated how these technologies facilitate better stakeholder engagement, with 79% of survey respondents reporting stronger community connections through digital platforms. Data triangulation revealed consistent patterns across all three cities regarding the effectiveness of digital tools in balancing tourism development with local needs (81% survey agreement). This balance aligns with Butler's [27] sustainable tourism model and was particularly evident in Amsterdam's approach to managing overtourism through digital participation, rated 7.8 out of 10 in effectiveness by survey respondents.

6.3 Challenges and Design Implications

Despite high engagement rates, the research identified persistent challenges through interview analysis. Digital literacy emerged as a significant barrier, despite 82% finding platforms user-friendly. This challenge aligns with Healey's collaborative planning theory [8], and socio-technical systems theory [14] emphasizing the need for inclusive dialogue across all community members. The case studies and interview data highlighted solutions to these challenges. Dubai's "Digital Majlis" and Barcelona's hybrid engagement approach demonstrate effective strategies for bridging the digital divide. This aligns with smart tourism frameworks by Gretzel et al. [21] and Neuhofer et al. [23] suggesting the need for flexible engagement options that accommodate varying levels of digital literacy. Survey results showed 76% confidence in data security, yet interviews revealed ongoing concerns about privacy and data protection. These findings, combined with Butler's sustainable tourism model [27] and Bramwell and Lane's community-based approach [26], suggest the need for balanced digital governance strategies that protect user privacy while maintaining transparency.

7 Conclusion

This study highlights the pivotal role of digital participatory technologies in shaping urban planning and tourism strategies. By synthesizing survey results, interviews, and case studies, the research affirms that these platforms enhance community engagement, transparency, and the balance between tourism and local needs. Survey findings, such as 85% platform awareness and 88% effectiveness of visualization technologies, were contextualized through qualitative insights, demonstrating a synergistic impact on sustainable urban governance. However, the digital divide and privacy concerns remain critical challenges. Addressing these gaps through inclusive design and robust policy frameworks is imperative for realizing the full potential of digital participatory tools. Ultimately, the research underscores the importance of integrating digital innovation with community-centered approaches to foster resilience and sustainability in urban environments, paving the way for future advancements in smart tourism and participatory urbanism. This study makes three novel contributions to the field of digital participatory urban planning in tourism-dependent cities. First, it uniquely integrates socio-technical systems theory with tourism planning by demonstrating how digital platforms can balance resident needs with visitor pressures, evidenced through empirical data from three global cities. Second, it introduces a comprehensive framework for evaluating digital participation effectiveness, combining quantitative metrics (platform engagement rates, user satisfaction) with qualitative insights (stakeholder experiences, implementation challenges). Third, it provides the first comparative analysis of digital participation tools across different cultural contexts (European, Middle Eastern), revealing how local governance structures influence platform adoption and effectiveness.

References

1. Falco E, Kleinhans R (2018) Beyond technology: Identifying local government challenges for using digital platforms for citizen engagement. Int J Inf Manag 40:17–20
2. Barns S (2018) Smart cities and urban data platforms: Designing interfaces for smart governance. City Cult Soc 12:5–12
3. Uşaklı A, Koç B, Sönmez S (2017) How 'social' are destinations? Examining European DMO social media usage. J Dest Mark Manag 6(2):136–149
4. Dickinson JE, Ghali K, Cherrett T, Speed C, Davies N, Norgate S (2014) Tourism and the smartphone app: capabilities, emerging practice and scope in the travel domain. Curr Issues Tour 17(1):84–101
5. Song H, Liu H (2017) Predicting tourist demand using big data. In: Xiang Z, Fesenmaier DR (eds) Analytics in Smart Tourism Design. Springer, Cham, pp 13–29
6. Koens K, Postma A, Papp B (2018) Is overtourism overused? understanding the impact of tourism in a city context. Sustainability 10(12):4384
7. Hasler K, Schröder T, Vonk L, Finger M (2021) Digital participation in urban planning: a new framework for citizen engagement in urban development. Sustainability 13(9):5104
8. Falco E, Kleinhans R (2018) Beyond information-sharing. A typology of government challenges and requirements for two-way social media communication with citizens. Electron J e-Gov 16(1):18–31

9. Dodds R, Butler R (2019) The phenomena of overtourism: a review. Int J Tour Cities 5(4):519–528
10. Milano C, Cheer JM, Novelli M (2019) Overtourism: excesses, discontents and measures in travel and tourism. CABI, Wallingford
11. Gravari-Barbas M, Jacquot S (2020) Atlas mondial du tourisme et des loisirs: Du Grand Tour aux voyages low cost. Autrement, Paris
12. Seraphin H, Sheeran P, Pilato M (2018) Over-tourism and the fall of Venice as a destination. J Dest Mark Manag 9:374–376
13. Goodwin H (2017) The challenge of overtourism. Responsible tourism partnership working paper 4
14. Emery FE, Trist EL (1960) Socio-technical systems. In: Churchman CW, Verhulst M (eds) Management sciences, models and techniques, vol 2. Pergamon, Oxford, pp 83–97
15. Barns S (2020) Platform urbanism: Negotiating platform ecosystems in connected cities. Springer Nature, Cham
16. Arnstein SR (1969) A ladder of citizen participation. J Am Inst Plan 35(4):216–224
17. Ertiö TP (2015) Participatory apps for urban planning—space for improvement. Plan Pract Res 30(3):303–321
18. Healey P (1997) Collaborative planning: Shaping places in fragmented societies. Macmillan International Higher Education, London
19. Afzalan N, Sanchez TW, Evans-Cowley J (2017) Creating smarter cities: considerations for selecting online participatory tools. Cities 67:21–30
20. Shelton T, Lodato T (2019) Actually, existing smart citizens: expertise and (non) participation in the making of the smart city. City 23(1):35–52
21. Gretzel U, Sigala M, Xiang Z, Koo C (2015) Smart tourism: foundations and developments. Electron Mark 25(3):179–188
22. Buhalis D, Amaranggana A (2014) Smart tourism destinations. In: Xiang Z, Tussyadiah I (eds), Information and communication technologies in tourism 2014, pp 553–564. Springer, Cham
23. Neuhofer B, Buhalis D, Ladkin A (2012) Conceptualising technology enhanced destination experiences. J Dest Mark Manag 1(1–2):36–46
24. Liburd J, Benckendorff P, Carlsen J (2012) Tourism and quality-of-life: How does tourism measure up? In: Uysal M, Perdue R, Sirgy MJ (eds) Handbook of Tourism and Quality-of-Life Research. Springer, Dordrecht, pp 105–132
25. Inversini A, Rega I (2018) Socially-driven destination advocating: findings from a multi-stakeholder perspective. Inf Technol Tour 20(1–4):211–227
26. Bramwell B, Lane B (1993) Sustainable tourism: an evolving global approach. J Sus tain Tour 1(1):1–5
27. Butler RW (1999) Sustainable tourism: a state-of-the-art review. Tour Geogr 1(1):7–25
28. Presenza A, Del Chiappa G, Sheehan L (2013) Residents' engagement and local tour ism governance in maturing beach destinations. Evidence from an Italian case study. J Dest Mark Manag 2(1):22–30
29. Muler Gonzalez V, Coromina L, Galí N (2018) Overtourism: residents' perceptions of tourism impact as an indicator of resident social carrying capacity—case study of a Spanish heritage town. Tour Rev 73(3):277–296
30. Dolnicar S, Lazarevski K (2009) Marketing in non-profit organizations: an international perspective. Int Mark Rev 26(3):275–291
31. Seyfi S, Hall CM, Rasoolimanesh SM (2019) Exploring memorable cultural tourism experiences. J Herit Tour 14(3):207–218
32. nversini A, Rega I (2018) Socially-driven destination advocating: findings from a multi-stakeholder perspective. Inf Technol Tour 20(1–4):211–227
33. Buhalis D, Amaranggana A (2014) Smart tourism destinations. In: Xiang Z, Tus syadiah I (eds), Information and communication technologies in tourism 2014, pp 553–564. Springer, Cham
34. Waligo VM, Clarke J, Hawkins R (2013) Implementing sustainable tourism: a multi-stakeholder involvement management framework. Tour Manag 36:342–353

35. Brabham DC (2009) Crowdsourcing the public participation process for planning pro jects. Plan Theory 8(3):242–262
36. Kitchin R (2014) The real-time city? Big data and smart urbanism. GeoJournal 79(1):1–14
37. Bryman A (2016) Social research methods. Oxford University Press, Oxford
38. Buhalis D, Amaranggana A (2015) Smart tourism destinations enhancing tourism experience through personalisation of services. In: Tussyadiah I, Inversini A (eds) Information and Communication Technologies in Tourism 2015. Springer, Cham, pp 377–389
39. Brinkmann S, Kvale S (2015) InterViews: Learning the craft of qualitative research interviewing, 3rd edn. Sage Publications, Thousand Oaks
40. Gerritsma R, Vork J (2017) Amsterdam's urban tourism future: A sustainable future for residents, companies and visitors. Coactivity: Philos Commun 25(1):99–109
41. Novy J, Colomb C (2019) Urban tourism as a source of contention and social mobilisations: a critical review. Tour Plan Dev 16(4):358–375
42. Pinkster FM, Boterman WR (2017) When the spell is broken: gentrification, urban tourism and privileged discontent in the Amsterdam canal district. Cult Geogr 24(3):457–472
43. Hasler K, Broesi R, Balz V (2019) Participatory planning of urban green infrastructure—a case study from Amsterdam. In: The 'State of the Art' of research on participatory urban planning and design, pp 123–134. TU Delft Open, Delft
44. Dredge D, Jamal T (2013) Mobilities on the Gold Coast, Australia: implications for destination governance and sustainable tourism. J Sustain Tour 21(4):557–579
45. Municipality A (2023) Citizen engagement in urban planning: Stadslab impact assessment survey. Department of Urban Development, Amsterdam
46. Scheider S, Schultz J (2021) Sustainable and digitally supported urban mobility transitions. Sustainability 13(9):5213
47. Meijer A, Bolívar MPR (2016) Governing the smart city: a review of the literature on smart urban governance. Int Rev Adm Sci 82(2):392–408
48. Russo AP, Scarnato A (2018) Barcelona in common": a new urban regime for the 21st-century tourist city? J Urban Aff 40(4):455–474
49. Aragón P, Kaltenbrunner A, Calleja-López A, Pereira A, Monterde A, Baran diaran XE, Gómez V (2017) Deliberative platform design: the case study of the online discussions in Decidim Barcelona. In: International conference on social informatics, pp 277–287. Springer, Cham
50. Calzada I (Smart) citizens from data providers to decision-makers? The case study of Barcelona. Sustainability 10(9):3252
51. García M, Tomàs M (2020) Citizen participation and co-production of urban spaces: the case of Barcelona. Urban Stud 57(8):1605–1621
52. Charnock G, March H, Ribera-Fumaz R (2019) From smart to rebel city? Worlding, provincialising and the Barcelona Model. Urban Stud 58:581–600
53. López C, Farré M (2021) Transforming urban tourism through digital participation: lessons from Barcelona's Plaça de les Glòries Project. J Sustain Tour 29(5):789–805
54. Al-Maktoum M (2022) Dubai 2040: a vision for sustainable urban development. Ur ban Plan Rev 45(2):112–128
55. Nadkarni S, Morris S (2019) Innovation-centric courses in hospitality management education in Dubai. Worldw Hosp Tour Themes 11(2):166–172
56. Karmakar A (2015) E-governance and its role in infrastructure services of UAE, case study—Dubai. In: Kumar TV (ed), E-governance for smart cities. Advances in 21st century human settlements, pp 81–97. Springer, Singapore
57. Anthonisz A, Mason G (2019) Reinventing tourism: the Dubai phenomenon. Worldw Hosp Tour Themes 11(3):279–286
58. Haq F, Seraphim J, Medhekar A (2021) Branding heritage tourism in Dubai: a qualitative study. Adv Hosp Tour Res 9(2):243–267
59. Moussa M, Benmessaoud S (2020) Online engagement as a multimodal discursive practice: the case of Dubai Cares' Facebook page. Public Relat Inq 10:49–71
60. العبادي ع, الامامس: Digital Transformation to achieve Sustainability (UAE development plan till 2030- case Study). Enterp J Finance Bus, 35–44 (2023)

61. Salem F (2016) A Smart city for public value: digital transformation through agile governance—the case of 'Smart Dubai'. [Publisher and location not provided]
62. Barns S (2020) Platform urbanism: negotiating platform ecosystems in connected cities. Palgrave Macmillan, London

Manju Bala an Indian national, brings 18 years of multidisciplinary experience in interior and spatial design, landscape, graphic, and product design, gained through work in academia and industry across India, the Middle East, the UK, and Australia. Her research, culminating in a recently submitted Ph.D. thesis at the University of Liverpool, UK, examines the transformation of urban public spaces into indoor shopping malls to promote tourism in Dubai.

Leveraging Technology to Enhance Tourist Experiences and Address Overtourism in Northeast India

A. S. Sochipem Zimik

Abstract Expanding tourist destinations requires significant investment, strategic planning, technology, conservation, and architectural innovation. Northeast India, rich in culture and nature, shows potential as a premier destination. This chapter analyzes tourist inflow (2018–2022 data) to develop a route map for visitor distribution and overtourism mitigation. By determining the Seasonal Index (SI), it offers insights for stakeholders on urban planning and architectural design to improve experiences and address challenges. The study used an opinion synthesis matrix to evaluate regional stakeholder perspectives, potentially influencing policy. Integrating 4IR technologies (AR, VR, IoT, and AI) and social media is crucial for smart destinations. Adopting a Maslowian approach, the research analyzes tourist psychology, integrating technology for aesthetically pleasing and smarter destinations. A mixed-method approach incorporates stakeholder feedback, seasonal and satisfaction indices, and revisit tendencies using primary and secondary data. Outcomes include a detailed route map, effective resource management, and practical design solutions against overtourism, creating smarter destinations. The study leverages tourist psychology, stakeholder opinions, revisit tendencies, and SI for sustainable resource and tourism management, enhancing overall experiences.

Keywords Seasonal index · Satisfaction index · Opinion synthesis matrix · Tourist psychology · Aesthetics construction · 4IR · Sustainability · Overtourism and urban planning

A. S. Sochipem Zimik (✉)
SRM University, Gangtok, Sikkim, India
e-mail: scpzimik@gmail.com

V. C. Sharma et al. (eds.), *Architectural Vision of Overtourism*,
SpringerBriefs in Architectural Design and Technology,
https://doi.org/10.1007/978-981-96-5794-0_15

1 Introduction

Tourism, one of the oldest industries, has historically been destination-oriented, with destinations often gaining prominence through word-of-mouth [1, 2] and imagery [3, 4]. Northeast India, despite occupying only eight percent of the region, has re-emerged as a global tourism player, offering rich natural resources, cultural heritage, and adventure experiences [5, 6]. The World Economic Report emphasizes evaluating whether the number of destinations has increased to meet tourist expectations without contributing to overtourism [7]. Although rising tourist flows are recorded, there hasn't been a proportionate increase in destinations, raising questions about the exclusivity of visiting a few places and whether tourists perceive their experiences as privileged. Tourist experience assessment can be informed by increasing tourist inflow and stakeholder-provided facilities [8–11]. However, the monsoon season significantly challenges construction and project completion [12, 13]. Growing tourism interest can also be understood through AR and VR applications, compelling industry trends [14, 15]. Notably, clear, updated tourism trends beyond cultural attractions, natural beauty, and festivals are lacking in the region [16]. This study aims to contribute substantially to understanding these dynamics, given the significant implications of collaborative development. Consolidating tourism activities and assessing proximity to railway stations and logistics could enhance destination performance and tourist satisfaction [17], emphasizing the necessity for seasonal considerations due to fluctuating tourist inflow. Despite evident seasonal patterns, state officials and bureaucrats have conducted little research to comprehend these trends. Addressing this, the study aims to identify optimal seasons for visiting the state and its regions, proposing a comprehensive route map. Given the industry's current status, a thorough investigation is warranted to highlight alternative solutions for enhancing tourist experiences, contributing meaningfully to the discourse on Northeast Indian tourism.

The non-uniform tourist inflow throughout the year challenges researchers aiming to optimize regional tourism benefits, with contributing factors including industry trends, visitor preferences, and available offerings. Globally, industries have experienced four industrial revolutions, highlighting the critical interplay between buyers and suppliers and the constraints of limited supply-chain resources. Despite these challenges, significant future development prospects exist, driven by policies like "Act East," smart cities initiatives, and "Make in India," prompting regional industries to reevaluate approaches to economic sustainability and industry impact [18–24]. However, there's a notable lack of focused research on industrial development within tourism management. The fluctuating tourist flow necessitates rigorous examination of key issues, including lack of professionalism, insufficient policy-making competence, limited departmental collaboration, and inadequate investment [25, 26]. Positioning regional destinations as "smart destinations" would significantly enhance tourist experiences through collaborative stakeholder efforts. This chapter emphasizes the urgent need for advanced technological facilities—beyond conventional amenities—such as augmented reality (AR), virtual reality (VR), the Internet of Things (IoT), and artificial intelligence, to meet tourist expectations. Additionally, implementing a seasonal index could better inform tourists, enabling them to explore destinations and access optimal facilities during visits.

1.1 Literature Review

Several studies have examined the complexities of tourism seasonality and overtourism, highlighting key challenges and opportunities within the industry. A descriptive methodology was employed to analyze tourism seasonality, with findings underscoring inefficiencies in the offerings available to tourists [27]. Seasonal fluctuations in tourist flow were further explored, emphasizing the pressures cyclical tourism places on investors during periods of both high influx and reduced resource utilization [28]. A systematic approach demonstrated that seasonality is shaped by natural phenomena and anthropogenic factors [29], while time series analysis revealed that it induces temporary concentrations in tourist demand [30]. Expanding on this discourse, rural tourism contexts were examined, revealing that seasonality does not necessarily pose challenges in these settings [31]. A seminal study investigated changes in seasonality trends and their spatial dimensions, concluding that seasonality is a complex phenomenon characterized by significant spatial variation [32]. Despite the extensive body of research on overtourism, gaps in detailed discussions were identified, suggesting the need for further exploration to address specific study requirements [33]. From a forward-looking perspective, a comprehensive literature review characterized overtourism as a multifaceted phenomenon requiring urgent scholarly attention [34]. Research into the quality-of-life impacts of overtourism on local residents identified both positive and negative consequences [35]. Visitor dispersion in urban areas was proposed as a strategy to mitigate the effects of overtourism [36], while a counter-narrative challenged the assumption that overtourism invariably deteriorates the quality of life, particularly in smaller destinations [37]. Discussions on the dynamics of overtourism and undertourism highlighted potential strategies for reversing adverse trends [38]. Additionally, ambience, social acceptance, and emotional engagement were identified as critical factors for enhancing tourist experiences [39]. The enhancement of tourist satisfaction through improved experiences was also noted as an effective strategy for addressing challenges in tourism [40–42]. While tourism experience and overtourism are distinct concepts, both significantly influence the trajectory of sustainable tourism. However, a critical gap exists in research focusing on the intersection of overtourism and seasonal trends within the tourism industry. Addressing this gap forms the basis for the present study, which seeks to contribute new insights to this underexplored area.

1.2 Research Questions

The identified research gaps inform the formulation of the following research questions (**RQ**$_n$) relevant to the region:

RQ$_1$ = How significant is the determination of the Seasonal Index (SI) for tourism development?

RQ$_2$ = In what ways do the Seasonal Index (SI) and stakeholder strategies enhance the tourist experience?

RQ$_3$ = What strategies are being adopted to address overtourism?

1.3 Objectives

For the study, the following objectives are formulated:

- To determine the Seasonal Index (SI) and propose a route map calendar for exploring the region.
- To analyze the role of stakeholders in enhancing the tourist experience in the region.
- To recommend strategies for managing overtourism in the region.

2 Methodology

To underscore the significance and seriousness of the study's objectives, background, and literature review, it is essential to recognize the existing research gaps in this domain. Given the limited available literature, the study will employ exploratory and descriptive methodologies, supported by robust and valid data to ascertain the Seasonal Index (SI) of industries in the region (Fig. 1).

To enhance the study's comprehensiveness, data were collected from both primary and secondary sources to thoroughly understand the study's content and generate rational insights. Tourist opinions play a pivotal role in promoting tourism in the region, and as such, data on tourist inflow from various states were collected through visits to state regional offices and relevant organizations to obtain official data. Unlike other industries, gathering tourist opinions using probability sampling poses a challenge. To address this, the study adopts Maslow's Hierarchy of Needs (Fig. 4) to interpret tourist opinions and expectations, ensuring that relevant policies align with these insights. The generalization of tourist opinions through this framework has been widely used by researchers in the field. Various stakeholders, including Civil Society Organizations (CSOs), policymakers, contractors, and engineers, were consulted to assess the status of competence, awareness, challenges, and collaborative efforts within their respective departments, in line with the study's objectives, research

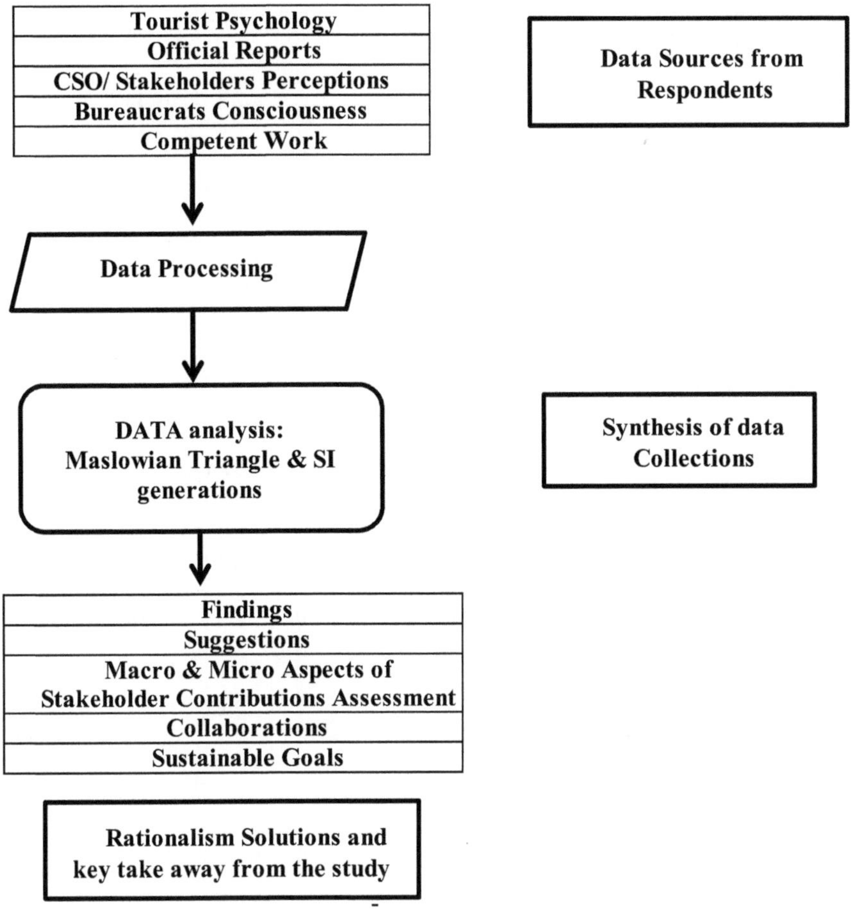

Fig. 1 Adopted research methodology flowchart. *Source* By Author

questions, and identified gaps. Stakeholder opinions are organized according to the Maslowian framework, with tourist inflow data serving as key factors that inform strategies to improve tourism performance in the region. To ensure a more accurate representation of tourist opinions and expectations, the data are analyzed using seasonal indices, as depicted in Figs. 2 and 3.

Given the complex, multifaceted nature of tourism, a holistic approach was adopted by synthesizing the perspectives of various stakeholders into a comprehensive tabular format (Table 2), highlighting their competencies and awareness. The study uses these findings to make practical, evidence-based recommendations to mitigate risks such as overtourism and resource exploitation, which could lead to the emergence of "dark tourism." By evaluating both macro and micro aspects of stakeholder involvement, the study offers a detailed approach to achieving tourism

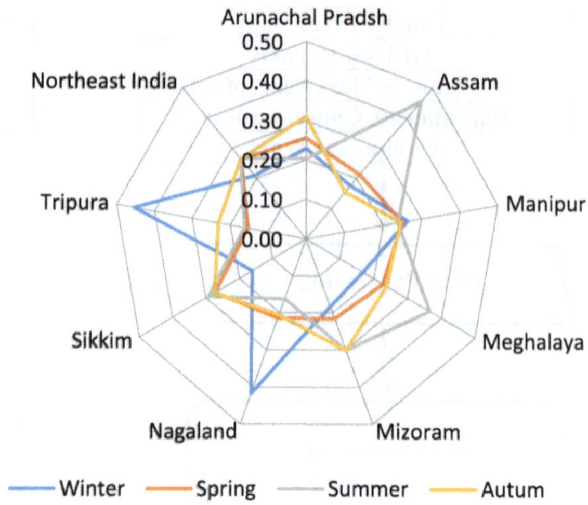

Fig. 2 SI in the region of 2019. *Source* By Author

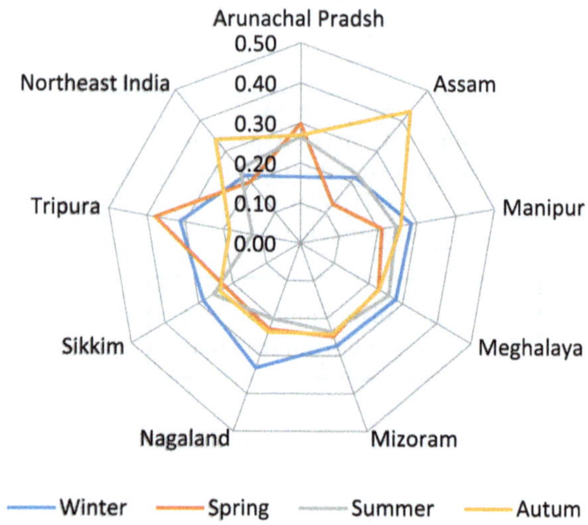

Fig. 3 SI in the region of 2022. *Source* By Author

excellence in the region. Technological advancements, aligned with the principles of sustainability and development, are emphasized as critical drivers of tourism success, as outlined in the study's objectives and further elaborated throughout the paper.

The determination of the Seasonality Index (SI) facilitates the identification of necessary actions that require rational and logical strategies in the domains of supply-chain management, transportation, offerings, and policy development. A strategic approach, considering both peak and off-peak seasons and the roles of stakeholders, is essential. This involves analyzing visitor psychology based on historical data and local community feedback, which can significantly enhance the formulation of policies aimed at improving the tourist experience in the region. Data necessary for the

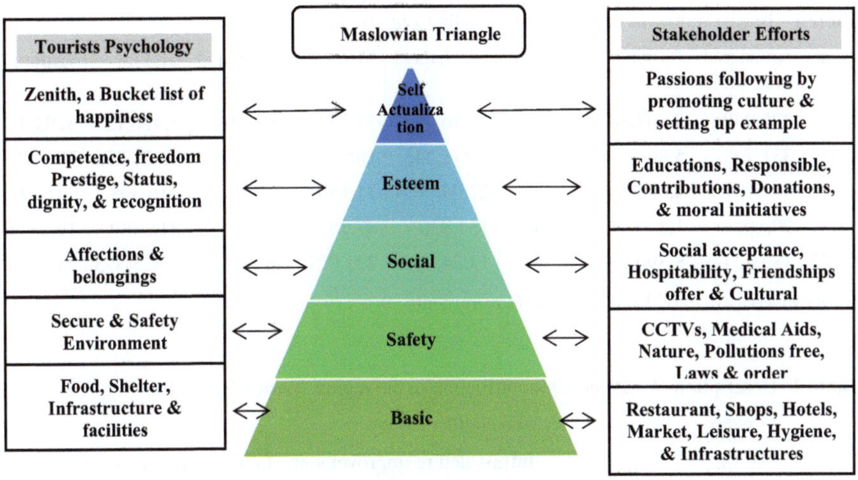

Fig. 4 Enhancing tourist experience in Northeast India with Maslow's Theory of Needs. *Source* By Author

calculation of the SI are sourced from official tourism department websites of the respective states, complemented by direct engagement with state officials to gather monthly visitor arrival data. This dual approach ensures the primary data is both robust and relevant for the study. Given the study's focus on tourist experiences, interaction with policymakers and stakeholders is critical. Discussions involve relevant departments, such as the Public Works and Tourism Departments, regarding development projects. Furthermore, insights from contractor and worker associations are solicited to assess competencies, technological capabilities, and resource allocation (including finances, personnel, materials, and machinery). This information, illustrated in Fig. 4, is examined from both macro and micro perspectives of stakeholder roles and responsibilities. In line with the study's objectives and scope, both primary and secondary data sources are utilized to provide a comprehensive understanding of the factors influencing the tourist experience. Additionally, secondary data will be collected from a range of sources, including books, articles, reports, and online platforms, and will be analyzed in the context of the study's objectives.

3 Tourist Inflow Trends in the Region from 2019 to 2022

Tourist inflow data for Northeast India in 2019 and 2022 is summarized in Table 1, revealing an upward trend in visitor numbers, coinciding with the region's recovery from the pandemic. The significant increase in tourist inflow observed in the region aligns with a global surge of 790 million tourists in the first seven months of 2024 [43], highlighting the potential of the area as a tourism hotspot. However, this situation underscores the challenges associated with limited resource allocation and

Table 1 Tourist inflow trend in Northeast India from 2019 to 2022

Year	Arunachal Pradesh	Assam	Manipur	Meghalaya	Mizoram	Nagaland	Sikkim	Tripura
2018	326,940	5,507,303	177,448	1,216,454	77,274	66,499	1,426,127	520,648
2019	331,735	5,804,684	52,862	1,271,446	165,040	72,927	1,421,823	705,052
2020	44,310	1,274,183	501 27	27,045	31,180	11,497	316,408	74,375
2021	90,100	1,470,913	151,934	154,820	87,388	9701	511,489	192,311
2022	185,474	8,390,962	61,281	1,024,864	221,021	50,915	1,625,573	343,494

Sources Department of Tourism, State Governments

Table 2 Description of stakeholder contribution in the region. *Source* By Author

SI	Levels	Stakeholder Efforts
1	Basic	Infrastructure improvements, facilities, partnerships
2	Safety	Community involvement, green tourism, security
3	Social	Cultural exchange, local hospitality, social acceptance
4	Esteem	Ecosystem balance, unique culture, visitor dignity
5	Self-Actualization	Data-driven strategies, visitor satisfaction, happiness

the region's capacity to effectively manage demand while meeting tourist expectations. The data presented in Table 1 indicate a natural and rapid recovery from the pandemic, marked by a sudden influx of tourists to the region. This trend raises significant concerns regarding the existing gaps in research, highlighting an urgent need for further investigation.

Understanding the seasonal patterns of tourist visits will be essential for effectively managing resources and the capacity of hosts to accommodate increased tourist inflow. Identifying these seasonal trends will provide valuable insights for both tourists and stakeholders in the regional tourism industry. Additionally, further research will facilitate resource allocation, capacity planning, and the formulation of relevant policies and strategies.

3.1 Seasonal Index (SI)

The analysis of the Seasonal Index (SI) has emerged as a critical topic of global discussion. It is imperative to address the seriousness of SI at both the state and regional levels, as this data is essential for planning effective policies. Engaging local communities in this conversation will foster awareness of the potential risks of overtourism in the region. As a developing nation, it is crucial for the region to remain informed—a key intention of this chapter, given the lack of thorough investigations and explorations conducted thus far. This study aims to fill that gap and contribute to the sustainable operation of the tourism industry.

To facilitate understanding, this chapter categorizes the year into four distinct seasons as follows:

- **Winter**: November, December, and January
- **Spring**: February, March, and April
- **Summer**: May, June, and July
- **Autumn**: August, September, and October.

3.2 *Calculations of SI*

SI of tourism industries visiting the region in the spring (February to April) of 2019:

SI = (Total number of tourists visiting the region from February to April in 2019)/ (Total number of tourists visiting the region in the year 2019)

$$SI = (792032 + 832624 + 1002372)/10006424$$

$$SI = 2627028/10006424$$

$$SI = 0.26$$

During the winter a high seasonal index of 0.45 is recorded for Tripura, indicating a strong seasonal influence. In contrast, the lowest seasonal index is observed in Sikkim at 0.16, suggesting minimal winter-related activity. Furthermore during the spring, a relatively balanced distribution is noted, with the highest seasonal index of 0.27 recorded in Sikkim. Assam shows a slightly lower index at 0.21, while Tripura has the lowest seasonal index of 0.15, reflecting less spring activity. In summer the highest seasonal activity is indicated by Assam, with a seasonal index of 0.46. Conversely, Sikkim and Tripura both exhibit lower seasonal indices of 0.16, suggesting reduced engagement during this season. Lastly, autumn records the highest seasonal index of 0.31 recorded for Arunachal Pradesh, indicating notable autumnal activity. The lowest seasonal index is found in Assam at 0.15, suggesting a weaker seasonal presence.

During winter the highest seasonal index is recorded by Nagaland (0.33), indicating a significant impact of winter activities, while the lowest index is observed in Arunachal Pradesh (0.16), suggesting a minimal seasonal influence. While in spring the highest level of seasonal index is indicated by Tripura (0.38), Assam (0.13) shows the least engagement during this season. In summer, a decrease in seasonal index is noted across most states, with Sikkim having the lowest seasonal index (0.20) and Arunachal Pradesh showing a relatively higher value (0.27). In autumn the highest seasonal index is observed in Assam (0.43), while Tripura is recorded with the lowest seasonal index (0.18), indicating reduced autumn-related activity. Overall, varying levels of seasonal influence are highlighted across the different states, with notable contrasts, particularly between Assam and Tripura.

The portrayed Figs. 2 and 3 reveal a significant irregularity in peak and off-peak seasons for industries in the region, highlighting an intriguing aspect of growth and sustainable development. Despite these fluctuations, it is evident that winter emerges as the most favorable season for tourism development. Following winter, summer also shows potential for attracting visitors. Conversely, autumn and spring are the least favorable seasons, likely due to monsoon rainfall and a scarcity of festivals during this time. Seasonal trends vary by year and region; notably, Nagaland, Sikkim, and Tripura excel during the winter months, while Assam and Meghalaya are more prominent in the summer. Arunachal Pradesh and Mizoram experience peak tourist activity in the spring, with autumn consistently recognized as the least favorable season for tourism.

4 Discussion

4.1 Tourist Experience

It is essential to emphasize that enhancing the tourist experience can be achieved by incorporating stakeholder opinions, which should align with an understanding of tourist psychology and awareness. Tourists' tastes and preferences can be categorized according to the various levels of Maslow's hierarchy of needs as illustrated in Fig. 4. Although the application of Maslow's hierarchy of needs to tourism management is a well-established concept, significant credit must be given to the rigorous research conducted by [6, 44, 45], which has further advanced this framework. This approach offers a particularly comprehensive and nuanced understanding of tourist psychology. To make the study more interesting, it has been asserted that there is a critical need for increased awareness and competence among stakeholders regarding tourists' expectations and requirements [45]. It is important to recognize that the tangible contributions of stakeholders diminish as one ascends the hierarchy. This model facilitates a more holistic assessment of the collective efforts of various stakeholders, ensuring that all aspects of the tourist experience are integrated into policy-making and service delivery. Furthermore, it provides a rational framework for bridging the gap between stakeholder perceptions and tourist expectations. Detailed descriptions of these stakeholder efforts are presented in Table 2, which supports the proposed framework. This work represents a significant advancement in the literature by applying Maslow's hierarchy in the context of tourism. Additionally, to enrich the study, the opinions of policymakers are incorporated, highlighting their role in managing resources effectively to align with tourists' needs, ultimately aiming to enhance satisfaction and foster loyalty. Given the region's mountainous location, it is crucial to address the challenges faced by different stakeholders from their unique perspectives. A distinctive and systematic approach involving the roles of various states in the region would be an ideal effort. To date, no comprehensive investigation has been conducted into the micro and macro aspects of stakeholder roles in the development of the tourism industry. This gap in the literature represents the unique contribution that the current study aims to address. Furthermore, to enhance the study's engagement, the roles of stakeholders in relation to tourists in the region are generalized and interpreted in Table 2.

Despite various policies aimed at overall development, the monsoon season poses significant challenges to project completion. Investment approvals and the necessary clearances often face delays, compounding issues with project timelines that require diligent follow-up. Additionally, inadequate location and poor connectivity to many destinations create logistical difficulties, hindering tourist accessibility. The region's technical and digital infrastructure must advance beyond basic requirements to enhance tourist satisfaction. Although the local population is known for its hospitality and social acceptance, the lack of fundamental services undermines the potential benefits for the tourism industry.

Despite strong advocacy for the digitization of transactions, rural destinations face challenges in providing adequate network services during sudden tourist influxes, making cash transactions more prevalent. While the state has introduced special transportation facilities, there is still no evidence of a solar transportation system in the region's tourist destinations.

From Fig. 4, it can be inferred that regional development can be aligned with tourist philosophy. The Maslowian triangle effectively reflects the expectations of tourists in light of industry trends. Addressing both micro and macro aspects of stakeholder roles and contributions is likely to result in positive impacts on improvement, which are outlined as follows.

4.1.1 Micro-Level Analysis of Stakeholder Roles in Northeast India

At the micro level, Northeast Indian states largely operate as independent tourism entities, creating significant barriers to collaborative infrastructure development due to distinct policies and budget allocations. There's a notable lack of both intra- and inter-departmental collaboration. Manipur, for instance, suffers from bureaucratic delays in budget and policy processes, hindering project completion. Inadequate infrastructure, particularly the absence of a robust railway network in Mizoram, Meghalaya, and Nagaland, leads to delays in goods receipt. Effective management of development during off-monsoon seasons and adherence to schedules are critical but often lacking. Manipur's ineffective policies contribute to its developmental lag compared to states like Assam and Arunachal Pradesh, which benefit from proximity to transport hubs. Micro-level analysis must prioritize policy formulation, focusing on design innovation and appealing outputs. National initiatives like Make in India, Digital India, Smart City, and Act East policies, alongside global platforms like G20 submissions, are crucial for enhancing the region's global tourism visibility. While Assam, Sikkim, and Arunachal Pradesh demonstrate competent workforces and appropriate fund allocation, other states exhibit disparities, raising questions about accountability.

Transparency in financial aid and budget allocations is vital but often scrutinized for effectiveness. Policymakers must incorporate tourist perspectives, especially in Meghalaya. Despite government funding, many internal development projects are locally driven. Academics note the absence of festival monuments for tourism promotion, and states like Mizoram and Tripura over-rely on external supplies. Addressing these implementation delays is crucial for aligning outcomes with expectations.

4.1.2 Macro-Level Analysis of Stakeholder Roles in Northeast India

The Northeast Indian tourism sector faces significant hurdles, primarily a pervasive lack of inter-state collaboration, hindering the development of a cohesive regional route map. A core objective must be to champion enhanced partnerships and collective efforts. Calculating a Seasonal Index (SI) for individual states and the region overall would provide vital data for policymakers, especially for addressing macro-level challenges like securing financial support. At the micro level, collaborative development and joint infrastructure improvements are urgently needed for interconnected destinations. Delays in state funding, notably in Manipur and Nagaland, impede progress despite contributions from organizations like the Asian Development Bank and state-led initiatives. Evidence of successful implementation of these efforts remains scarce. Local communities, such as those in Mizoram, are actively engaging with new policies, signaling a desire for practical application. There's strong regional interest among scholars, tourists, and local entrepreneurs (e.g., in Meghalaya) in adopting contemporary technologies. The integration of Fourth Industrial Revolution (4IR) technologies like VR, AR, IoT, and AI presents significant opportunities for innovation, potentially bolstering regional policies and providing a competitive edge. However, the region often adheres to a classical development approach, making it resistant to proposed modifications, as highlighted by the Contractors Association of Manipur. Despite this, tourists in Meghalaya envision AI applications replacing traditional guides, transforming destinations into "smart" and well-organized locales. Ultimately, serious investment from both micro and macro perspectives is essential to redefine destinations, improve leisure activities, and enhance safety measures. A synchronized and collaborative approach among all stakeholders is paramount for enhancing the region's appeal and establishing it as a smart, sustainably developed, and innovative destination.

5 Findings

Systematic Prevention is Key: Addressing unpredictable overtourism and dark tourism requires prompt, effective, and preventive actions. The Seasonal Index (SI) analysis (2019, 2022) provides a systematic framework to anticipate and mitigate these issues, revealing significant regional variations in tourist inflow.

Overtourism's Dual Nature: Though rare, overtourism's negative impacts are severe, akin to a pandemic, influencing tourist perception. However, it also signals rising demand and market expansion opportunities. The challenge is to improve the tourist experience and leisure activities amid these contrasting effects.

Infrastructure Gaps and Connectivity: Tourists report poor telecommunication and network issues, impeding online interactions. Sudden tourist surges worsen these problems, causing social network instability and reduced revisit intentions, mirroring dark tourism challenges.

Economic and Environmental Strain: Unmanaged tourist influxes create significant marketing risks and ecological threats, potentially displacing locals. Economic benefits often favor the industry, at the cost of communities and increased environmental degradation (pollution, resource management).

Community and Resource Shortfalls: Overtourism can cause tension with local communities and compromise tourist expectations due to limited carrying capacity. Resource and service shortages (e.g., transport, rising prices) during peak seasons necessitate vital community involvement and acknowledgment of local contributions, especially for sensitive areas.

6 Suggestions

Based on the discussions presented in this chapter, the following suggestions are proposed to effectively address overtourism and enhance the tourism experience in the region (Table 3).

By adopting these strategies, the region can better manage tourism challenges, enhance visitor experiences, and ensure sustainable development for local communities.

7 Conclusion

This research utilized Figs. 2 and 3 to determine the Seasonal Index, culminating in a new tourism chart for the Northeast Indian region. This is significant as previous studies lacked adequate identification of optimal tourist seasons, focusing instead on

Table 3 Tourism improvement recommendations. *Source* By Author

Category	Recommendation	Description
Managing tourism flow	Seasonal Tourism Routes	Develop maps for optimal visitation times, distributing tourist flow evenly
	Off-Peak Travel Promotion	Encourage travel to alternative destinations/events outside peak season to reduce pressure
	Temporary Destination Closures	Implement closures during peak seasons to protect environment, considering local livelihoods
	Off-Season Event Organization	Organize events during off-season to balance visitor distribution and protect natural habitats
Sustainable development	Infrastructure Investment (4IR)	Invest in tourism infrastructure, incorporating Fourth Industrial Revolution principles for efficiency
	Promote Sustainable Practices	Implement sustainable tourism with local and government support for community/environmental benefit
	Utilize Underused Land	Explore barren land (riverbanks, shorelines) for parking and temporary homestays
	Wastewater/Emission Controls	Introduce policies for wastewater treatment and carbon emission control (e.g., secondary water use)
	Plastic Use Abolition	Implement measures to eliminate plastic usage, promoting sustainable practices
Collaboration & data	Inter-Departmental Collaboration	Foster collaboration between government departments and neighboring states for cohesive strategies
	Data-Driven Planning	Conduct comprehensive research to inform effective tourism planning strategies
	Spiritual Destinations Promotion	Highlight lesser known pilgrimage sites to attract visitors during off-peak periods
	Social Media Engagement	Utilize social media to raise awareness about overtourism and foster responsible travel

economic importance and stakeholder collaboration. Interpreting Figs. 2 and 3, this chapter reveals critical insights into tourist flow trends, illustrating potential overflow and identifying specific seasons prone to overtourism, along with management strategies. The radar data in Figs. 2 and 3 signify a substantial breakthrough in identifying the most favorable tourist seasons, aligning with the study's second and third objectives by emphasizing stakeholder roles and suggesting strategies to mitigate sudden tourist influxes. Unlike other destinations, the study areas have not recorded any incidents of dark tourism, offering valuable insights for policymakers.

Collaborative efforts among state departments, encompassing micro and macro perspectives, are vital for strengthening stakeholder roles and responsibilities. Assam emerges as a promising tourism corridor, benefiting from a strong railway network that could elevate the region's profile. The study underscores the SI's importance in mitigating dark tourism risks stemming from overtourism, offering a proactive management approach. Variations in tourist inflow highlight the necessity for effective supply-chain management. While definitive solutions to overtourism are lacking, it can be managed through careful carrying capacity assessments, focusing on community sentiment, promoting collaborations, transport management, and event management. Special emphasis should be placed on promoting untapped locations. Overtourism challenges may necessitate adaptations in local cuisine and the promotion of foreign language learning and cultural exchange, contributing to a more sustainable tourism model. Enhancing the tourist experience while fostering sustainability requires strong partnerships among local communities, stakeholders, tourists, and industry players. As highlighted in Table 2, local hosts must take serious responsibility for management efforts, particularly during peak times, focusing on infrastructure development, technology advancement, and investment in critical areas to mitigate negative economic impacts. This research provides valuable insights and recommendations for establishing a more sustainable and responsible tourism framework.

References

1. Xu F, Niu W, Li S, Bai Y (2020) The mechanism of word-of-mouth for tourist destinations in crisis. Sage Open 10(2). https://doi.org/10.1177/2158244020919491
2. Goyal C, Taneja U (2023) Electronic word of mouth for the choice of wellness tourism destination image and the moderating role of COVID-19 pandemic. J Tourism Futures. https://doi.org/10.1108/JTF-08-2022-0207
3. Zimik S, Barman A (2022) Cross-national state of affairs on performance of tourists attraction a comparison among India, Singapore, and Thailand. Interdis Res J AIIAS Graduate School 10(2):813–837
4. Zimik S, Barman A (2024) Emerging competencies for tourism management. In: NA, futuristic trends in management; Vol 3, Book 24 (pp 208–217). NA: IIP Series
5. Barman A, Zimik S (2021) Developing tourism in Northeast India- need for a censure of its policy & governance. World Education Organisation, NA
6. Zimik S, Barman A (2021) Constructed environment as unkempt issue in north East Indian tourism: a review based triangulation for research. Int J Multidis Educ Res 10(5):137–153

7. WEF (2023) World Economic Forum. Retrieved October 4, 2024, from World Economic Forum Web site. https://www.weforum.org/agenda/2023/10/what-is-overtourism-and-how-can-we-overcome-it/
8. Jangra R, Kaushik S, Saini S (2021) An analysis of tourist's perceptions toward tourism development: study of cold desert destination. India. Geogr Sustain 2(1):48–58
9. Beritelli P, Reinhold S, Laesser C (2020) Visitor flows, trajectories and corridors: planning and designing places from the traveler's point of view. Ann Tour Res 82. https://doi.org/10.1016/j.annals.2020.102936
10. Achmad F, Prambudia Y, Rumanti A (2023) Improving tourism industry performance through support system facilities and stakeholders: the role of environmental dynamism. Sustainability 15(5). https://doi.org/10.3390/su15054103
11. Wani G (2024) Stakeholder perspectives on tourism infrastructure development in Kashmir Valley. Int J Hosp Tourism Manag 8(2):39–51. https://doi.org/10.11648/j.ijhtm.20240802.13
12. Kirtane & Pandit (2024) Kirtane & Pandit. Retrieved October 4, 2024, from Kirtane & Pandit Web site: https://www.kirtanepandit.com/blog-detail/37/the-ripple-effect-of-monsoon-on-the-construction-industry-and-related-financial-nuances
13. Guha H, Biswas P (2008) Monsoon risks for construction sites in India. Indus Eng Eng Manag. https://doi.org/10.1109/IEEM.2008.4738167. NA: IEEM
14. Zimik S, Barman A (2024) Next dynamics in designing artificial intelligence to support tourism development. Int J Eng Technol Manag Res 6(11):44–57. https://doi.org/10.29121/ijetmr.v11.i6.2024.1465
15. Zhao X, Ren Y, Cheah K (2023) Leading virtual reality (VR) and augmented reality (AR) in education: bibliometric and content analysis from the web of science (2018–2022). SAGE Open 13(3). https://doi.org/10.1177/21582440231190821
16. Zimik S, Barman A (2021) Whether Northeast India wishes to exploit its Tourism Potential? Some Doubts. Int J Creative Res Thoughts 9(5):836–841
17. Zimik S, Barman A (2024) Sustainable tourism through novel approach. Int J Res Rev 11(5):252–266. https://doi.org/10.52403/ijrr.20240530
18. Kesavan K (2020) Observer research foundation. Retrieved October 2024, 2024, from Observer Research Foundation Web site: https://www.orfonline.org/expert-speak/indias-act-east-policy-and-regional-cooperation-61375
19. PTI (2024) Business standard. Retrieved October 4, 2024, from Business Standard Web site: https://www.business-standard.com/india-news/govt-working-to-make-ne-gateway-of-act-east-policy-increased-fund-allocation-over-10-years-prez-124062700571_1.html
20. Satghare H (2022) Role of smart city mission in the development of smart tourism destinations in India: a need of an hour after the COVID-19 pandemic. Perspect Bus Manag Econ 7:49–58
21. Maitra R (2017) The development of smart cities in India: a smart notion of promoting travel and tourism. Int J Res Tourism Hosp 3(1):38–49
22. Habeeb N, Weli S (2020) Relationship of smart cities and smart tourism: an overview. High Tech Innov J 1. https://doi.org/10.28991/HIJ-2020-01-04-07
23. Yuvaraj S, Suganthiya M (2016) Make in India policy on tourism towards sustainable development. Int J Bus Administr Res Rev 1(1):467–470
24. Kumar R, Pathania K (2022) Make In India scheme for tourism and hospitality sector—a case study of Himachal Pradesh. J Emerg Technol Innov Res 9(6):708–715
25. Zimik S, Barman A (2024) Role of constructed environment for tourism development: a global trend analysis base on triangulated review. J Bus Manag Econ Dev 2(1):159–190
26. Zimik S, Barman A (2023) Global trend and roles of constructed environment in tourism development. Europ Econ Lett 13(3):783–793
27. Corluka G (2019) Tourism seasonlity- an overview. J Bus Paradigms 4(1):21–40
28. Alshuqaiqi A, Omar S (2019) Causes and implication of seasonality in Tourism. J Adv Res Dyn Control Syst 11:1480–1486
29. Zvaigzne A, Litavniece L, Dembovska I (2022) Tourism seasonality: the causes and effects. Worldwide Hosp Tourism Themes 14(5):421–430. https://doi.org/10.1108/WHATT-07-2022-0080

30. Duro J, Prats J (2019) Tourism seasonality worldwide. Tourism Manag Perspect 31:38–53
31. Martín J, Fernández J, Martín J, Rey M (2019) Analysis of tourism seasonality as a factor limiting the sustainable development of rural areas. J Hosp Tourism Res 44(1). https://doi.org/10.1177/1096348019876688
32. Butler R (1998) Seasonality in tourism: issues and implications. Tourist Rev 53(3):18–24. https://doi.org/10.1108/eb058278
33. Rojo C, Amorós M, García J (2023) Overtourism and sustainability: a bibliometric study (2018–2021). Technol Forecast Soc Chang 188. https://doi.org/10.1016/j.techfore.2022.122285
34. Capocchi A, Vallone C, Pierotti M, Amaduzzi A (2019) Overtourism: a literature review to assess implications and future perspectives. Sustainability 11(12). https://doi.org/10.3390/su11123303
35. Mihalic T, Kuščer K (2022) Can overtourism be managed? Destination management factors affecting residents' irritation and quality of life. Tourism Rev 77(1):16–34. https://doi.org/10.1108/TR-04-2020-0186
36. Vagena A (2021) Overtourism: definition and Impact. Academia Lett. https://doi.org/10.20935/AL1207
37. Krajickova A, Hampl F, Lancosova E (2022) Visitors' perception of overtourism impacts in a small destination. Int J Tourism Hosp Res 33(2):236–246
38. Mihalic T (2020) Conceptualising overtourism: a sustainability approach. Ann Tour Res 84. https://doi.org/10.1016/j.annals.2020.103025
39. Coelho M, Gosling M, Almeida A (2018) Tourism experiences: core processes of memorable trips. J Hosp Tour Manag 37:11–22. https://doi.org/10.1016/j.jhtm.2018.08.004
40. Ma S (2024) Enhancing tourists' satisfaction: leveraging artificial intelligence in the tourism sector. Pacific Int J 7(3):89–98
41. Crames A, Supriadi B, Setyadi S (2023) Enhancing tourist loyalty through destination attributes and the mediating role of tourist satisfaction. Cross Curr Int J Econ Manag Media Studies 5(4):73–79
42. Bagheri F, Guerreiro M, Ghaderi Z (2023) From tourist experience to satisfaction and loyalty: exploring the role of a sense of well-being. J Travel Res 63(8). https://doi.org/10.1177/00472875231201509
43. UNWTO (2024) UN Tourism barometer. Madrid: UNWTO: United Nation World Tourism Organization
44. Zimik S, Barman A (2023) Emergence of constructed environment concept for tourism: a study base on triangulation review. 10th International Baskent congress on humanities and social sciences, pp 381–386. Baskent: NA
45. Zimik S, Barman A (2023) Investment in constructed environment for tourism development. Formosa J Multidis Res 2(11):1799–1814. https://doi.org/10.55927/fjmr.v2i11.6883

A. S. Sochipem Zimik is an Assistant Professor at SRM University, Sikkim, with a Ph.D. in Tourism Industries domain from Assam Central University, India. An accomplished researcher, he has published 22 papers in reputed journals, including those indexed in Scopus and ABDC.

The Future of Tourism

Part-IV The Future of Tourism, looks ahead to emerging solutions and technologies, including the potential of the metaverse in promoting sustainable travel practices and the role of technology in cultural regeneration. Through two distinct but complementary perspectives, this part explores how emerging technologies might transform tourism experiences while addressing the persistent challenges of overtourism. The first contribution examines the intersection of technological innovation and cultural regeneration in Indian heritage precincts, from the bustling streets of Ahmedabad and Mumbai to the historic landscapes of Hampi and Mysore. The second chapter ventures into the virtual realm, investigating how Metaverse marketing could revolutionize sustainable travel practices and offer alternative ways to experience destinations without physical presence. Together, these chapters paint a picture of tourism's potential future—one where physical and virtual experiences coexist, where technology serves as a tool for both preservation and access, and where innovative solutions help balance the competing demands of tourism growth and heritage conservation. This part offers valuable insights for architects, urban planners, and policymakers working to shape the future of tourist destinations in an increasingly digital world.

Reimagining Heritage Spaces: The Role of Technology in Cultural Regeneration and Overtourism

Srinidhi D. Acharya and Amit Kinjawadekar

Abstract Overtourism presents an increasingly relevant threat to historic public spaces, damaging their cultural and architectural heritage. Hence, this study focuses on how technological innovation can regenerate such places from the influence of mass tourism while providing sustainable urban development. Using a comparative case study, this work reflects on efforts toward cultural regeneration in the Ahmedabad Heritage Precinct and the Mumbai Fort Precinct in light of development pressures, stakeholder engagement, and heritage preservation. It further extends an analysis to smaller towns, such as Hampi and Mysore, where the unique scale of the towns can offer particular insight into the management of tourism and preservation. This study took a more expansive view of overtourism as it assessed the capabilities of emerging technologies and their applications in managing overcrowding, raising infrastructure levels, regulating tourist flows, and encouraging greater community participation. Focus areas involved innovative architectural interventions, adaptive reuse strategies, and digital platform use for participatory urban design. It contributes to Sustainable Development Goal 11, which underscores the need for building resilient and sustainable urban environments that protect cultural heritage. The study provides design strategies that should promote a balanced approach to tourism but harmonize local communities with cultural preservation and urban growth since it is promoted by tourism. The work combines tradition and technology to build futures in cultural regeneration for historic public spaces.

Keywords Overtourism · Technological Innovations · Sustainable tourism · Heritage conservation · Cultural heritage

S. D. Acharya (✉) · A. Kinjawadekar
Manipal School of Architecture and Planning, Manipal Academy of Higher Education, Manipal, Karnataka, India
e-mail: srinidhi.msapmpl2023@learner.manipal.edu

A. Kinjawadekar
e-mail: amit.ck@manipal.edu

V. C. Sharma et al. (eds.), *Architectural Vision of Overtourism*,
SpringerBriefs in Architectural Design and Technology,
https://doi.org/10.1007/978-981-96-5794-0_16

1 Introduction

1.1 *The Rise of Overtourism and Its Challenges*

Tourism has become an essential world sector, contributing to GDP, employment, and development in various regions. Rapid tourism growth, however, especially in most culturally rich and historic public spaces, has triggered concerns over overtourism: overcrowding visitors that disrupt the local community and jeopardize the preservation of cultural and historical landmarks [1]. The most dramatic effect of overtourism on heritage sites is the physical strain and wear associated with continuous high foot traffic, which fast-tracks deterioration and requires repeated efforts to restore the sites. Such restoration may damage the site's authenticity, destroying its intangible cultural value. Not only that, but even cultural events and practices will get commodified in deference to tourist tastes, losing their intrinsic meaning and value [2]. Overtourism can also have profound social implications. That is to say, increased property values and cost of living with tourism often push the residents out, and when you take long-standing communities and split them up, it does not just rupture the social tissue but also diminishes local interest in preserving and maintaining cultural sites [3, 4]. It further complicates residents' association with their heritage and handicaps the site's protection of authenticity and significance.

Tackling overtourism is very delicate in planning and requires sustainable tourism strategies to balance its benefits with the need to protect cultural heritage and the community within its presence. It can only be ensured that such sites are supported rather than threatened if tourism is approached with mindful management [5]. This paper focuses on the issues of overtourism and how they can be countered by exploring technological development in the direction of sustainable regeneration in historic public spaces. This study has case studies from two Indian cities, namely Ahmedabad and Mumbai, through which it assesses the usage of technology in managing tourist flows, safeguarding cultural heritage, and enriching community engagement. Among the smaller towns are Hampi and Mysore, which possess uniqueness along with the challenges of scarce infrastructure and growing demand for tourism. In line with the UN's Sustainable Development Goal 11, which stresses an inclusive, safe, resilient, and sustainable city, this research examines how techniques to preserve heritage are being balanced with sustainable urban growth strategies. This paper stresses community involvement in long-term technological solutions to preserve culture.

2 Research Methodology

This paper adopts a qualitative comparative approach to explore how technological innovations address overtourism, allowing sustainable cultural regeneration. The investigation examines four case studies—Ahmedabad, Mumbai, Hampi, and

Mysore—which reflect diverse urban scales, heritage importance, and scales of technological integration (Fig. 1).

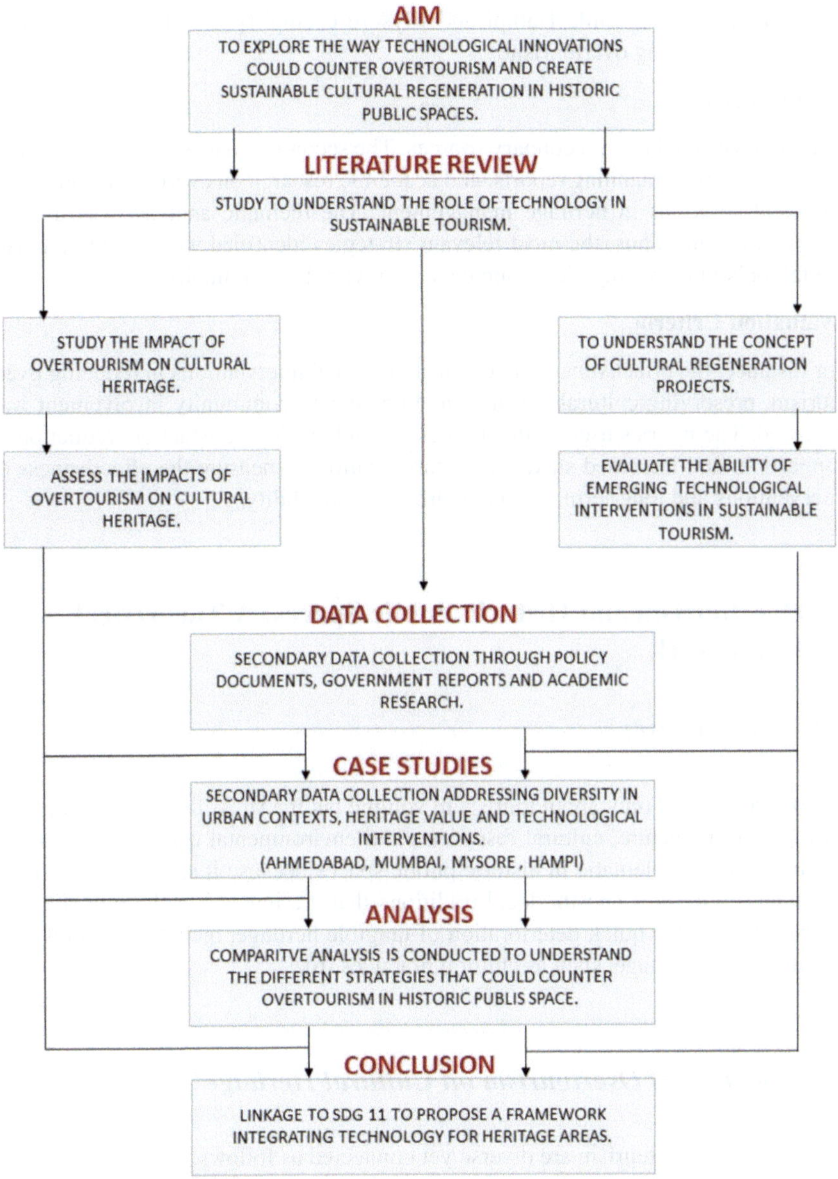

Fig. 1 Methodical approach to the study

Case Study Selection

To ensure diversity in the choice of case studies, each was selected based on different urban contexts, heritage values, and the varying nature of technological adoption. Ahmedabad and Mumbai represent large metropolitan cities with established heritage conservation efforts, while Hampi and Mysore exemplify the challenges smaller towns face regarding overtourism.

Data Collection

Data was collected from secondary sources. The secondary data sources were policy documents, urban planning reports, and academic research on overtourism and technological solutions in heritage management. The thematic analysis was used to analyze the data. Thus, the most relevant strategies identified were adaptive reuse, digital tools in managing cities, and engaging with the communities.

Evaluation Criteria

For instance, the effectiveness of such technological interventions in reducing overtourism, preserving cultural heritage, and enhancing community involvement was observed. The metrics used in this process include visitor satisfaction, reduction in congestion, and increased stakeholder participation to measure the effectiveness of interventions and long-term effects on urban sustainability.

3 Overtourism and Historic Public Spaces: A Theoretical Framework

3.1 Overtourism

Overtourism is not only about tourists in volume but the stress that the influx places on local infrastructure, cultural resources, and environmental capacity. This stress is particularly problematic in historic public spaces because it endangers the physical integrity of sites and the local traditions that are integral to these areas. This further leads to the quick deterioration of tangible heritage, including monuments, and intangible heritage, such as cultural practices [6].

3.2 Impacts of Overtourism on Cultural Heritage

The effects of overtourism are diverse yet connected as follows.

Structural Destruction The congestion that usually characterizes these regions hastens the deterioration of monuments and buildings. The repeated restoration process may distort or permanently damage historical features, thus losing their authenticity and worth [7].

Cultural Erosion Folk festivals, customs, and rites are commercialized so that the tourists can receive their expectations, diluting meaning and emotions. Those activities are more likely to lose their original spirituality and culturally pertained significance by becoming performative.

Displacement of Communities Tourism increased value with increased property prices, displacing local communities against community attachment to the place. This kind of action through displacement does not support social capital that would otherwise be needed for effective, sustained heritage conservation measures [8, 9].

3.3 Cultural Regeneration and Technological Innovation: A Conceptual Understanding

Cultural regeneration is restoring historical spaces that respect their cultural and architectural value [10]. Technological innovation leads toward modern tools that offer new management for tourism, thereby improving visitor experience and local community participation. They range from digital heritage platforms and smart city infrastructure to advanced conservation techniques that effectively respond to the challenges of balancing tourism with heritage preservation and community development for urban centers [11].

Digital technologies allow culture to be documented and shared but do not burden the historic sites: Visitors and locals can access historical information without burdening the cities where historic sites are located [12]. The implementation of intelligent technologies in towns allows for real-time monitoring whereby data is gathered to manage the flow of visitors better, thus preventing overcrowding and degradation on the part of historic sites. Advanced means of conservation also help preserve the originality of the historic sites despite the increased access.

Introducing such technologies in a city will not only safeguard the cultural heritage but also improve responsible tourism that significantly supports its people's local economic and social growth. The approach makes it possible to make historic areas remain functional while still meaning so much to future years, catering to modern demands for tourism [13].

4 Emerging Technologies in Sustainable Tourism and Heritage Conservation

4.1 Architectural Interventions and Smart Heritage

Innovative architectural strategies are now critical in preserving historic buildings, mainly through adaptive reuse and sustainable building technologies. Adaptive reuse refers to giving an old structure a new function, such as turning historic structures into museums or galleries or utilizing them as cultural spaces. Not only does this extend the lives of these heritage structures, but it also prevents the environmental implications associated with new constructions. Adaptive reuse preserves the cultural values attached to historic buildings without compromising their usability for current urban activities. This is done by restoring an old building into a useful one for current usage [12].

Another sustainable building technology that leads toward protecting heritage sites includes energy-efficient materials and techniques so that heritage site conservation serves both purposes: preserving heritage sites and, at the same time, promoting sustainability in urban development [14].

Aside from these techniques, innovative heritage systems have recently emerged as one of the main innovations in heritage conservation. These systems utilize sensors, data analytics, and real-time monitoring to assess the structural health of heritage buildings. For instance, the embedded sensors within the building will track anomalies in temperature, humidity, or even stress levels in the structure, revealing possible deterioration. These technologies, in turn, provide permanent data regarding the state of the building and allow preventive maintenance. This reduces the need for more invasive and potentially harmful restoration work. Adaptive reuse, sustainable technologies, and intelligent heritage systems comprehensively preserve historical architecture while considering new urban development needs [15].

4.2 Digital Platforms for Community Participation

Digital participatory platforms for designing towns revolutionize the process of cultural regeneration, making it more democratic and interactive. Such platforms present an open space for diversified stakeholders, from residents, tourists, architects, and other city authorities, who can concretely and actively take part in determining how heritage should be conserved and developed. The core objective of these websites is to ensure that the plans and strategies for heritage are adequately considered through community needs and priorities. This can be achieved by bringing into the activity those otherwise impacted by tourism and urban development.

The most significant benefit of participation in urban design is the ability to empower the local community, often putting them in decision-making roles taken directly from external planners or government officials. Hence, this democratized approach encourages residents to have more vital ownership and augments community participation in heritage conservation. When local knowledge, cultural values, and everyday experiences are considered, heritage conservation projects are more likely to succeed in the protection of historical sites while maintaining their relevance to the people who live and work in those areas.

These platforms allow the meaningful involvement of heritage sites in the minds of tourists because they can provide feedback or suggestions on how to make cultural experiences better without sacrificing the integrity of the location. The collaborative framework thereby seeks a way to find the intrinsic values of cultural heritage while still developing tourism.

The digital platforms also help reduce the potential clash between development and conservation when promoting urban development since they involve various stakeholders in their decision-making process. Therefore, by involving residents and local businesses, tourism strategies are not merely a means of amassing economic advantages but also sustaining the area's cultural and social fabric. In this case, the impact of these platforms is to ensure sustainable tourism management by cities and towns to preserve the culture for future users while promoting the well-being of the present residents.

4.3 Virtual and Augmented Reality

Virtual reality (VR) and augmented reality (AR) technologies have dramatically changed the world of sustainable tourism because they tackle many salient problems of heritage sites with innovative solutions. One of the significant problems in tourism is overcrowding at spectacular historical locations; this is prone to the physical destruction of these fragile heritage sites. This is because VR allows tourists to see heritage sites from a closer perspective while sitting at home or in their office. It enables consumers to enjoy historical landmarks' beauty and cultural significance through immersive virtual tours without necessarily viewing them live. This thus reduces pressure on the actual sites while making such cultural heritage accessible to more people, including those who may not have the capacity for traveling.

On the other hand, AR enables bringing digital elements to the physical world to make an experience more attractive to a visitor's site. During their visits to heritage sites, AR might superimpose real-time information such as historical facts or what a site of interest must have looked like in the past over the actual landmarks. This virtual engaging experience promotes intense cultural interaction with visitors to comprehend the background history at the site without exposing the region to additional stress and strain [12]. A tourist visiting a ruined palace can apply AR to watch a digital reconstruction of how the building looked at its peak and then make sense of its former power as they learn about it in real time.

Heritage sites can now give everybody immersive, educational experiences that engage modern tourists in new ways while preserving the physical integrity of those sites through VR and AR. They represent the forward-thinking approach to the growth of tourism and cultural preservation so that heritage sites are well protected for future generations [12].

5 Case Studies: Cultural Regeneration in India's Historic Public Spaces and Tourism

5.1 *The Ahmedabad Heritage Precinct*

One of India's first city listings as a UNESCO World Heritage City, Ahmedabad, shows how technology, policy, and community engagement can support heritage conservation. The heritage precinct that characterizes the city has few things more characteristic than its Pol houses, historical mosques, and lively markets. Since it achieved the UNESCO World Heritage status, the number of visitors there has multiplied [16] (Fig. 2).

This growth underscores the need to marry technological innovation with local action while conserving cultural sites and engaging visitors. This success of Ahmedabad has demonstrated how heritage can be well maintained using intelligent strategies and active involvement in a community.

Technological Interventions

To confront the issues of overtourism over its heritage, Ahmedabad has brought in technology-enabled solutions that help enhance this city's heritage efficiently. Virtual Heritage Platforms using VR and AR: Such a novelty will enable virtual exploration by visitors at significant cultural and historical sites without actually visiting those places, thus ensuring these places are not overcrowded. Besides the ease of pedestrian movement, these platforms provide rich educative experiences by telling the visitor in detail about the significance of many landmarks. It encourages more responsible forms of tourism and reduces overcrowding at sensitive heritage

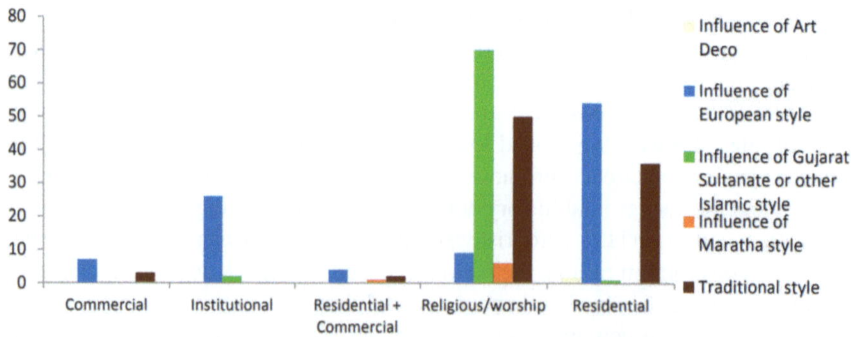

Fig. 2 Use of buildings by architectural style in Ahmedabad Heritage Precinct [17]

sites [18]. In addition, Ahmedabad applies Smart Traffic Systems to enhance the flow of tourists and residents. These use real-time data tracking pedestrian and vehicle flows around the city. This data analysis allows the authorities to divert tourists away from heavily congested areas; therefore, heritage sites are not damaged through wear and tear. This means that better traffic flow will offer tourists and residents a more enjoyable visit and living experience because it will prevent overcrowding and reduce damage to historic buildings [18].

Innovative Methods of Restoration form another integral part of Ahmedabad's heritage preservation policy. Advanced 3D scanning and 3D printing technologies permit the accurate restoration of complex architectural details. These methods will help conserve the historic look and feel of heritage sites housed by the city with reduced aggressive restoration practices. For example, the worn or damaged elements can be restored using 3D scanning to record all the details accurately, and 3D printing enables them to replicate all these features with minimal intervention to guarantee as much authenticity as possible in their architectures. Through such collaboration, Ahmedabad effectively deals with overtourism and keeps its cultural wealth for later generations.

Community Engagement

The great success of Ahmedabad in heritage conservation is owed to the community-based approach. All residents actively participate in the restoration works with even formal training in traditional construction techniques to not deplete the city's heritage. Heritage walks led by the community members are also organized, enabling them to share their knowledge about the legacy and bond it to themselves more. This helps in the equitable dispersal of economic advantages of tourism so that it is felt positively by the community and allows local pride in preserving their historic surroundings [16].

5.2 *The Mumbai Fort Precinct*

Despite Mumbai's Fort being a unique combination of Victorian Gothic and Art Deco architecture, difficulties in preserving heritage and urban commercial activity exist. As one of the major tourist centers and a busy business district, the precinct feels the burden of hosting visitors on top of the daily needs of its urban workforce. The cultural and architectural value of the area is attributed to most historic buildings, which have been put to new use but demand high conservation to maintain their integrity. It requires innovative strategies such as adaptive reuse of heritage structures and intelligent urban planning to ensure the area remains functional for residents and tourists. Technological intervention through digital mapping and traffic management has also been proposed to reduce congestion, enhance heritage conservation, and foster the coexistence of tourism and commerce in the precinct [19].

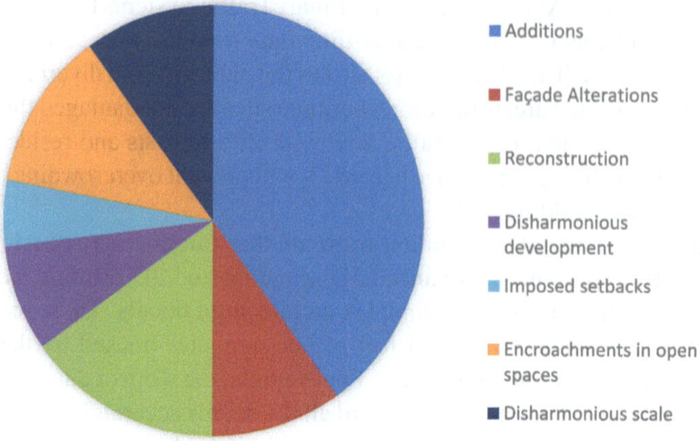

Fig. 3 Physical transformation over the years in Mumbai Fort Precinct [19]

Adaptive Reuse of Heritage Buildings

Many of the old colonial structures in Mumbai have successfully been repurposed to serve as museums, galleries, and cultural spaces, thereby helping relieve pressure on heritage structures like these. Thus, it is through adaptive reuse that these old buildings are given such new, sustainable functions that stretch their lifespan while being culturally relevant and architecturally valued. In doing so, the city protects its treasure trove of the historical past and satisfies increasing demands regarding development and public spaces. This is appreciably in sync with sustainable urban design principles: it ensures that new uses are compatible with modern needs and heritage conservation principles. The adaptive reuse of Mumbai enables the city to balance its need for growth with the protection of its historical assets, thereby creating spaces suitable for contemporary urban life without compromising the integrity of its past [20] (Fig. 3).

Digital Tools for Urban Management

Mumbai's participatory urban design initiatives involve several key stakeholders, beginning with the Municipal Corporation of Greater Mumbai (MCGM), which oversees urban planning and management. Urban Centre Mumbai (UCM) plays a vital role in the engagement and participatory planning process, and so does the academic institution in terms of research and innovative methodologies. Lastly, without local community organizations, it is impossible to ensure resident participation in the design process [21]. All these collaborative efforts toward enhancing urban sustainability, community ties, and protecting cultural heritage finally lead to better urban development strategies. This, therefore, means that the efforts by the city toward conservation are aligned with the different needs and values of its community. This can be achieved by incorporating group feedback to balance heritage preservation with modern urban requirements better and set up a shared vision for the city's future.

Table 1 Comparative analysis of Ahmedabad heritage precinct and Mumbai fort precinct

Parameters	Ahmedabad Heritage Precinct	Mumbai Fort Precinct	References
Technological interventions	Digitized heritage portals and intelligent traffic management reduce congestion and promote responsible tourism	Digital tools for participatory urban design ensure community input and adaptive reuse of colonial structures	[18, 19]
Community engagement	Social involvement: Capacity building and heritage walks enhance ownership significantly in the local community	Digital platforms allow inclusive planning to empower residents to participate in conservation efforts	[16, 19]
Preservation efforts	Restoration practices are also minimalistic in newer approaches, and authenticity becomes the highest priority in heritage conservation efforts	Adaptive reuse maintains the historical value of buildings while performing a contemporary function	[19, 22]
Economic impact	Increased heritage tourism benefits the local economies and acts as a source of employment while reinforcing cultural identities	Economic vitality through tourism and commerce interplay increases property value, challenging the residents	[7, 16]

Using digital tools has been worth it as they allow local participation in planning rather than pouring decisions down from top management. Residents and other stakeholders are now involved in restoring and repurposing heritage buildings and public spaces. Therefore, ownership and responsibility among the community members are being fostered for more sustainable and culturally sensitive outcomes.

Besides, digital platforms provide real-time data and interactive mapping to visualize proposed projects' impact and adapt strategies accordingly.

Making local voices part of the decision-making process, the models of participatory urban design in Mumbai guarantee that development initiatives would not solely be for tourism or commercial interests but would look toward meeting the long-term well-being needs of its residents and protecting the vast cultural heritage of this place. Through collaborative efforts, this approach will help strengthen the balance of growth and preservation within the city and facilitate a development that may be directed to serve a shared future for the community as well [19] (Table 1).

5.3 *Small Towns: Hampi and Mysore*

Major cities have better infrastructure and resource bases, thus putting them in a stronger position to deal with tourists. Hampi and Mysore, being small towns, have great cultural and architectural heritage but face the challenge of managing tourism. As the number of tourists increases, it becomes difficult for these small towns to preserve their heritage sites. Small towns do not have as much support for the influx of tourists as big cities do. This hampers effective crowd management, thereby maintaining the quality of tourist experiences by providing heritage site protection. Hampi, a UNESCO World Heritage Site, boasts ancient temples and ruins but finds its narrow roads and inadequate facilities struggling to cope with periodic surges in tourism. This overload is considerably burdening the historic buildings as well as the quality of life of the inhabitants. Similar is the case with Mysore—its royal palaces and cultural festivals have attracted a large throng, especially during mega events such as an annual Dussehra celebration—hence generally straining local infrastructure and resources [23]. The towns require innovative dovetailing approaches for managing tourism sustainably. The reduction of stress put on those towns will be realized through digital crowd management tools, real-time visitor monitoring, and off-peak tourism. Consultation with local communities in tourism management and preservation will better align with their needs and values of heritage conservation. By emphasizing sustainable tourism practices and upgrading infrastructural setup, smaller towns like Hampi and Mysore can better preserve their cultural assets without losing the economic benefits of tourism.

Hampi: Managing Tourist Flows Through Technology

Hampi, a UNESCO World Heritage Site, has witnessed unprecedented growth in tourism in the last few years, with its age-old temples and monumental ruins earning colossal acclamation [23] (Fig. 4).

With the growing crowd of people, it presented dilemmas concerning the balance between tourism and heritage. To minimize the congestion, Hampi has explored innovative technological solutions for handling the flow of visitors to the historical sites more thoughtfully.

One such initiative would be mobile applications that show the real-time levels of crowds at any given location in the town. These mobile applications help the tourists schedule their visits according to real-time conditions, thereby keeping most of the crowded sites off the tour and trying to distribute foot traffic more evenly around the area [23]. This will also enhance the visitor experience while reducing the pressure on

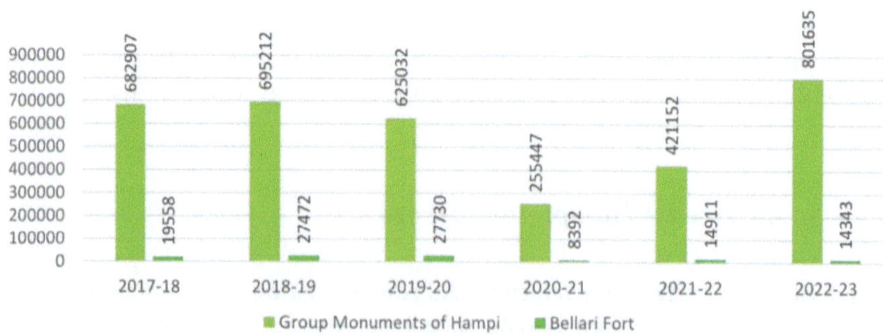

Fig. 4 Tourist footfall at major tourist destinations in Hampi [23]

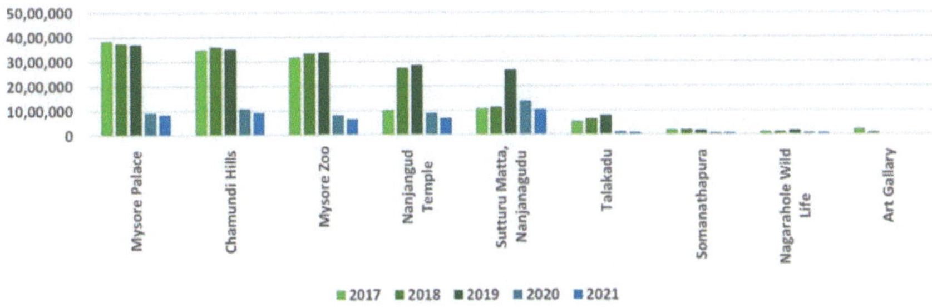

Fig. 5 Tourist footfall at major tourist destinations in Mysore [24]

the particularly fragile or popular sites for better preservation for future generations. Apart from its application through mobile apps, it has been noted that the organization uses drones to monitor every visitor activity, especially at sensitive archeological sites. Using drones ensures that the tourists remain on the given routes and paths and do not unknowingly damage the fragile ruins or invade the restricted areas. It is a proactive surveillance measure taken to preserve Hampi's historical monuments even as it encourages considerable tourist activities. Leverage these technologies so that Hampi has taken a step forward in managing the impacts of the rising tourism while protecting its invaluable heritage. The strategy also significantly contributes toward a model of sustainable tourism, not only for the community but also for preserving cultural heritage in the town (Fig. 5).

Mysore: Integrating Tradition with Technology

Mysore is a city that has a royal heritage, palaces, and life-vibrant cultural festivals. The city has now embraced AR technology to improve this tourism experience. With such a rich history and architectural landmarks, this city faces the challenge of handling a large number of tourists while ensuring the preservation of cultural assets. To meet this need, the city has innovatively merged AR into its tourist attractions, allowing visitors to experience some of the outstanding historical and cultural resources innovatively preserved in the city [25]. With AR technology, one can explore the virtual reconstruction of significant historical events and how architecture transforms. For example, through AR technology, visitors can view how the famous Mysore structures, like the Mysore Palace, have evolved, or they can attend one of the most popular cultural events in the city, the Mysore Dasara festival—virtually [24]. This visit is more exciting as it offers thorough knowledge and appreciation of the city's history and heritage beyond what is observable by the naked eye just by viewing the physical structures as they are today.

Mysore's introduction of AR in its tourism industry has had tremendous success as it improved visitor interactions through experiences such as at Mysore Palace and Chamundi Hills. Cultural heritage is also not left behind, as historical elements are recreated digitally, and tourist inflow is highly witnessed on such sites, mainly by those with the spirit of exploring uncommon experiences [24]. However, there are still some limitations within the use of AR, such as accessibility, as not all tourists may have the appropriate devices. Moreover, the development of high-quality AR content might be very expensive, and what is more, the experience depends on the technologies familiar to the user. Ongoing tasks for the local tourism authorities are also regular maintenance and updates of AR content.

6 Results and Discussion

Analyzing historic Indian public spaces like Ahmedabad, Mumbai, Hampi, and Mysore indicates that proper harnessing of technological advancement is an indispensable tool for mitigating the impacts of overtourism. At the same time, it is achievable through sustainable urban development. Ahmedabad and Mysore have shown successful integration, but other cities reflect varying degrees of success, and critical trends suggest a need for balanced integration (Table 2).

6.1 Managing Overtourism Through Technology

Technological interventions turned out to be essential tools in tackling overtourism. In Ahmedabad, virtual heritage platforms and intelligent traffic management systems effectively reduced overcrowding at heritage sites while enhancing visitor experiences. Similarly, Mumbai utilized digital mapping and participatory urban design tools to balance tourism and urban commerce. These strategies align with Goodwin's emphasis on using technology to minimize physical strain on cultural assets [6]. However, for smaller towns like Hampi and Mysore, the adaptation of technologies in such a manner depends on the infrastructure, which necessitates custom solutions such as real-time crowd monitoring and augmented reality (AR) experiences.

Table 2 Comparative analysis of technological interventions and cultural regeneration strategies in selected case studies

Study Location	Technological Interventions	Cultural Regeneration strategies
Ahmedabad	Smart city infrastructure	Adaptive reuse
	Digital heritage platforms	Community engagement
Mumbai	Real-time monitoring systems	Stakeholder participation
	Smart transport apps	Cultural events management
Hampi	Virtual Reality Applications	Heritage Tourism Management
	QR Codes	
Mysore	Augmented Reality tools	Local community involvement
	Digital ticketing	
	e-guides	

6.2 Cultural Regeneration and Community Engagement

Community involvement was an essential component of the success of heritage conservation initiatives. In Ahmedabad, training residents in traditional construction methods and holding heritage walks were some of the activities that empowered local communities. These activities support Salazar's suggestion that community-based cultural tourism promotes sustainable heritage management [13]. On the other hand, reliance on technological solutions in Mumbai indicates a missed opportunity for harvesting community involvement, thus highlighting a point of failure that could be improved to ensure long-term sustainability.

6.3 Adaptive Reuse and Heritage Preservation

Adaptive reuse techniques in Mumbai reveal how colonial buildings can be redeployed as workable spaces, providing an example of how the architectural character is maintained while meeting contemporary objectives. The literature illustrates this kind of intervention in reconciling heritage preservation with urbanization [20]. In medium-sized towns like Mysore, AR technology can facilitate virtual cultural festivals and architectural transformations, reducing physical strain on heritage structures and extending their lifespan.

6.4 Broader Implications

The case study interventions support Sustainable Development Goal 11 on sustainable urban development. These case studies demonstrate scalable models that balance tourism growth with cultural preservation by applying digital tools, adaptive reuse, and engagement with the community. However, various contextual challenges, such as resource constraints in smaller towns, underscore the need for adaptable solutions that take into account local capacities.

7 Conclusion: A Balanced Approach to Cultural Regeneration

Overtourism has become one of the most critical challenges for historic public spaces, threatening the cultural and architectural heritage. The study examines how technological innovations can regenerate these sites while pursuing sustainable urban development. It shows various cases in Ahmedabad and Mumbai, as well as small towns like Hampi and Mysore, in which the need to balance growth in tourism with heritage preservation and community involvement came to the forefront. Key findings suggest emerging technologies are crucial in managing tourist flows, enhancing infrastructure, and fostering community involvement. Some strategies suggested in the study to reduce overtourism include the adaptive reuse of historic structures, innovative heritage systems, and digital participatory platforms. For example, adaptive reuse converts older buildings into valuable spaces such as museums or cultural centers, retaining their historical value but answering the needs of modern cities. Innovative heritage systems monitor the state of heritage sites through sensors and data analytics to ensure that preventive maintenance can minimize physical stress from high traffic volume.

Community participation in decisions made regarding heritage conservation is also highlighted within this research. This engagement is possible through digital platforms: several stakeholders, including residents, tourists, and urban planners, are granted the power to participate in planning and site management. In doing so, the local communities are empowered while the conservation efforts align with the values of the community's culture and needs. Future studies should explore more virtual and augmented reality unification to integrate it further into tourism management. Virtual and augmented reality technologies can create immersive experiences that reduce the pressure of physical contact on heritage sites while making cultural heritage accessible to a broader audience. One example is that Virtual Reality allows a home-bound user to visit historical landmarks, reducing congestion at tourist destinations. AR can thus be used to improve site visitation by layering information over physical objects, such as landmarks, enhancing visitors' knowledge without damaging the integrity of the sites in question. This study significantly contributes to Sustainable Development Goal 11: It advocates for inclusive urban development that protects cultural heritage amid the challenges of overtourism. Cities can, therefore, be created in a resilient environment, honoring their cultural legacies while promoting sustainable tourism practices. The findings thus underscore the urgent need to continue researching innovative solutions that balance tourism growth with safeguarding historic public space so these precious cultural resources do not crumble for future generations.

References

1. Verissimo M, Moraes M, Breda Z, Guizi A, Costa C (2020) Overtourism and tourismphobia: a systematic literature review. Tourism 68(2). https://doi.org/10.37741/T.68.2.4
2. Smith M, Robinson M (2006) Cultural tourism in a changing world: politics, participation and (re)presentation. Ann Tour Res 34(3):813–815. https://doi.org/10.1016/j.annals.2007.01.009
3. Singh B, Chhering M (2024) Overtourism in India: impacts, challenges, and solutions for sustainable destinations. Int J Recent Res Commer Econ Manag (IJRRCEM) 11:92–98. https://doi.org/10.5281/zenodo.13472122
4. Scott D (2014) The tourist: a new theory of the leisure class. Ann Lei-sure Res 17(1):113–114. https://doi.org/10.1080/11745398.2014.890513
5. Santos-Rojo C, Llopis-Amorós M, García-García JM (2023) Overtour-ism and sustainability: a bibliometric study (2018–2021). https://doi.org/10.1016/j.techfore.2022.122285

6. Goodwin H (2018) Responsible tourism and the conservation of heritage in Asian Urban Areas. https://doi.org/10.23912/978-1-911396-58-1-4048
7. Robinson M, Picard D (2006) Tourism, culture and sustainable development. https://unesdoc.unesco.org/ark:/48223/pf0000147578
8. Tosun C (2006) Expected nature of community participation in tourism developement. Tour Manage 27(3):493–504. https://doi.org/10.1016/j.tourman.2004.12.004
9. Su MM, Wall G (2014) Community participation in tourism at a world-heritage site: Mutianyu great wall, Beijing. China. Int J Tourism Res 16(2):146–156. https://doi.org/10.1002/jtr.1909
10. Harrison R (2015) Heritage: critical approaches. Norwegian Archaeological Rev 49(1):90–92. https://doi.org/10.1080/00293652.2015.1126632
11. Lotfi S (2011) Culture-led regeneration: a reflection upon cultural funda-ments and the act of regeneration. Retrieved from https://www.researchgate.net/publication/323105311
12. Guttentag DA (2010) Virtual reality: applications and implications for tourism. Tour Manage 31(5):637–651. https://doi.org/10.1016/j.tourman.2009.07.003
13. Salazar NB (2012) Community-based cultural tourism: issues, threats, and opportunities. J Sustain Tour 20(1):9–22. https://doi.org/10.1080/09669582.2011.596279
14. Peeters P, Dubois G (2010) Tourism travel under climate change mitigation constraints. J Transp Geogr 18(3):447–457. https://doi.org/10.1016/j.jtrangeo.2009.09.003
15. Jacob AM (2023) Urban regeneration for historic commercial centers. IOSR J Eng (IOSRJEN) 13. Retrieved from www.iosrjen.org
16. Ahmedabad Municipal Corporation (2016) Historic City of Ahmedabad India heritage management plan. Retrieved from https://whc.unesco.org/en/list/1551/
17. Indian National Trust for Art and Culture (INTACH) (2014) Listing and grading of heritage structures (excluding walled city) Volume I: Listing database of heritage structures in AMC jurisdiction. Ahmedabad Municipal Corporation. Retrieved from www.umcasia.org.
18. Solanki AS, Patel C, Doshi N (2019) Smart cities-a case study of Porto and Ahmedabad. Procedia Comput Sci 160:718–722. https://doi.org/10.1016/j.procs.2019.11.021
19. Urban Design Research Institute (2010) Fort-Management-Plan-Report. https://www.udri.org/pdf/02-Fort-Management-Plan-Report.pdf
20. Garrod B, Fyall A (2024) Managing heritage tourism. Retrieved from www.elsevier.com/locate/atoures
21. Sarode AV (2023) A study on challenges and opportunities for sustainable tourism in Mumbai. https://www.researchgate.net/publication/370111695
22. UNESCO World Heritage Committee (2017) Convention concerning the protection of the world cultural and natural heritage. World Heritage 41 COM WHC/17/41.COM/18 Krakow
23. Deloitte Touche Tohmatsu India LLP in association with (2023) Destination master plan, strategy and action plan for Hampi project development and management consultants under Swadesh Darshan 2.0 Scheme for Karnataka. Karnataka Tourism Infrastructure Limited (KTIL), Government of Karnataka
24. Deloitte Touche Tohmatsu India LLP in association with (2023) Destination master plan, strategy and action plan for Mysuru project development and management consultants under Swadesh Darshan 2.0 Scheme for Karnataka. Karnataka Tourism Infrastructure Limited (KTIL), Government of Karnataka
25. Chiu YH, Lee MS, Wang JW (2019) Culture-led urban regeneration strategy: an evaluation of the management strategies and performance of urban regeneration stations in Taipei City. Habitat Int 86:1–9. https://doi.org/10.1016/j.habitatint.2019.01.003

Srinidhi D. Acharya is a postgraduate student in Urban Design at the Manipal School of Architecture and Planning, Manipal. Her academic and research interests lie in the fields of history, heritage, cultural regeneration, and sustainable urban environments.

Amit Kinjawadekar is a Professor at the Manipal School of Architecture and Planning, MAHE, Manipal, India. Their research focuses on urban mobility and accessibility studies to improve societal well-being, particularly within sustainable urban environments. Their broader research interests include universal design and the relationship between the built environment and health, as evidenced by publications.

Exploring the Potential of Metaverse Marketing to Promote Sustainable Travel Practices and Reduce Overtourism

Mirjana Maksimović and Zvjezdana Krstić

Abstract Tourism is a fast-growing global sector crucial in boosting a host country's economy. Sustainable tourism aims to enhance benefits while minimizing negative effects associated with tourism activities. It encompasses economic, social, and environmental dimensions and promotes a more responsible way of traveling to protect the environment, preserve local cultures, and support communities. Ecological degradation, cultural loss, and strain on local resources often result from overtourism. This chapter examines the potential of Metaverse marketing to promote sustainable travel practices and reduce overtourism in popular destinations, highlighting its advantages, challenges, and limitations. The Metaverse, a virtual environment, allows travelers to explore destinations without visiting them in person. Implementing marketing strategies in the Metaverse can significantly influence tourist behavior, engagement, and interactions, raise awareness, promote sustainable tourism practices, and lead to better management of tourist numbers while preserving the cultural and environmental integrity of well-known destinations. However, virtual travel should not replace physical travel but serve as an additional tool for sustainable development and economic growth in local communities. This chapter contributes to the existing literature by examining the effects of Metaverse marketing on sustainable travel practices through a survey that explores travelers' behaviors, experiences with overtourism, and perceptions of technology's role in fostering sustainable choices.

Keywords Metaverse · Marketing · Sustainability · Tourism

M. Maksimović (✉)
Faculty of Electrical Engineering, University of East Sarajevo, East Sarajevo, Bosnia and Herzegovina
e-mail: mirjana.maksimovic@etf.ues.rs.ba

Z. Krstić
Independent Researcher (Previously affiliated with Faculty of Business Economics Bijeljina, University of East Sarajevo), Bijeljina, Bosnia and Herzegovina
e-mail: zvjezdana.gavrilovic@gmail.com

V. C. Sharma et al. (eds.), *Architectural Vision of Overtourism*,
SpringerBriefs in Architectural Design and Technology,
https://doi.org/10.1007/978-981-96-5794-0_17

1 Introduction

According to the United Nations (UN) World Tourism Barometer [1], around 790 million tourists traveled internationally in the first seven months of 2024, bringing international tourism back to 96% of what it was before the pandemic. The international tourism export revenues hit USD 1.8 trillion, almost matching the levels seen before the pandemic. The direct contribution of tourism to Gross Domestic Product (GDP) has also recovered, estimated at USD 3.4 trillion, which represents approximately 3% of the global GDP. The UN Panel of Tourism Experts reported that the primary challenge facing the industry is inflation in travel and tourism, especially the rising costs of transportation and accommodation [1].

An increasing number of tourists has resulted in greater effects from tourism, but it has also caused popular destinations to become so crowded that the capacity of tourist attractions and the residents' tolerance for receiving visitors are exceeded. Overtourism can disrupt the local community because of inadequate transportation options, limited communication services, and concerns regarding safety and security [2]. McKinsey reports that 80% of travelers visit only 10% of the world's tourist destinations, causing local infrastructure issues, harming natural and cultural sites, and disturbing local communities [3]. The United Nations World Tourism Organization (UNWTO) defines overtourism as the effect of tourism on a destination, which adversely affects the quality of life for residents and the experiences of visitors [4]. In numerous popular tourist destinations, overtourism is starting to pose a serious problem.

The tourism sector attracts visitors through smart and sustainable tourism practices, focusing on creating environmentally safe, recyclable, and biodegradable products, energy-efficient operations, waste management, and preserving cultural and natural resources [5]. Sustainable tourism considers economic, social, and environmental effects while meeting the needs of travelers, the industry, the environment, and local communities [6]. The hospitality industry is increasingly adopting sustainable practices by minimizing waste, conserving energy and water, sourcing local and organic products, and encouraging responsible tourism. Booking.com's 2023 Sustainable Travel Report reveals that 76% of people want sustainable leisure trips, but 49% find them too expensive due to increasing living expenses [7].

The smart tourism ecosystem consists of customers (tourists), businesses, and destinations, with customers benefiting from personalized support and businesses promoting collaboration and resource sharing. It aims to improve competitiveness and enhance the quality of life for all stakeholders, including residents and visitors, at the destination level [8]. Technology is essential in promoting sustainable and socially responsible practices in travel and tourism. The Metaverse, a virtual platform, allows tourists to be completely submerged in a destination's culture and beauty, reducing overtourism and promoting environmental sustainability [9]. It has potential to revolutionize travel and tourism management and marketing, offering immersive and interactive experiences that improve destination awareness, branding, coordination, and overall management. Consequently, it will facilitate trip planning, interaction, and engagement [10, 11]. The World Economic Forum (WEF) estimates that by 2030, over 700 urban areas will adopt Metaverse platforms, with Seoul, Santa Monica, and Dubai being leaders in embracing such technologies [12]. These cities use the Metaverse not only as a marketing tool but also to drive improvements in various sectors such as transportation systems, education, healthcare services, urban planning, mobility initiatives, sustainability efforts, and economic strategies [13].

Metaverse is expected to transform travel and tourism management and marketing, fostering a more sustainable and socially responsible tourism industry. It provides immersive and interactive experiences and reduces overtourism, but it should not replace physical travel as it can harm local economies. Virtual tourism should support sustainable development and economic growth in local communities [9]. It is beneficial for disabled individuals, disadvantaged individuals, and busy professionals, making tourism more inclusive and accessible. Additionally, it reduces travel costs, boosts efficiency, lowers carbon emissions, and enhances sustainability by eliminating in-person travel for activities [11].

1.1 Motivation for Research, Chapter Organization, and Contributions

The Metaverse can enhance marketing strategies for smart tourist destinations, increasing the city's brand awareness and product sales [13]. Marketing in Metaverse can also promote sustainable tourism, address overtourism, and improve visitor experience by transforming destination promotion. The purpose of this chapter is to clarify the idea of Metaverse marketing and reflect on its significance, potential advantages, challenges, and limitations linked to its adoption in the travel and tourism sector. Our previous studies [5, 10] served as the foundation and starting point for this research.

Therefore, the rest of this chapter is structured as follows (Fig. 1). The concept of Metaverse, and its role in marketing are presented in Sect. 2. Metaverse-associated technological tools and innovations in the sustainability and tourism industry are shown in Sect. 3. Section 3 is enhanced with a subsection examining survey results on the influence of Metaverse marketing on sustainable tourism. Challenges and ethical considerations are discussed in Sect. 4. Section 5 concludes the chapter.

The key contributions of this research are

- Investigation of the synergy between Metaverse and marketing strategies.
- Comprehensive analysis of Metaverse marketing implementations in the tourism industry.
- Provision of valuable insights on how marketing in the Metaverse can promote sustainable travel practices and address overtourism.
- Providing insights into travel behavior and people's attitudes toward immersive technologies and tourism through survey analysis.
- Identification of challenges and limitations associated with implementing Metaverse marketing in the tourism industry.

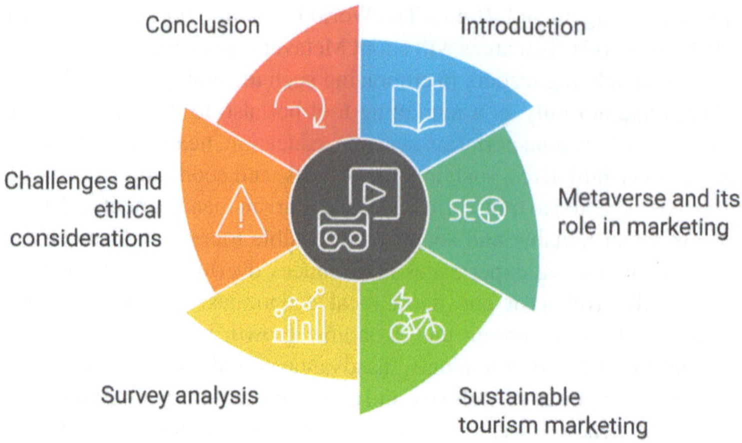

Fig. 1 The structure of the study

2 Metaverse and Its Role in Marketing

The term Metaverse describes a future iteration of the Internet consisting of persistent, shared, 3D virtual spaces linked into a perceived virtual universe. Its design fuses technology (e.g., Virtual Reality (VR), Augmented Reality (AR), Mixed Reality (MR), Extended Reality (XR), Artificial Intelligence (AI), Internet of Things (IoT), Digital Twins (DTs), 5G/6G, blockchain, and holography) and creativity to deliver immersive and captivating experiences. Important aspects include virtual spaces, user engagement, and data management. Metaverse allows users to interact with each other and a computer-generated environment, by representing avatars and simulating presence in space, with communication possible through text, voice, and more. Creating a Metaverse requires setting clear goals, identifying the target audience, and utilizing advanced 3D software to incorporate interactivity. Some challenges include ensuring compatibility across various platforms, scaling services to accommodate a growing user base, and keeping users engaged with interesting experiences [14, 15].

Marketing in the Metaverse or Metaverse marketing can be viewed as the integration of Metaverse technologies with marketing strategies to create immersive and interactive brand experiences (Fig. 2). In other words, in marketing, the Metaverse serves as a fresh platform for brands to connect with consumers and offer experiences that captivate users in innovative ways, providing a strong sense of realism and immersion that enriches their online interactions, ultimately enhancing engagement and loyalty. Enabling brands to collect user data—such as preferences, behaviors, and interests—can enhance the personalization of marketing messages and strategies within the Metaverse. Hence, Metaverse as a virtual environment promotes consumer-brand interactions, shaping how consumers process, perceive, and respond to advertising, unlike traditional marketing which relies more on passive, one-way communication (Fig. 3) [16].

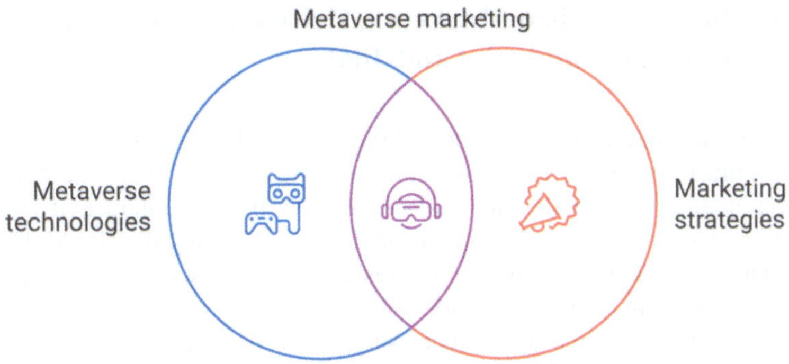

Fig. 2 Metaverse marketing: Synergy of Metaverse technologies and marketing strategies

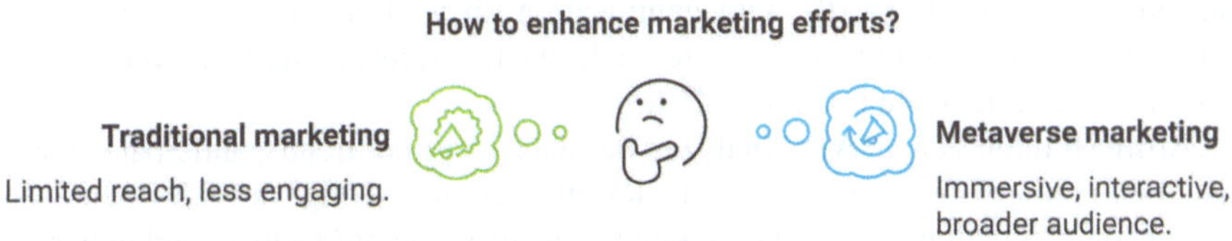

Fig. 3 From traditional marketing to Metaverse marketing

Metaverse marketing has the potential to significantly impact the tourism industry by promoting sustainable practices. By utilizing immersive digital experiences, businesses can engage with environmentally conscious travelers, showcase indigenous cultures, and promote environmental preservation. This strategy not only enhances customer experiences but also aligns with global sustainability goals. Since Metaverse allows destinations and attractions to be presented in a virtual environment, it allows travelers to explore them without visiting them in person. Implementing Metaverse marketing strategies can significantly influence tourist behavior, engagement, and interactions, providing new opportunities for trip planning. This approach can raise awareness and promote sustainable tourism practices, leading to better management of tourist numbers and preservation of cultural and environmental integrity. It is important to emphasize that virtual travel should complement physical travel by promoting sustainable development and economic growth, while still preserving the advantages of in-person tourism for local communities.

3 Sustainable Tourism Marketing Driven by Metaverse's Underlying Technologies and Tools

Destination marketing organizations (DMOs) and hospitality/tourism service providers must efficiently present offerings to attract visitors [17]. The Metaverse enhances intangible, experiential tourism elements through interactivity, immersion, and visualization. Its underlying technologies—AR/VR/MR/XR, AI, DT, IoT, Blockchains, and 5G/6G—improve both resident and tourist experiences. Benefits to tourism marketing include the following.

AR/VR/MR/XR Technologies: Increasingly used for immersive tourist experiences. VR fully immerses; AR overlays information onto physical space; MR blends virtual/physical with smart glasses [18]. These boost engagement, confidence, and satisfaction by offering destination previews, controlling tourist flow, and aiding informed decisions [17–20]. With gamification, they offer innovative marketing, transform consumer interactions, increase destination interest, manage overtourism, and promote sustainable travel [21, 22].

Artificial Intelligence (AI): Analyzes vast data, identifies trends, anticipates travel patterns, predicts tourist behavior, creates strategic marketing, provides personalized recommendations, supports bookings, manages pricing/revenue, and enhances security [23–27]. AI tailors travel experiences, refines destination management, and streamlines customer service. It creates marketing campaigns aligned with preferences, crafts personalized virtual tours, aids resource management, and reduces carbon footprint, contributing to sustainable tourism [9, 28, 29].

Digital Twins (DT): Transforms tourism marketing by producing accurate digital replicas of places (destinations, attractions). DTs create virtual city models, capturing architecture, culture, and atmosphere, continuously updated in real time [30]. This immersive experience allows pre-visit exploration and preserves historical sites [9]. DTs also forecast traffic, manage visitor flow, enhance transport/security, optimize resource utilization, and contribute to sustainable tourism goals. Immersive virtual tours attract tourists, personalize recommendations, and improve business processes and communications [8, 9, 31, 32].

Blockchain: Open-source code improving tourism marketing transparency, security, and efficiency. It establishes loyalty, claims digital ownership, enables secure transactions, automates agreements, and removes third-party intermediaries, lowering costs [33]. Benefits include secure bookings, payments, reservations, rewards, and loyalty programs, minimizing errors/delays and deterring hackers. NFTs in the Metaverse represent transferable digital asset rights [34], used in marketing (merchandise, collectibles, lotteries) [35]. Travel companies leverage NFT-based services for passenger monitoring and service quality. Blockchain's decentralized communication system facilitates direct tourist-local community interaction, tackling overtourism and fostering sustainable practices [9].

Internet of Things (IoT): Greatly influences tourism by integrating intelligent devices/systems. It collects/analyzes customer data for personalized marketing and smart tourism. IoT helps managers track/enhance resource utilization [36]. IoT integration with blockchain manages tourist flows, predicts movement, and enhances infrastructure/services, minimizing tourism's impact on communities/environment [9]. In summary, IoT improves customer service, increases revenue, fosters loyalty, and contributes to tourism sustainability.

Communication Technologies (5G/6G): Improve connectivity and immersive experiences in tourism marketing. 5G/6G networks meet Metaverse demands, enabling AR/VR, boosting tourist engagement/satisfaction. They enhance check-in, provide smart in-room services, and improve travel experience accuracy/security. Real-time data sharing enhances navigation/recommendations. 5G/6G may enable new business models like remote tourism. Effects are assessed via user feedback and performance indicators (visitor numbers, user interaction) [37, 38].

Ubiquitous Computing: Connects customers to information infrastructures consistently, enabling anytime/anywhere services. This is via real-time user/environment data, location-based services, and smooth interaction [39, 40]. Tourism marketers enhance visitor experience and destination management by providing more connected, personalized, and efficient travel.

The use of the above-presented technologies enables Metaverse marketing's significant influence on the tourism industry. The impact of Metaverse marketing before, during, and after travel is presented in Table 1.

Metaverse offers tourism and travel companies a powerful sale and marketing tool, enhancing their offerings, and saving time and money. It digitizes historical artifacts, ensuring their preservation for future generations [45]. Metaverse aligns tourism with Sustainable Development Goals (SDGs) (Fig. 4), introducing virtual eco-destinations, enhancing travel planning, preserving endangered destinations, fostering cultural exchange, and supporting local economies [46].

Having in mind that Metaverse is being utilized in sustainable tourism efforts to improve visitor experiences and encourage sustainability, numerous companies are crafting digital replicas of tourist destinations usually adopting a comprehensive

Table 1 Metaverse marketing's effects on the travel sector [9, 11, 31, 41–44]

Before Travel	During Travel	After Travel
Immersive exploration and virtual tours aid travel choices with previews	Interactive local experiences and chatbots offer real-time guidance and support	Digital memory preservation fosters brand loyalty; virtual content reduces overtourism
Personalized recommendations tailor marketing based on preferences	Interactive itinerary updates adapt to location/preferences	Gamified loyalty programs motivate engagement; social sharing attracts new visitors
Balanced tourism promotes off-peak travel and lesser known areas for sustainability	Eco-friendly travel uses AR for sustainable education	Feedback via virtual interactions boosts service quality and reputation
Social engagement via metaverse events promotes destinations	Virtual communities enable traveler interaction	Focusing on ongoing community

Fig. 4 Metaverse contribution to the achievement of SDGs in the tourism industry

approach that involves technology partnerships, a focus on user-centered design, and strong privacy protections. Table 2 illustrates some key strategies.

Some examples of Metaverse tourism where the strategies mentioned above are applied include the following [45, 51–54]:

- Cuenca, the first World Heritage City in Spain, has introduced “Cuenca Land,” a virtual environment that allows avatars to explore and engage with different areas of the city’s Old Town.
- The Castle of Montilla in Spain is being digitally recreated as a 3D virtual museum through photogrammetry, enabling visitors from around the world to experience its rich heritage without the need for physical travel.
- The Great Tapestry of Scotland has been transformed into a Metaverse tourism experience by New Frontier. This initiative showcases digital representations of diverse themes, including battles, traditions, and royalty, with plans for future interactive games and expansion.
- The Seoul Metropolitan Government has introduced Metaverse Seoul, a virtual replica of the actual city. This initiative enables users to create avatars and engage with the digital environment.
- ZEPETO World is a smartphone app that lets users create their avatars and explore Korea, featuring a detailed interactive map of Han River Park. Users can also chat with one another, shop, and enjoy various performances.
- The BCB Group, a prominent player in crypto banking, has developed a Metaverse city that includes famous landmarks such as the Great Wall of China and the Statue of Liberty.

Table 2 Strategies for developing immersive, sustainable, and responsible Metaverse experiences [11, 44, 46–50]

Creating immersive experiences	• Focus on high-definition graphics, spatial audio, and haptic technology to enhance realism • Tailor experiences to user preferences with interactive storytelling and personalized features • Introduce social components and reward systems to foster community and engagement • Organize exclusive, limited-time events to generate excitement and promote repeat visits
Collaborating with technology providers	• Collaborate with startups, telecommunications companies, communication service providers (CSPs), technology companies, and IoT providers • Form partnerships with firms that focus on VR/AR, AI, and 3D design • Team up with analytics companies to customize experiences for each user • Join forces with VR headset manufacturers or Cloud computing services to enhance performance and accessibility
Implementing data privacy measures	• Clearly explain how user data is collected, stored, and used in the Metaverse • Employ encryption and data anonymization techniques to safeguard personal information • Provide users with control over their data • Comply with the General Data Protection Regulation (GDPR) and other relevant data protection regulations
Encouraging sustainable practices	• Create environments that prioritize environmental sustainability • Motivate sustainable actions through gamified rewards • Develop educational materials focused on sustainable tourism and conservation
Integrating physical and virtual experiences	• Create hybrid events and promotions that blend real-world advantages with virtual engagement • Enable users to explore destinations virtually before making a physical visit

- Saudi Arabia's Royal Commission for AlUla has revealed that the ancient city of Hegra is now part of the Metaverse, marking it as the first UNESCO World Heritage Site to enter this digital realm.
- The Coachella festival has launched the Coachellaverse digital experience, enabling those who cannot attend in person to engage with various event features from home.
- Rendezverse is a Metaverse tourism and hospitality platform designed to replicate the experience of events and conferences held in hotels.

- Mövenpick Hotel Amsterdam City Centre is leveraging the Rendezverse service to improve pre-event venue inspections. This Metaverse tourism solution enables event organizers to virtually explore the venue and discover the hotel's amenities in different areas, facilitating event planning with no need for an in-person visit. This innovative approach saves time, effort, and money for organizing events.
- Hotel Riu Plaza España in Madrid has made its debut in the Metaverse, featuring a blend of real-world elements and digitally recreated spaces.
- Qatar Airways has effectively introduced Metaverse tourism with QVerse, a VR experience built on Unreal Engine. This platform allows users to take a virtual tour of both Business Class and Economy Class cabins.
- Spanish low-cost carrier Vueling plans to enable customers to visualize and book their travels through Metaverse distribution channels.
- Emirates has incorporated VR experiences into its Metaverse tourism offerings, providing a 360-degree virtual tour of aircraft cabins on its website, a VR app for exploring the interiors of Airbus A380 and Boeing 777, and an interactive 3D seat map that allows customers to select their preferred seats in a virtual setting.

3.1 Survey Analysis on Metaverse Marketing for Sustainable Tourism

To assess the impact of Metaverse marketing on sustainable tourism, we conducted a survey whose process flowchart is shown in Fig. 5, and whose methodology is presented in Table 3.

Regarding leisure travel frequency, a survey revealed that 6.7% travel less than once a year, 50% travel 1–2 times annually, 38.3% travel 3–5 times, and 5% travel more than 6 times. For destination selection, 10% choose based on fame, 13.3% prefer less well-known places, and 76.7% explore a combination. On overtourism, 86.7% confirmed experiencing it, while 13.3% had not. Those who experienced it highlighted concerns about lengthy wait times, overcrowding at popular attractions, rising costs for accommodation and transportation, and adverse environmental effects. Furthermore, 73.3% indicated overtourism might shape their future travel preferences depending on the destination, 15% affirmed no effect, and 11.7% would actively evade such locations. When asked about factors promoting sustainable travel, participants emphasized consumer education (61.4%), increased awareness through marketing (50.9%), more investment in technologies (43.9%), government policy changes (38.6%), and increased funding. On the influence of technology like the Metaverse, VR, and AR on travel sustainability and overtourism, most suggested reduced congestion at airports and bus stations and improved accessibility to destinations. The importance of sustainability in travel decision-making, such as choosing eco-friendly accommodations and minimizing carbon footprint, was deemed important by 43.4% of respondents, neutral by 33.3%, very important by 13.3%, and not important by 10%. Most respondents experienced overtourism, and enhancing awareness through consumer education and marketing is considered essential for

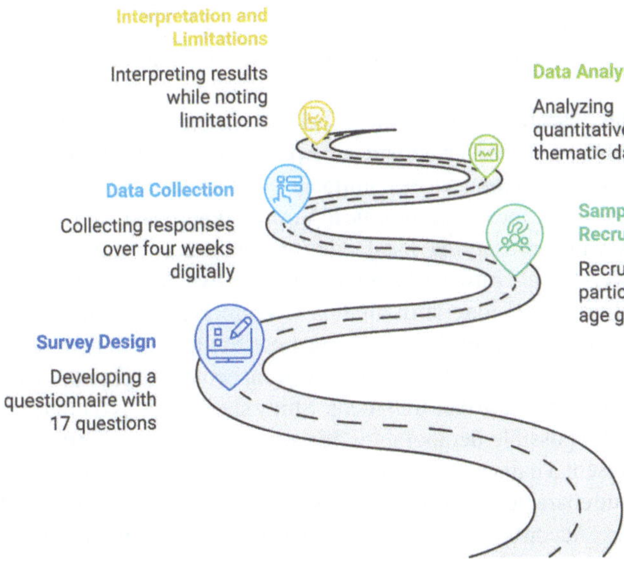

Fig. 5 Survey process flowchart

Table 3 Survey's methodology

Survey design	17 questions, including the Likert scale, and dichotomous, closed, and open-ended questions to collect comprehensive data on travel behavior and attitudes related to immersive technologies and tourism
Sample and participants	60 individuals aged • under 20: 16.7%, • 21–30: 23.3% • 31–40: 8.3% • 41–50: 38.3% • 51–60: 6.7% • over 61: 6.7% aiming to understand generational perspectives on sustainable tourism and Metaverse experiences
Justification for the survey instrument	Collecting quantitative and qualitative data on travel habits, preferences, concerns, experiences with overtourism, and Metaverse technology perceptions in tourism
Mixed-methods approach	Quantitative analysis of close-ended responses and thematic analysis of open-ended responses to explore how Metaverse technologies influence travel sustainability, providing context to data on overtourism, travel frequency, and sustainable travel preferences
Limitations	A small sample size of 60 respondents, potential demographic biases, and the reliance on self-reported data, which could affect data reliability because of personal recall and subjective interpretation

promoting sustainable travel. A majority reported no experience with VR/AR platforms or the Metaverse, implying these technologies need greater inclusion to assist travelers.

In summary, this research explores travel behavior, immersive technology attitudes, and sustainable tourism/overtourism perceptions. It indicates immersive technologies can enhance destination access, mitigate overtourism, and encourage eco-friendly travel, benefiting local communities. Virtual experiences may ease overtourism pressures while preserving local environments and cultures. The study suggests the Metaverse could raise sustainable practices awareness, supporting local economies and cultural sites' long-term vitality. It's important to highlight that this research does not directly and in detail explore how Metaverse tourism can positively impact local communities regarding job creation, economic growth, or cultural preservation; this will be a future research focus. Study shortcomings include reliance on self-reported data, possible demographic biases, and a small sample size of 60 respondents. To present a balanced viewpoint, a variety of age groups were included to understand unique characteristics and develop effective profiles for potential Metaverse platform users. According to [13, 55–57], Gen Y is drawn to immersive experiences and loyalty programs, Gen Z prioritizes social elements in games like Roblox, Minecraft, and Fortnite, while Millennials, particularly in tourism, would utilize the Metaverse and social media for better travel decisions.

4 Challenges and Ethical Considerations

The Metaverse is transforming the tourism industry by offering immersive and interactive experiences, therefore enhancing the discovery of destinations, promoting lesser known locations, and helping reduce overtourism and environmental degradation. Marketers in the tourism industry can tailor experiences by using user information, hence creating more impactful campaigns, providing greater engagement, as well as encouraging sustainable tourism practices [31, 42].

Apart from the exciting opportunities that Metaverse marketing creates within the tourism industry, there are also associated challenges and ethical concerns. Dealing with these issues is essential for the achievement of effective, responsible, and sustainable marketing practices within the Metaverse tourism. The following are some of the key challenges being faced in implementing Metaverse marketing in the tourism [9–11, 18, 42, 44, 49, 58–60]:

- Technology and equipment: The Metaverse relies on cutting-edge technologies and software to ensure effective communication and user interface standards. Unfortunately, not everyone has access to advanced technologies and reliable Internet connectivity, which excludes certain groups from participating. Marketers need to focus on inclusivity in their strategies to prevent disparities in access to travel experiences.

- Data privacy and security: The Metaverse's marketing faces challenges in data privacy and security due to user interactions and preferences. Tourists may not understand data collection, storage, and usage, making them vulnerable to breaches. Companies may also face complexities due to data vulnerabilities.
- Identifying the customers: When performing marketing strategies in the Metaverse, it's essential to take into account the target audience and how they connect with other platforms.
- Metaverse products versus real-life products: Companies should also think about whether they are using the Metaverse to market virtual products or physical ones and reflect on the potential implications of Metaverse products surpassing their real-world equivalents.
- Ethical considerations: Virtual environments raise ethical and legal concerns, including harassment, assaults, cyberstalking, and child exploitation [11, 61]. Despite the potential to reduce carbon emissions, the Metaverse also has environmental impacts due to energy consumption in data centers, further requiring ethical practices in digital operations. Ethical issues from a user perspective and associated concerns in the Metaverse are presented in Table 4.

Establishing best practices can help guide ethical decision-making, which includes respecting consumer data, effectively responding to misinformation attacks, and selecting suitable technologies and experiences. Proactively addressing safety, privacy and ethical concerns is mandatory.

Governments progressively incorporate social, environmental, and economic goals into their tourism and environmental management strategies to promote long-term sustainability. Therefore, national and regional governments should prioritize fostering the development and utilization of the Metaverse in the tourism industry

Table 4 Ethical issues and associated concerns in the Metaverse [49, 62–64]

Privacy	• Concerns about managing personal information, data collection, cybersecurity, and hacking risks • Access through headsets and smart glasses enables the monitoring of movements and physiological reactions
Societal control	• Concerns about constant tracking and control by elites or regulators
Health	• Concerns about physical and psychological health issues • Virtual reality can affect memory and create a sense of déjà vu
Social dynamics	• Concerns about exclusion due to lack of technological skills, affordability, and elderly individuals
User misbehavior	• Concerns about user anonymity and challenges in meeting new people
Globalization	• Concerns about the impact on culture because of global user gatherings
Uncertainty	• Concerns about the Metaverse's potential for significant growth

to fully maximize economic and social benefits from it. By offering tax incentives, implementing regulations, and investing in infrastructure, they can attract eco-conscious businesses and investors. Establishing minimum environmental standards for energy efficiency, data privacy, and ethical content is essential for creating a digital ecosystem that honors social and environmental considerations. Public–Private Partnerships (PPPs) can collaborate to develop sustainable Metaverse platforms and applications, while grants and funding aimed at sustainable innovation can assist startups and developers dedicated to green technologies. Education and awareness initiatives can inform the public about low-impact tourism options, and eco-certifications can motivate developers to achieve sustainability goals. Digital inclusion efforts can guarantee fair access to Metaverse tourism, especially for underserved communities. Additionally, promoting local culture and heritage through virtual cultural tourism can help mitigate overtourism and encourage a sustainable, inclusive, and responsible tourism ecosystem in the Metaverse [13, 65–67].

5 Conclusion

Smart tourism addresses economic, socio-cultural, and environmental challenges by creating enjoyable and responsible travel experiences. It promotes sustainable travel practices through shared platforms. The Metaverse can enhance sustainable tourism by offering immersive virtual experiences, reducing overtourism, enabling digital access to cultural and environmental sites, and providing a lower carbon alternative to traditional travel. Metaverse's long-term impact on the tourism industry comprises the transformation of travel habits, the evolution of virtual tourism, attracting a global audience, and generating new revenue streams for local communities. However, for its successful integration, it is essential to have dependable infrastructure, technical expertise, compatibility with current systems, and a strong focus on security and privacy. Successfully incorporating the Metaverse into the tourism industry involves finding the right balance between technological progress, educating customers, and implementing focused marketing strategies. A complete implementation of the Metaverse in tourism necessitates cooperation among various stakeholders, such as technology providers, tourism operators, DMOs, government agencies, and data privacy advocates.

The goal of our research was to contribute to the current understanding of how the Metaverse can be used in tourism, specifically examining the impact of Metaverse marketing on promoting sustainable tourism practices. To accomplish this, we also conducted a survey aimed at collecting insights into the travel habits and experiences of respondents, along with their views on the potential integration and influence of the Metaverse in tourism. This survey enabled us to investigate public awareness, interest, and concerns related to the Metaverse. Based on the results, it can be concluded that educating consumers and focusing marketing strategies on increasing awareness of sustainable development are of immense importance.

Metaverse marketing has the potential to encourage responsible travel by highlighting sustainable practices and informing travelers about responsible behaviors. While the Metaverse is a valuable marketing tool for tourist attractions, tour operators, hotels, and activity providers, it's crucial to understand that it cannot fully replicate the genuine experiences of real-world travel. By offering a mix of virtual and physical tourism packages, hybrid tourism experiences can be created, helping balance visitor distribution and manage demand during peak seasons. Destinations can enhance their eco-friendly branding by presenting themselves as leaders in sustainability. Tailoring data-driven sustainability initiatives to address both environmental concerns and consumer preferences can further enhance these efforts. Ultimately, these strategies can help establish a tourism model that focuses on long-term sustainability, supports the conservation of destinations, and promotes a culture of responsible global travel.

References

1. World Tourism Organization (2024) World tourism barometer. UN Tourism, Madrid 22(3). https://doi.org/10.18111/wtobarometereng
2. Menon A, Jayawant S (2021) Designing sustainable tourism experiences for the tourists of tomorrow.https://doi.org/10.1007/978-981-16-0119-4_66
3. McKinsey & Company (2024) Destination readiness: preparing for the tourist flows of tomorrow. Available at: https://www.mckinsey.com/industries/travel-logistics-and-infrastructure/our-insights/destination-readiness-preparing-for-the-tourist-flows-of-tomorrow
4. World Tourism Organization (UNWTO) (2018) 'Overtourism'?—understanding and managing urban tourism growth beyond perceptions. Executive Summary. Available athttps://doi.org/10.18111/9789284420070
5. Gavrilović Z, Maksimović M (2018) Green Innovations in the Tourism Sector. 41 Strategic Manag 23(1):36–42
6. World Tourism Organization (UNWTO) (2019) Sustainable development. Available at: https://www.unwto.org/sustainable-development
7. World Economic Forum (WEF) (2023) Survey of travellers finds 76% want more sustainable options. Available at: https://www.weforum.org/agenda/2023/04/survey-of-travelers-finds-76-want-more-sustainable-options/
8. Rahmadian E, Feitosa D, Virantina Y (2023) Digital twins, big data governance, and sustainable tourism. Ethics Inf Technol 25:61. https://doi.org/10.1007/s10676-023-09730-w
9. Kouroupi N, Metaxas T (2023) Can the metaverse and its associated digital tools and technologies provide an opportunity for destinations to address the vulnerability of overtourism? Tourism Hosp 4(2):355–373. https://doi.org/10.3390/tourhosp4020022
10. Krstić Z, Maksimovic M (2023) The Metaverse: a game-changing opportunity for marketing. International scientific conference strategic management and decision support systems in strategic management, 533–540. https://doi.org/10.46541/978-86-7233-416-6_68
11. Buhalis D, Leung D, Lin M (2023) Metaverse as a disruptive technology revolutionising tourism management and marketing. Tour Manage 97. https://doi.org/10.1016/j.tourman.2023.104724
12. World Economic Forum (WEF) (2023) These 3 cities already have their own metaverse. Available at: https://www.weforum.org/agenda/2023/11/metaverse-digital-cities-urban/
13. Florido-Benítez L (2024) Metaverse cannot be an extra marketing immersive tool to increase sales in tourism cities. Int J Tourism Cities 10(3):974–994. Emerald Publishing Limited. https://doi.org/10.1108/IJTC-01-2024-0001
14. Weinberger M (2022) What is metaverse?—a definition based on qualitative meta-synthesis. Future Internet 14:310. https://doi.org/10.3390/fi14110310

15. Huynh-The T, Pham QV, Pham XQ, Nguyen TT, Han Z, Kim DS (2023) Artificial intelligence for the metaverse: a survey. Eng Appl Artif Intell 117:105581
16. Rosarió AT, Lopes PR, Rosarió FS (2023) Metaverse in marketing: challenges and opportunities. In: Handbook of research on AI-based technologies and applications in the era of the metaverse. IGI Global. pp 51–82. https://doi.org/10.4018/978-1-6684-8851-5.ch003
17. Lim WM, Jasim KM, Das M (2024) Augmented and virtual reality in hotels: Impact on tourist satisfaction and intention to stay and return. Int J Hosp Manag 116. https://doi.org/10.1016/j.ijhm.2023.103631
18. Dwivedi YK et al (2023) Metaverse marketing: How the metaverse will shape the future of consumer research and practice. Psychol Mark 40:750–776. https://doi.org/10.1002/mar.21767
19. Bretos MA, Ibañez-Sanchez S, Orús C (2024) Applying virtual reality and augmented reality to the tourism experience: a comparative literature review. Spanish J Market 28(3):287–309. https://doi.org/10.1108/SJME-03-2023-0052
20. Nayyar A, Mahapatra B, Le D, Suseendran G (2018) Virtual Reality (VR) & Augmented Reality (AR) technologies for tourism and hospitality industry. Int J Eng Technol 7(2.21):156–160
21. Calisto ML, Sarkar S (2024) A systematic review of virtual reality in tourism and hospitality: the known and the paths to follow. Int J Hosp Manag 116. https://doi.org/10.1016/j.ijhm.2023.103623
22. Yung R, Khoo-Lattimore C, Potter LE (2020) Virtual reality and tourism marketing: conceptualizing a framework on presence, emotion, and intention. Curr Issue Tour 24(11):1505–1525. https://doi.org/10.1080/13683500.2020.1820454
23. Majan K, Mishra A, Tiwari V, Prasad Vemuri V (2024) Artificial intelligence: the next frontier for marketing in the tourism industry. Acad Market Studies J 28(1):1–12
24. Bulchand-Gidumal J, Secin EW, O'Connor P, Buhalis D (2023) Artificial intelligence's impact on hospitality and tourism marketing: exploring key themes and addressing challenges. Curr Issue Tour. https://doi.org/10.1080/13683500.2023.2229480
25. Moilanen T, Laatikainen E (2023) Opportunities and challenges of artificial intelligence in the tourism industry. Glob J Tourism Leisure Hosp Manag 1(1):555553. https://doi.org/10.19080/GJTLH.2024.01.555553
26. García-Madurga M-A, Grilló-Méndez A-J (2023) Artificial intelligence in the tourism industry: an overview of reviews. Administr Sci 13:172. https://doi.org/10.3390/admsci13080172
27. Ku ECS, Chen C-D (2024) Artificial intelligence innovation of tourism businesses: from satisfied tourists to continued service usage intention. Int J Inf Manage 76. https://doi.org/10.1016/j.ijinfomgt.2024.102757
28. Xiang Z, Fuchs M, Gretzel U, Höpken W (2020) Handbook of e-tourism. Springer Cham. https://doi.org/10.1007/978-3-030-05324-6_110-1
29. Bulchand-Gidumal J (2020) Impact of artificial intelligence in travel, tourism, and hospitality. Springer Nature Switzerland AG
30. Gallist N, Hagler J (2024) Tourism in the Metaverse: digital twin of a City in the Alps. MUM '23: Proceedings of the 22nd international conference on mobile and ubiquitous multimedia, pp 568–570. https://doi.org/10.1145/3626705.3631880
31. Sanchez-Amboage E, Crespo-Pereira V, Membiela-Pollan M, Jesus Faustino JP (2024) Tourism marketing in the metaverse: a systematic literature review, building blocks, and future research directions. PLoS ONE 19(5):e0300599. https://doi.org/10.1371/journal.pone.0300599
32. Doğan S, Bayar SB (2024) Hospitality and tourism marketing in an artificially intelligent world. Ondokuz Mayıs Üniversitesi Karadeniz Turizm Araştırmaları Dergisi 2(1):Ocak: 51–72
33. Rashideh W (2020) Blockchain technology framework: current and future perspectives for the tourism industry. Tour Manage 80. https://doi.org/10.1016/j.tourman.2020.104125
34. Ozdemir Ucgun G (2024) The Effects of metaverse on the tourism industry. J Metaverse 4(1):71–83. https://doi.org/10.57019/jmv.1466997
35. Liu YY, Lace N, Chen LL (2023) Metaverse for destination marketing in the Era of great changes. Proceedings of the 14th international multi-conference on complexity. Informatics and Cybernetics (IMCIC). Pp 156–161. https://doi.org/10.54808/IMCIC2023.01.156

36. Ordóñez MD et al (2020) IoT technologies and applications in tourism and travel industries. In: Internet of things—the call of the edge. River Publishers, pp 341–360
37. Santos JD, Rodrigues L, Lourenço JM (2023) The potential of 5G technology applied to tourism marketing. In: Santos J, Sousa B (Eds). Promoting organizational performance through 5G and agile marketing. IGI Global, pp 28–41. https://doi.org/10.4018/978-1-6684-5523-4.ch002
38. Vignaroli et al (2020) The touristic sector in the 5G technology Era: the 5G-TOURS project approach. 2020 IEEE Globecom workshops, Taipei, Taiwan, pp 1–6. https://doi.org/10.1109/GCWkshps50303.2020.9367418
39. Rokhsaritalemi S et al (2022) Ubiquitous tourist system based on multicriteria decision making and augmented reality. Appl Sci 12:5241. https://doi.org/10.3390/app12105241
40. Gökalp E, Eren PE (2014) Improvement of service quality by using ubiquitous computing in hospitality. In: Mola L, Carugati A. Kokkinaki A, Pouloudi N (eds) Proceedings of the 8th mediterranean conference on information systems. Verona, Italy
41. Ioannidis S, Kontis AP (2023) Metaverse for tourists and tourism destinations. Inf Technol Tourism 25:483–506. https://doi.org/10.1007/s40558-023-00271-y
42. Monaco S, Sacchi G (2023) Travelling the Metaverse: potential benefits and main challenges for tourism sectors and research applications. Sustainability 15:3348. https://doi.org/10.3390/su15043348
43. Gursoy D, Malodia S, Dhir A (2022) The metaverse in the hospitality and tourism industry: an overview of current trends and future research directions. J Hosp Market Manag 31(5):527–534. https://doi.org/10.1080/19368623.2022.2072504
44. Chen Z (2024) Beyond boundaries: exploring the Metaverse in tourism. Int J Contemp Hosp Manag. https://doi.org/10.1108/IJCHM-06-2023-0900
45. Triviño-Tarradas P, Mohedo-Gatón A, Carranza-Cañadas P, Hidalgo-Fernandez RE (2024) The application of Metaverse in the tourism sector as a tool for enhancing sustainability—case study: a medieval 'Perfume Burner' of the local historical museum of Montilla (Cordoba, Spain). Sustainability 16:6966. https://doi.org/10.3390/su16166966
46. Adnan N, Rashed MF, Ali W (2024) Embracing the metaverse: cultivating sustainable tourism growth on a global scale. Curr Issue Tour. https://doi.org/10.1080/13683500.2024.2390678
47. Buhalis D, Karatay N (2022). Mixed reality (MR) for generation Z in cultural heritage tourism towards Metaverse. In: Stienmetz JL, Ferrer-Rosell B, Massimo D (eds) Information and communication technologies in tourism 2022. ENTER 2022. Springer, Cham. https://doi.org/10.1007/978-3-030-94751-4_2
48. Das SD, Bala PK, Behera RK (2023) Data privacy in the metaverse ecosystem. In: Khang A, Shah V, Rani S (Eds), Handbook of research on AI-based technologies and applications in the era of the metaverse. IGI Global, pp 381–394. https://doi.org/10.4018/978-1-6684-8851-5.ch020
49. Murti K, Darma G, Mahyuni L, Gorda AN (2023) Immersive experience in the metaverse: implications for tourism and business. International J Appl Bus Res 5(02):187–207
50. Shin H, Kang J (2024) How does the metaverse travel experience influence virtual and actual travel behaviors? Focusing on the role of telepresence and avatar identification. J Hosp Tour Manag 58:174–183. https://doi.org/10.1016/j.jhtm.2023.12.009
51. Ahlgren L (2022) Try before you fly: vueling launches metaverse ticket sales experience. Simple Flying. Available at: https://simpleflying.com/vueling-metaverse-ticket-sales/
52. Constantin M, Genovese G, Munawar K, Stone R (2023) Tourism in the metaverse: can travel go virtual? McKinsey & Company. Available at: https://www.mckinsey.com/industries/travel-logistics-and-infrastructure/our-insights/tourism-in-the-metaverse-can-travel-go-virtual
53. Telefonica (2022) Tourism is joining the metaverse. Available at: https://www.telefonica.com/en/communication-room/news/tourism-is-joining-the-metaverse/
54. Barten M (2024) Metaverse tourism: overview, benefits, examples and more… REVFINE. Available at: https://www.revfine.com/metaverse-tourism/
55. Lee YC, Nguyen MN, Yang Q (2023) Factors influencing Vietnamese generation MZ's adoption of metaverse platforms. Sustainability 15(20):14940. https://doi.org/10.3390/su152014940

56. Lamba SS, Malik R (2022) Into the metaverse: Marketing to gen Z consumers. Applying metalytics to measure customer experience in the metaverse. India: IGI Global, pp 92–98. https://doi.org/10.4018/978-1-6684-6133-4.ch008
57. Corbisiero F, Monaco S, Ruspini E (2022) Millennials, generation Z and the future of tourism. Bristol: Channel View Publications. https://doi.org/10.21832/9781845417628
58. Shi F et al (2023) A new technology perspective of the Metaverse: Its essence, framework and challenges. Digital Commun Networks. https://doi.org/10.1016/j.dcan.2023.02.017
59. Rawat DB, El Alami H (2023) Metaverse: requirements, architecture, standards, status, challenges, and perspectives. arXiv preprint, arXiv:2302.01125v1
60. Saneinia S, Zhai X, Zhou R et al (2024) Beyond virtual boundaries: the intersection of the metaverse technologies, tourism, and lifelong learning in China's digital discourse. Humanit Soc Sci Commun. 11:1287. https://doi.org/10.1057/s41599-024-03624-y
61. Qin HX, Wang Y, Hui P (2022) Identity, crimes, and law enforcement in the metaverse. arXiv preprint arXiv:2210.06134
62. Okkerman M (2023) Holding up a black-mirror: exploring ethical issues of the Metaverse from a user perspective. Master Thesis. University of Twente. Available at: http://essay.utwente.nl/95127/1/Okkerman_MA_BMS.pdf
63. Hackl C (2020) Now is the time to talk about ethics and privacy in the metaverse. Forbes. Available at: https://www.forbes.com/sites/cathyhackl/2020/08/02/now-is-the-time-to-talk-about-ethics--privacy-in-the-metaverse/
64. Petr C, Caudan P (2024) Ethical marketing framework for metaverse simulated experiences of tourism (SET): an exploration of consumers' aspirations and fears. J Retail Consum Serv 79. https://doi.org/10.1016/j.jretconser.2024.103785
65. OECD (2010) Tourism 2020: policies to promote competitive and sustainable tourism. OECD Publishing. Paris, OECD Tourism Trends and Policies. https://doi.org/10.1787/tour-2010-4-en
66. Arbanas J et al (2023) Considerations for regulating the metaverse: new models for content, commerce, and data. Deloitte Insights. Available at: https://www2.deloitte.com/us/en/insights/industry/technology/emerging-regulations-in-the-metaverse.html
67. De Giovanni P (2023) Sustainability of the metaverse: a transition to industry 5.0. Sustainability 15(7):6079. https://doi.org/10.3390/su15076079

Mirjana Maksimović is an Associate Professor at the Faculty of Electrical Engineering, University of East Sarajevo, Bosnia and Herzegovina. She specializes in telecommunications, automation, and information technology, focusing on 5G networks and IoT applications. Dr. Maksimović holds a Ph.D. in Electrical Engineering and has authored three books and over 100 scientific papers. Recognized among the top 2% of most-cited researchers globally (Stanford University, 2020–2024), she actively participates in international conferences and standardization committees, Erasmus+ projects and COST actions.

Zvjezdana Krstić is an expert in marketing, specializing in digital transformation, consumer behavior, and artificial intelligence. She holds a Ph.D. in Marketing and has authored over 50 academic works, including books and book chapters, contributing significantly to marketing strategies. In 2023, she was appointed as an Associate Professor for a six-year term. Dr. Krstić has taught courses such as Consumer Behavior, Service Marketing, Digital Economy, and Marketing in Banking. She has actively participated in Erasmus+ projects and professional collaborations. In November 2024, she resigned from her position at the Faculty of Business Economics in Bijeljina and relocated to Munich, Germany, for new professional opportunities.

MIX
Papier aus verantwortungsvollen Quellen
Paper from responsible sources
FSC® C105338

If you have any concerns about our products,
you can contact us on
ProductSafety@springernature.com

In case Publisher is established outside the EU,
the EU authorized representative is:
Springer Nature Customer Service Center GmbH
Europaplatz 3, 69115 Heidelberg, Germany

Printed by Libri Plureos GmbH
in Hamburg, Germany

SpringerBriefs in Architectural Design and Technology

Series Editor

Thomas Schröpfer, Architecture and Sustainable Design, Singapore University of Technology and Design, Singapore, Singapore

Indexed by SCOPUS

Understanding the complex relationship between design and technology is increasingly critical to the field of Architecture. The *Springer Briefs in Architectural Design and Technology* series provides accessible and comprehensive guides for all aspects of current architectural design relating to advances in technology including material science, material technology, structure and form, environmental strategies, building performance and energy, computer simulation and modeling, digital fabrication, and advanced building processes. The series features leading international experts from academia and practice who provide in-depth knowledge on all aspects of integrating architectural design with technical and environmental building solutions towards the challenges of a better world. Provocative and inspirational, each volume in the Series aims to stimulate theoretical and creative advances and question the outcome of technical innovations as well as the far-reaching social, cultural, and environmental challenges that present themselves to architectural design today. Each brief asks why things are as they are, traces the latest trends and provides penetrating, insightful and in-depth views of current topics of architectural design. *Springer Briefs in Architectural Design and Technology* provides must-have, cutting-edge content that becomes an essential reference for academics, practitioners, and students of Architecture worldwide.

Vikas Chand Sharma · Sandeep Dua ·
Pankaj Kumar Tyagi · Elena Tarsi
Editors

Architectural Vision of Overtourism

Volume 1

Springer

Editors
Vikas Chand Sharma
Liberal Arts and Humanities Cluster,
University Institute of Architecture
Chandigarh University
Mohali, Punjab, India

Pankaj Kumar Tyagi
University Institute of Tourism
and Hospitality Management
Chandigarh University
Mohali, Punjab, India

Sandeep Dua
Department of Architecture
Guru Nanak Dev University
Amritsar, Punjab, India

Elena Tarsi
Department of Architecture
University of Florence
Florence, Italy

ISSN 2199-580X ISSN 2199-5818 (electronic)
SpringerBriefs in Architectural Design and Technology
ISBN 978-981-96-5793-3 ISBN 978-981-96-5794-0 (eBook)
https://doi.org/10.1007/978-981-96-5794-0

This Springer imprint is published by the registered company Springer Nature Singapore Pte Ltd.
The registered company address is: 152 Beach Road, #21-01/04 Gateway East, Singapore 189721, Singapore

Preface

In an era of unprecedented global mobility and digital connectivity, overtourism has emerged as one of the most critical challenges facing destinations worldwide. The phenomenon has evolved from a mere management issue to a complex crisis that threatens the very essence of our cultural heritage sites, historic cities, and local communities. While tourism continues to be a vital economic driver, generating approximately 10% of global GDP and supporting millions of jobs, its unchecked growth has created an unsustainable burden on many destinations. From the crowded streets of Venice to the delicate ecosystems of the Himalayas, from ancient religious monuments to modern urban centers, the impacts of overtourism are increasingly visible and concerning. This crisis demands a fundamental rethinking of how we approach tourism development, particularly from an architectural and urban planning perspective. The intersection of architecture, technology, and sustainable tourism practices offers promising pathways forward, yet also raises critical questions about preservation, authenticity, and the future of travel itself. This book, presents a comprehensive examination of these challenges and opportunities, exploring how architectural interventions and technological innovations can address the complex challenges posed by overtourism while promoting sustainable urban development.

The volume is structured in four interconnected parts, each addressing distinct yet interrelated aspects of the overtourism phenomenon. The first part, *Understanding Overtourism*, establishes the theoretical foundation by examining fundamental concepts and presenting case studies from various global contexts, Tourism, once celebrated primarily for its economic benefits, has evolved into a complex phenomenon that challenges the very fabric of our architectural and urban spaces. This part presents five distinct perspectives on how excessive tourism impacts our built environment and cultural heritage, from religious monuments in Florence to the intricate canals of Venice. Authors examine how overtourism manifests across diverse settings: historic religious sites in Florence grappling with preservation and accessibility; traditional villages in Turkey and Indonesia balancing cultural authenticity with tourist demands; Venice's unique urban fabric facing pressures that threaten its community identity; and the emerging role of smart cities in managing tourist flows.

Through these varied lenses, authors examine innovative solutions that integrate strategic planning, technological innovation, and community engagement.

This part underscores a crucial message: the architectural response to overtourism must extend beyond physical infrastructure to encompass the social, cultural, and technological dimensions of our evolving cities and heritage sites. The authors present frameworks and strategies that aim to transform overtourism's challenges into opportunities for more sustainable and resilient urban environments, while preserving the authentic character that makes these destinations worth visiting in the first place.

The second part, *Technology and Overtourism*, explores the dual nature of technology in tourism management. While acknowledging technology's potential contribution to overtourism, this part also investigates how digital solutions, artificial intelligence, and smart infrastructure can be harnessed to mitigate its negative impacts. The chapters in this part present we delve into the complex relationship between technology and overtourism, exploring both its promises and pitfalls in shaping our architectural and urban landscapes. The chapters in this part present a critical examination of how technological solutions, particularly artificial intelligence and smart infrastructure, are being deployed to address the challenges of excessive tourism.

From the dark side of technology's influence on tourism trends to the practical application of AI in managing tourist flows in Nainital's lake town, these contributions reveal the multifaceted nature of technological interventions. Through case studies spanning from Florence's Historic Centre to Slovakia's High Tatras, our authors examine how smart tourism infrastructure and digital innovations are reshaping destination management strategies. However, this part goes beyond mere technological optimism. It critically evaluates whether technology truly offers solutions to overtourism or potentially exacerbates the problem. The authors present both the opportunities and limitations of digital tools in preserving architectural heritage, managing tourist flows, and maintaining community balance. Their insights suggest that while technology can be a powerful ally in addressing overtourism, it must be implemented thoughtfully and in conjunction with sound urban planning principles.

The third part, *Architectural Responses to Overtourism*, focuses on concrete design interventions and urban planning solutions. Through case studies from diverse geographical contexts, the chapters explore innovative architectural and urban planning solutions that respond to the challenges of excessive tourism, with a particular focus on emerging economies and cultural heritage sites. Through diverse case studies spanning from the historic Shahjahanabad in Delhi to the sensitive ecosystems of the Himalayas, and from traditional Asian markets to the urban centers of Northeast India, this part examines how architectural interventions can help balance tourism demands with community needs. Authors present thoughtful analyses of how sustainable design, smart technologies, and community engagement can work together to address overtourism's impacts. The chapters explore various approaches: the integration of digital technologies in traditional markets, the use of smart urban design in mountain destinations, the role of community participation through digital platforms, and the importance of understanding tourist psychology in architectural solutions.

A common thread running through these contributions is the emphasis on combining technological innovation with sensitive architectural design to preserve cultural heritage while managing tourist flows. The authors demonstrate how architectural solutions, when properly conceived and implemented, can help distribute tourism pressure, enhance visitor experiences, and protect local communities and their cultural assets. This part emphasizes that successful architectural responses to overtourism must be holistic, considering not just physical infrastructure, but also the social, cultural, and environmental dimensions of tourism development.

The final part, *The Future of Tourism*, looks ahead to emerging solutions and technologies, including the potential of the metaverse in promoting sustainable travel practices and the role of technology in cultural regeneration. Through two distinct but complementary perspectives, this part explores how emerging technologies might transform tourism experiences while addressing the persistent challenges of overtourism. The first contribution examines the intersection of technological innovation and cultural regeneration in Indian heritage precincts, from the bustling streets of Ahmedabad and Mumbai to the historic landscapes of Hampi and Mysore. The second chapter ventures into the virtual realm, investigating how Metaverse marketing could revolutionize sustainable travel practices and offer alternative ways to experience destinations without physical presence. Together, these chapters paint a picture of tourism's potential future—one where physical and virtual experiences coexist, where technology serves as a tool for both preservation and access, and where innovative solutions help balance the competing demands of tourism growth and heritage conservation. This part offers valuable insights for architects, urban planners, and policymakers working to shape the future of tourist destinations in an increasingly digital world.

This book serves as a resource for urban planners and architects seeking strategies to mitigate the negative impacts of excessive tourism. Tourism management professionals and policymakers responsible for developing sustainable tourism strategies will find guidance on best practices. Academics and researchers in urban studies, tourism, and sustainable development will benefit from the book's theoretical framework and empirical evidence. Technology professionals developing smart city solutions and digital tourism management tools will find relevant explorations of innovative technologies and their applications. Finally, the book offers a comprehensive overview of sustainable urban tourism for graduate students in architecture, urban planning, tourism management, and related disciplines. By bringing together perspectives from scholars and practitioners across multiple disciplines, this volume contributes to the growing body of knowledge on sustainable tourism development while offering practical solutions to one of the most significant challenges facing urban centers today. The book aligns with the United Nations Sustainable Development Goal 11, which aims to make cities inclusive, safe, resilient, and sustainable, while addressing the specific challenges posed by mass tourism.

As tourism continues to grow globally, the need for innovative solutions to manage its impact becomes increasingly urgent. This book provides a timely contribution to this critical discourse, offering both theoretical frameworks and practical approaches to creating more resilient and sustainable urban environments in the face of growing tourism pressures.

Mohali, India Vikas Chand Sharma
Amritsar, India Sandeep Dua

Contents

Editors and Contributors

About the Editors

Vikas Chand Sharma is an Associate Professor at Chandigarh University and a researcher at Guru Nanak Dev University, Amritsar, brings 13 years of experience in architectural design, urban planning, and research. He specializes in sustainable urban design strategies for historic cities, combining phenomenological approaches with digital mapping. His foundational research on religious spatial patterns and new urbanism was published in 2016. He serves on the Scientific Committee for Contesti. Città, Territori, Progetti and is a member of the Editorial Board for several international journals. His work appears in the International Journal of Construction Education and Research, Journal of Architecture & Urbanism, and Landscape Research.

Sandeep Dua is a Professor in the Department of Architecture at Guru Nanak Dev University, Amritsar. He holds a PG Diploma in Urban Design from CEPT University, Ahmedabad, and a PhD in Architecture from IIT Roorkee. His research interests encompass mathematics in architecture and urban form, history and theory of architecture and urban design. With a 30-year experience, Dr. Dua has published extensively on topics such as medieval domestic architecture, colonial and post-colonial architectural styles, and urban form and architecture. He has also undertaken research project funded by AICTE, New Delhi.

Pankaj Kumar Tyagi is an accomplished academician, researcher, and consultant. He is presently working as a professor at Chandigarh University, Mohali, India. He is an alumnus of Hemwati Nandan Bahuguna Garhwal University (Central University) and Kurukshetra University, India. He has more than 22 years of experience in the industry, teaching, training, consultancy, and expertise in destination management, travel process and disruptive technologies. He has published over 40 research papers in international journals indexed in the quality database and authored/edited

20 books. He is a reviewer for many international journals in tourism, management, and technology.

Elena Tarsi is an architect and has done Ph.D. in urban territorial and environmental planning. She is currently a researcher at the Department of Architecture of University of Florence where she is the coordinator of LabPSM—Laboratory City and territory in the Global South. She is also "collaborator researcher" at the Centre for Social Studies of University of Coimbra. Her research interests focus on informal settlements dynamics both in the Global South and the Global North. Recently, she has dealt with the impacts of climate change in cities and the role of cities and territories in the European Green Deal policies.

Contributors

Srinidhi D. Acharya Manipal School of Architecture and Planning, Manipal Academy of Higher Education, Manipal, Karnataka, India

Sobia Ahsan Department of Architecture, Jamia Millia Islamia, New Delhi, India

Bassam Samir Al-Romeedy Tourism Studies Department, Faculty of Tourism and Hotels, University of Sadat City, Sadat City, Egypt

Rashmi Ashtt Indira Gandhi Delhi Technical University for Women, Delhi, India

MinJoo Baek Department of Architecture, Xi'an Jiaotong-Liverpool University, Suzhou, China

Vanshika Bajaj Department of Architecture, Jamia Millia Islamia, New Delhi, India

Manju Bala University of Liverpool, Liverpool, UK

Sriya Banerjee Department of Architecture and Regional Planning, Indian Institute of Technology, Kharagpur, West Midnapore, India

Gianluca Belli University of Florence, Florence, Italy

Giada Cerri Dipartimento di Architettura, Università degli Studi di Firenze, Florence, Italy

Ana Rita Conde University Lusófona, Porto, Portugal

Marta Conte University of Florence, Florence, Italy

Giuseppe De Luca University of Florence, Florence, Italy

Carlo Francini Municipality of Florence, Florence, Italy

Martina Franco University of Florence, Florence, Italy

Tomáš Gajdošík Faculty of Economics, Matej Bel University in Basnká Bystrica, Banská Bystrica, Slovakia

Abraham George Department of Architecture and Regional Planning, Indian Institute of Technology, Kharagpur, West Midnapore, India

Shivani Goel Indira Gandhi Delhi Technical University for Women, Delhi, India

Ricardo Jorge da Costa Guerra Polytechnic University of Guarda, Seia, Portugal

Jit Kumar Gupta College of Architecture, IET Bhaddal Technical Campus, Ropar, Punjab, India

Prasoon Gupta Indira Gandhi Delhi Technical University for Women, Delhi, India

Zeeshan Ibrar Department of Architecture and Planning, Netaji Subhas University of Technology (NSUT), New Delhi, India

Nisar Khan Department of Architecture, Jamia Millia Islamia, New Delhi, India

Amit Kinjawadekar Manipal School of Architecture and Planning, Manipal Academy of Higher Education, Manipal, Karnataka, India

Maximiliano E. Korstanje University of Palermo, Buenos Aires, Argentina

Zvjezdana Krstić Independent Researcher (Previously affiliated with Faculty of Business Economics Bijeljina, University of East Sarajevo), Bijeljina, Bosnia and Herzegovina

Ahmet Kurnaz Istanbul Sabahattin Zaim University, Istanbul, Turkey

Jose Antonio Lara-Hernandez Auckland University of Technology, Auckland CBD, Auckland, New Zealand

Luísa Lopes Instituto Politécnico de Bragança, Mirandela, Portugal

Shem Wambugu Maingi Kenyatta University, Nairobi, Kenya

Mirjana Maksimović Faculty of Electrical Engineering, University of East Sarajevo, East Sarajevo, Bosnia and Herzegovina

Marco Martins CiTUR, Coimbra, Portugal

Iasef Md. Rian Department of Architecture, Xi'an Jiaotong-Liverpool University, Suzhou, China

Lara Santos Polytechnic University of Cávado and Ave, Barcelos, Portugal

Vikas Chand Sharma University Institute of Architecture, Chandigarh University, Mohali, 140413 Punjab, India

A. S. Sochipem Zimik SRM University, Gangtok, Sikkim, India

Pankaj Kumar Tyagi University Institute of Tourism and Hospitality Management, Chandigarh University, Mohali, 140413 Punjab, India

Monali Wankar Indira Gandhi Delhi Technical University for Women, Delhi, India

Understanding Overtourism

Part-I Understanding Overtourism, establishes the theoretical foundation by examining fundamental concepts and presenting case studies from various global contexts, Tourism, once celebrated primarily for its economic benefits, has evolved into a complex phenomenon that challenges the very fabric of our architectural and urban spaces. This part presents five distinct perspectives on how excessive tourism impacts our built environment and cultural heritage, from religious monuments in Florence to the intricate canals of Venice. Authors examine how overtourism manifests across diverse settings: historic religious sites in Florence grappling with preservation and accessibility; traditional villages in Turkey and Indonesia balancing cultural authenticity with tourist demands; Venice's unique urban fabric facing pressures that threaten its community identity; and the emerging role of smart cities in managing tourist flows. Through these varied lenses, authors examine innovative solutions that integrate strategic planning, technological innovation, and community engagement. This part underscores a crucial message: the architectural response to overtourism must extend beyond physical infrastructure to encompass the social, cultural, and technological dimensions of our evolving cities and heritage sites. The authors present frameworks and strategies that aim to transform overtourism's challenges into opportunities for more sustainable and resilient urban environments, while preserving the authentic character that makes these destinations worth visiting in the first place.

Managing Overtourism: An Antithesis to Responsible Tourism Issues and Options

Vikas Chand Sharma and Jit Kumar Gupta

Abstract Considering the role, relevance, and importance of tourism in promoting economic development and cultural exchange, this paper examines overtourism as an antithesis to responsible tourism, exploring how unmanaged mass tourism contradicts principles of sustainability and community well-being. Through analysis of case studies and existing literature, the research identifies key factors contributing to overtourism and evaluates their multifaceted impacts across physical, social, cultural, economic, and environmental dimensions. The study proposes a comprehensive management framework integrating strategic planning, operational management, stakeholder engagement, and technological innovation, supported by institutional architecture and resource allocation mechanisms. This framework contributes to transforming overtourism into responsible tourism by emphasizing the need for context-specific solutions while maintaining alignment with broader sustainability principles. The research highlights the critical balance between economic benefits and destination sustainability, offering a structured yet flexible approach for policymakers and tourism stakeholders to achieve more sustainable tourism practices.

Keywords Overtourism · Responsible tourism · Sustainable destination management · Tourism policy · Community impact

1 Introduction

Tourism, long celebrated as a driver of economic growth and cultural exchange, has in many destinations evolved into its antithesis—overtourism. This phenomenon, characterized by number of visitors exceeding a destination's carrying capacity, stands in stark contrast to the principles of responsible tourism. While responsible tourism aims

V. C. Sharma (✉)
University Institute of Architecture, Chandigarh University, Mohali 140413, Punjab, India
e-mail: ar.vikassharma@gmail.com; vikas.uia@cumail.in

J. K. Gupta
College of Architecture, IET Bhaddal Technical Campus, Ropar, Punjab, India

V. C. Sharma et al. (eds.), *Architectural Vision of Overtourism*,
SpringerBriefs in Architectural Design and Technology,
https://doi.org/10.1007/978-981-96-5794-0_1

to create better places to live and visit, overtourism often results in environmental degradation, cultural erosion, and marginalization of local communities. The concept of overtourism, while relatively recent in academic discourse, has gained prominence in tourism studies. Goodwin (2017) defines overtourism as a situation where, "hosts or guests, locals or visitors, feel that there are too many visitors and that the quality of life in the area or the quality of the experience has deteriorated unacceptably" [1]. This definition emphasizes the subjective nature of overtourism, focusing on perception rather than absolute numbers. The United Nations World Tourism Organization offers a more comprehensive definition, describing overtourism as "the impact of tourism on a destination, or parts thereof, that excessively influences perceived quality of life of citizens and/or quality of visitors experiences in a negative way" [2]. This definition broadens the scope to include both residents and visitors, highlighting the dual impact of excessive tourism. Peeters et al. (2018), in a study for the European Parliament, conceptualize overtourism as "the situation in which the impact of tourism, at certain times and in certain locations, exceeds physical, ecological, social, economic, psychological, and/or political capacity thresholds" [3]. This definition introduces the notion of capacity thresholds across various dimensions, providing a more holistic view of the phenomenon. Milano et al. (2019) argue that overtourism is not merely a numbers game but a complex socio-spatial phenomenon. They suggest that it "should be analysed as the excessive growth of visitors leading to overcrowding in areas where residents suffer the consequences of temporary and seasonal tourism peaks, which have enforced permanent changes to their lifestyles, access to amenities and general well-being" [4]. Interestingly, Boissevain's earlier work (1996) on mass tourism in Europe foreshadowed many of the issues now associated with overtourism, highlighting the long-standing nature of these concerns despite the relatively recent coinage of the term [5]. The diverse definitions in the literature reflect the multifaceted nature of overtourism, encompassing quantitative measures of visitor numbers, qualitative assessments of experience, and the broader socio-ecological impacts on destinations. However, Goodwin enumerates the impact of overtourism in terms of volume, linking it with conflict in behaviour, privacy, and experience of both visitors and residents. He says, "Overtourism is not only a function of visitor volumes, but also of conflicting behaviours, crowding in inappropriate places and privacy" [1]. Overtourism occurs when too many visitors flock to the same place, causing detrimental impacts on the environment, local communities, and visitor experience. This complexity underscores the challenge in developing universally applicable strategies to mitigate overtourism's effects. This paper examines the paradox of overtourism as a direct challenge to responsible tourism practices. It aims to address the critical need for transforming mass tourism into sustainable visitation by analysing the issues arising from overtourism and exploring management options that align with responsible tourism principles.

1.1 Causes of Overtourism

Study and analysis of the literature identify multiple interrelated factors contributing to the emergence and persistence of overtourism. These causes are often rooted in broader socio-economic trends and technological advancements. Globalization and economic liberalization have played a significant role in facilitating increased international travel. Dodds and Butler (2019) argue that the ease of cross-border movement and trade has made tourism more accessible to a wider population [6]. This trend is further amplified by the growing influence of global agencies like UNDP in promoting tourism as a development strategy [2]. The rapid growth of the aviation sector has been a key driver of overtourism. Gössling and Peeters (2015) note that increased competition and technological advancements have led to more affordable air travel, with global passenger numbers nearly tripling over two decades [7]. Rapid growth in air travel capacity has significantly reduced barriers to international tourism. Economic factors also play a crucial role. The expansion of the global middle class, estimated at 140 million new entrants annually as of 2017, has created a vast new market for tourism [8]. This demographic shift, coupled with increasing disposable incomes in many countries, has fuelled demand for travel experiences. Prioritizing travel as an essential activity of human living and dedicating greater proportion of discretionary income to travel with spending on foreign travel increasing more than four times faster than physical goods, between 2014 and 2016, highlight the shift in priorities in the developing countries like the USA.

Technological advancements, particularly in the realm of online booking and sharing platforms, have revolutionized the tourism industry. Guttentag (2015) highlights how platforms like Airbnb have disrupted traditional accommodation models, often leading to overtourism in residential areas, not traditionally equipped for high tourist volumes [9]. Social media has emerged as a powerful force in shaping travel trends. Jansson (2018) argues that the desire for shareable experiences drives tourists to specific, often already popular, destinations, creating a self-reinforcing cycle of overtourism [10]. The rise of the "digital nomad" phenomenon, accelerated by the COVID-19 pandemic, has blurred the lines between tourism and residency. Nash et al. (2021) suggest that this trend is reshaping urban dynamics in popular destinations like Lisbon and Mexico City. Inadequate tourism regulation and planning at the destination level have exacerbated overtourism [10]. Oklevik et al. (2019) point out that many destinations lack the governance structures and policies necessary to manage rapid increase in visitor numbers effectively [11].

The promotion of tourism as an economic development strategy by national and local governments has often prioritized quantity over quality. Seraphin et al. (2018) argue that this approach, focused on increasing visitor numbers without adequate consideration of carrying capacity, has contributed significantly to overtourism in many destinations [12]. Lastly, the phenomenon of "bucket list" tourism, driven by media and popular culture, has concentrated a large number of tourists in a relatively small number of iconic destinations. Zygmunt (2019) suggest that this trend has led to the overexploitation of cultural and natural heritage sites [13]. Understanding these multifaceted causes is crucial for developing effective strategies to mitigate overtourism and promote more responsible forms of tourism.

1.2 Impact of Overtourism

The manifestation of overtourism's impacts became particularly evident during and after the COVID-19 pandemic. While the tourism industry experienced unprecedented disruption with international arrivals declining by 72% in 2020 [2], the post-pandemic recovery has highlighted the persistent challenges of unsustainable tourism practices. The rapid rebound in international travel, with tourist numbers doubling in the first quarter of 2023 compared to 2022, coupled with projections of tourism contributing 11.6% to the global economy by 2033, signals potential intensification of overtourism issues [3]. This surge in tourism, concentrated in select destinations, presents significant environmental challenges. The UNWTO (2018) projects that transport-related carbon emissions from tourism will increase by 25% between 2016 and 2030, elevating tourism's contribution to global emissions from 5 to 5.3%. This trend aligns with Milano et al.'s (2019) observation that overtourism's environmental impact extends beyond local degradation to contribute significantly to global climate challenges [4].

The socio-economic implications of overtourism are equally concerning. As Dodds and Butler (2019) note, the growing disparity between locals and visitors often manifests in social tensions and community displacement [6]. This phenomenon is exacerbated by tourist behaviours that Seraphin et al. (2018) identify as particularly problematic: disrespect for local customs, vandalism of cultural sites, and intrusion into residential areas [12]. The physical manifestations of overtourism, such as noise pollution from tour groups and vehicular congestion in residential areas, further contribute to what Goodwin (2017) describes as a deterioration in residents' quality of life [1]. These impacts, representing overtourism's antithetical relationship to responsible tourism principles, necessitate a comprehensive analytical framework. Following Milano et al.'s (2019) approach, the adverse consequences can be systematically categorized into physical, social, cultural, economic, and environmental dimensions, providing a structured basis for understanding and addressing these challenges (Table 1).

The key impacts of overtourism demonstrate its pervasive influence across physical, social, cultural, economic, and environmental dimensions. Notably, the analysis reveals that these impacts often manifest as long-term and potentially irreversible changes, affecting not only the physical environment and infrastructure but also the social fabric, cultural heritage, economic stability, and overall well-being of local communities. This underscores the urgency for sustainable tourism management practices.

Table 1 Socio-cultural, economic, and environmental impacts of tourism

Impact category	Specific consequence/ manifestation	Key affected areas	Severity of impact
Physical	Overcrowding and congestion in urban spaces, leading to long queues and wait times	Tourists, local residents, infrastructure	Long-term, potentially severe, impacting quality of experience and daily life
	Deterioration and damage to heritage sites due to increased visitor traffic and wear and tear	Historical and cultural sites, cultural heritage, future generations	Long-term, potentially irreversible, impacting cultural preservation
Social	Displacement of local communities due to rising housing costs and short-term rentals	Local residents, low-income communities, community fabric	Long-term, potentially severe, leading to social disruption and loss of community identity
	Increased social stratification and dilution of community values due to mixing of diverse tourist populations	Local communities, social cohesion, cultural identity	Long-term, potentially moderate to severe, impacting social harmony and cultural integrity
Cultural	Dilution, damage, and destruction of local culture and traditions due to commercialization and exposure to external influences	Local culture, traditional practices, cultural heritage	Long-term, potentially irreversible, leading to loss of cultural identity and authenticity
	Rise in cultural clashes and conflicts between locals and tourists due to differing values and behaviours	Local residents, tourists, intercultural relations	Short- to long-term, potentially moderate to severe, impacting social harmony and tourism experience
Economic	Increased cost of living and economic marginalization of local residents due to rising prices and competition for resources	Local residents, low-income communities, local businesses	Long-term, potentially severe, impacting economic stability and social equity
	Siphoning off of economic benefits by external businesses and corporations, with limited benefits for local communities	Local economy, local businesses, local communities	Long-term, potentially moderate to severe, impacting local economic development and sustainability

(continued)

Table 1 (continued)

Impact category	Specific consequence/ manifestation	Key affected areas	Severity of impact
Environmental	Increased waste generation, pollution (air, water, noise), and strain on local resources due to increased tourism activity	Local environment, ecosystems, public health, natural resources	Long-term, potentially severe, impacting environmental sustainability and public well-being
	Increased pressure on natural resources and ecosystems, including damage to biodiversity, flora, and fauna	Local ecosystems, biodiversity, natural environment	Long-term, potentially irreversible, impacting ecological integrity and environmental sustainability

Source By Authors

1.3 Principles for Promoting Responsible Tourism

The preceding discussion of overtourism's impacts clearly demonstrates the need for a paradigm shift in tourism management. Rather than focusing solely on maximizing visitor numbers and economic returns, critical would be to have a more responsible approach that prioritizes the well-being of local communities, ensures the protection of cultural and natural heritage, and promotes long-term sustainability of destinations. The core principles guiding this shift towards responsible tourism can be enumerated as follows (Table 2).

2 Discussion

2.1 Current Research on Managing Overtourism

The growing challenges of overtourism have impelled concerns among local communities in popular tourist destinations worldwide. Barcelona and Venice, two of the most visited destinations globally, exemplify the critical state of mass tourism, where both residents and visitors experience disillusionment and marginalization. The fundamental question emerging from this crisis is whether destinations will use tourism or be used by it, leading cities and nations to develop various management strategies for addressing unsustainable tourism practices. Amsterdam's response to overtourism demonstrates the complexity of local interventions. The city launched a comprehensive citizen-driven movement focusing on reducing congestion, regulating tourist numbers, and improving visitor behaviour while prioritizing local residents' quality of life [6]. Similarly, smaller destinations like the Austrian lakeside village of Hallstatt, with just 800 residents facing approximately 10,000 daily visitors, have

Table 2 Principles for promoting responsible tourism

Category	Principles for promoting responsible tourism	Source
Understanding and assessment	Understanding the causes of unsustainable tourism Evaluating the social, economic, environmental, cultural, and physical impacts of unsustainable tourism	[14, 15]
Local community focus	Making tourism supportive and empowering of local communities (socially, physically, economically) Involving local communities as co-partners in addressing tourism issues and planning Avoiding marginalization of local communities in accessing basic amenities Improving the quality of life of local residents Avoiding tourism in areas of special significance to the local community	[16, 17]
Sustainable practices and development	Rationalizing tourism with a focus on promoting the local economy, environment, employment, and culture Focusing on quality rather than quantity and promoting sustainable tourism Making tourism supportive of the local culture, environment, and ecology Reinventing tourism for value addition to places and public spaces Making tourism a promoter of sustainable and planned development Ensuring the induction of sustainable practices in all tourism-related goods and services Creating a sustainable visitor economy that balances quality of living and working	[18, 19]
Environmental and resource management	Minimizing air and noise pollution Rationalizing traffic and transportation Maintaining the sanctity of public spaces by eliminating overcrowding Ensuring appropriate maintenance and upkeep of local heritage and tourist destinations	[20–22]

Source By Authors

implemented strict visitor caps through tour coach restrictions [12]. Venice residents campaigned for cruise ship restrictions, while the Marseille Provence Cruise Club introduced an innovative flow management system around popular attraction. Orkney, Scotland, limited cruise ship docking, and destinations like Venice and Barcelona implemented tourist taxes based on accommodation type [2]. Rome focused on behavioural regulations, restricting activities at popular heritage sites like the Trevi Fountain and Spanish Steps to preserve their integrity.

The mounting challenges of overtourism have led cities worldwide to explore and implement various management strategies to mitigate its adverse impacts. Research has shown that several approaches have emerged as particularly effective in addressing the influx of visitors. A significant trend identified in recent studies involves the implementation of radical policy measures, with destinations taking decisive action by restricting travel modes and closing terminals that traditionally brought a large number of tourists [11]. Some destinations have embraced the concept of degrowth, as documented by Milano et al. (2019), by intentionally reducing tourist numbers through economic deterrents such as increased fees, charges, and bureaucratic requirements at the local level [4]. This approach aligns with the broader movement towards sustainable tourism management. Similarly, the strategy of desaturation has gained traction, with local tourism industries being encouraged to shift their focus from quantity to quality, prioritizing higher yield tourism segments rather than mass tourism [12].

Research by Dodds and Butler (2019) highlights the effectiveness of dispersal strategies, where destinations work to redistribute tourist flows across wider areas to prevent concentration in specific locations [6]. However, they emphasize the importance of careful implementation to avoid merely shifting overtourism problems from one area to another. In parallel, demarketing has emerged as a long-term management tool, with varying degrees of success. This approach includes strategic price increases, reduced promotional activities, and transparent communication about deteriorating visitor experiences [1]. Studies have also documented the implementation of differential entry fees as an economic tool for managing visitor flows. According to Peeters et al. (2018), this approach operates on the assumption that visitors to overcrowded destinations will accept higher charges, with the generated revenue being reinvested in sustainable tourism infrastructure and management solutions [3]. Complementing these structural approaches, research has demonstrated the effectiveness of advocacy and awareness campaigns in modifying tourist behaviour. These initiatives focus on promoting responsible tourism practices and encouraging visitors to take accountability for their impact on local communities and cultures [2].

These diverse management approaches reflect the complex nature of overtourism and the need for multifaceted solutions. Their implementation and effectiveness vary across destinations, highlighting the importance of context-specific strategies in sustainable tourism management.

2.2 *Options for Rationalizing Overtourism*

The global proliferation of overtourism demands systematic and strategic interventions that align with the principles of responsible tourism while addressing the complex challenges faced by destinations worldwide. Drawing from successful case studies, we propose a comprehensive framework of options for rationalizing overtourism, organized around four key pillars: strategic planning, operational management, stakeholder engagement, and technological innovation.

- Strategic Planning and Policy Framework

Category	Strategies/actions
Destination capacity assessment	Evaluate carrying capacities: physical, social, ecological
	Develop evidence-based visitor thresholds
	Implement dynamic pricing based on demand and seasons
	Establish destination hierarchies (cultural, environmental, infrastructure)
Spatial planning and distribution	Create tourism master plans integrating urban strategies
	Establish dedicated tourism zones with infrastructure
	Develop visitor dispersal strategies with alternative attractions
	Create tourism circuits to distribute visitor loads

- Operational Management Systems

Category	Strategies/actions
Visitor flow management	Use timed entry and advanced bookings
	Deploy flexible pricing mechanisms
	Establish dedicated transport and access systems
	Monitor visitors through feedback systems
Quality control and standards	Develop multi-criteria destination rating systems
	Certify tourism service providers
	Monitor environmental and social impacts
	Ensure quality visitor experiences

- Stakeholder Engagement and Community Partnership

Category	Strategies/actions
Community integration	Facilitate community participation in planning
	Ensure equitable revenue-sharing programs
	Foster dialogue between residents and stakeholders
	Use community-based monitoring systems
Industry collaboration	Build partnerships with tour operators
	Create standards for responsible tourism

(continued)

(continued)

Category	Strategies/actions
	Launch collaborative sustainable marketing efforts
	Jointly train and build capacity

- Technological Innovations and Digital Solutions

Category	Strategies/actions
Smart tourism infrastructure	Introduce digital visitor management
	Develop real-time monitoring and crowd solutions
	Create integrated booking and capacity platforms
	Offer digital wayfinding systems
Virtual and augmented experiences	Enable virtual tourism for sensitive sites
	Enhance understanding through augmented reality
	Implement digital interpretation tools
	Provide online educational platforms

The effective operationalization of these management options necessitates a robust implementation framework anchored in two critical pillars: institutional architecture and resource allocation. The institutional component requires establishing dedicated tourism management authorities that can facilitate cross-departmental coordination while implementing comprehensive monitoring and evaluation frameworks. This organizational structure must be complemented by adaptive management systems that can respond to changing tourism dynamics. Equally crucial is the strategic allocation of resources, encompassing sustainable funding mechanisms, infrastructure investment programs, and capacity-building initiatives. As Peeters et al. (2018) emphasize, the successful implementation of technological solutions further demands structured adoption strategies and sustained resource commitment [3]. This dual focus on institutional capacity and resource management creates a comprehensive framework that allows destinations to customize solutions while adhering to sustainable tourism principles. The framework's adaptability enables destinations to address their unique challenges while maintaining alignment with broader sustainability goals, as demonstrated in cities like Barcelona and Venice. This integrated approach, combining strategic planning, operational excellence, stakeholder engagement, and technological innovation, provides a structured pathway for transforming overtourism into responsible tourism practices.

3 Conclusion

This study has critically examined overtourism as an antithesis to responsible tourism, demonstrating the complex interplay between tourism growth and destination sustainability. The analysis reveals that overtourism is not merely a product of increased visitor numbers but emerges from the intersection of multiple factors including globalization, technological advancement, and inadequate tourism governance. The comprehensive examination of impacts across physical, social, cultural, economic, and environmental dimensions underscores the urgent need for transformative approaches to tourism management.

The research contributes significantly to the existing body of knowledge by synthesizing diverse management approaches and presenting an integrated framework that combines strategic planning, operational management, stakeholder engagement, and technological innovation. This framework, supported by the proposed institutional architecture and resource allocation mechanisms, offers a structured yet flexible approach to addressing overtourism challenges. As demonstrated through various case studies, successful management of overtourism requires a delicate balance between economic benefits and destination sustainability. The findings emphasize that transforming overtourism into responsible tourism is not a one-size-fits-all proposition but rather a context-specific process that demands careful consideration of local conditions, community needs, and destination capacity [12]. The proposed management framework provides destinations with the tools to develop tailored solutions while adhering to broader principles of sustainable tourism development. Future research could valuably explore the long-term effectiveness of these integrated approaches and their adaptation across different destination types and scales.

References

1. Goodwin H (2017) The challenge of overtourism. Responsible tourism partnership working paper 4
2. UNWTO (2018) 'Overtourism'? Understanding and managing urban tourism growth beyond perceptions. UNWTO. https://www.unwto.org/global/publication/overtourism-understanding-and-managing-urban-tourism-growth-beyond-perceptions-volume-0
3. Peeters PM, Gössling S, Klijs J, Milano C, Novelli M, Dijkmans CHS, Eijgelaar E, Hartman S, Heslinga J, Isaac R, Mitas O, Moretti S, Nawijn J, Papp B, Postma A (2018) Research for TRAN committee—overtourism: impact and possible policy responses. European parliament, directorate general for internal policies, policy department B: structural and cohesion policies, transport and tourism. https://doi.org/10.2861/919195
4. Milano C, Cheer JM, Novelli M (2019) Overtourism: excesses, discontents and measures in travel and tourism. CABI. https://doi.org/10.1079/9781786399823.0000
5. Boissevain J (ed) (1996) Coping with tourists: European reactions to mass tourism. Berghahn Books. https://doi.org/10.3167/9781571818782
6. Dodds R, Butler R (2019) The phenomena of overtourism: a review. Int J Tour Cities 5(4):519–528. https://doi.org/10.1108/IJTC-06-2019-0090
7. Gössling S, Peeters P (2015) Assessing tourism's global environmental impact 1900–2050. J Sustain Tour 23(5):639–659. https://doi.org/10.1080/09669582.2015.1008500

8. Kharas H (2017) The unprecedented expansion of the global middle class: an update. Glob Econ Develop Brookings 1–29. https://www.brookings.edu/articles/the-unprecedented-expansion-of-the-global-middle-class-2/
9. Guttentag D (2015) Airbnb: disruptive innovation and the rise of an informal tourism accommodation sector. Curr Issues Tour 18(12):1192–1217.https://doi.org/10.1080/13683500.2013.827159. https://haroldgoodwin.info/wp-content/uploads/2020/08/rtpwp4overtourism012017.pdf
10. Jansson A (2018) Rethinking post-tourism in the age of social media. Ann Tour Res 69:101–110. https://doi.org/10.1016/j.annals.2018.01.005
11. Nash C, Jarrahi MH, Sutherland W, Phillips G (2018) Digital nomads beyond the buzzword: defining digital nomadic work and use of digital technologies. In: Chowdhury G, McLeod J, Gillet V, Willett P (eds) Transforming digital worlds. Springer, Cham, pp 207–217. https://doi.org/10.1007/978-3-319-78105-1_25
12. Oklevik O, Gössling S, Hall CM, Steen Jacobsen JK, Grøtte IP, McCabe S (2019) Overtourism, optimisation, and destination performance indicators: a case study of activities in Fjord Norway. J Sustain Tour 27(12):1804–1824. https://doi.org/10.1080/09669582.2018.1533020
13. Seraphin H, Sheeran P, Pilato M (2018) Over-tourism and the fall of Venice as a destination. J Destin Mark Manag 9(000):374–376. https://doi.org/10.1016/j.jdmm.2018.01.011
14. Zygmunt K (2019) 1 Sposoby przeciwdziałania negatywnym skutkom overtourismu w atrakcjach i destynacjach turystycznych. https://doi.org/10.17951/B.2019.74.0.45-57
15. World Tourism Organization (UNWTO) (2005) Tourism highlights, 2004 edition [eISBN 978-92-844-0791-0]. https://doi.org/10.18111/9789284407910
16. Honey M (2008) Tourism: preventing conflict, promoting peace. Paper commissioned for USIP, 2. https://www.usip.org/sites/default/files/tourism_developing_world_sr233_0.pdf
17. Scheyvens R (1999) Ecotourism and the empowerment of local communities. Tour Manage 20(2):245–249. https://doi.org/10.1016/S0261-5177(98)00069-7
18. Tosun C (2006) Expected nature of community participation in tourism development. Tour Manage 27(3):493–504. https://doi.org/10.1016/j.tourman.2004.12.004
19. Swarbrooke J (1999) Sustainable tourism management. CAB International. https://doi.org/10.1079/9780851993140.0000
20. Butler R (ed) (2006) The tourism area life cycle, vol 1. Channel view publications. https://doi.org/10.21832/9781845410278
21. Hunter C (1997) Sustainable tourism as an adaptive paradigm. Ann Tour Res 24(4):850–867. https://doi.org/10.1016/S0160-7383(97)00036-4
22. Gössling S (2002) Global environmental consequences of tourism. Glob Environ Chang 12(4):283–302. https://doi.org/10.1016/S0959-3780(02)00044-4
23. Łapko A, Panasiuk A, Strulak-Wójcikiewicz R, Landowski M (2020) The state of air pollution as a factor determining the assessment of a city's tourist attractiveness—based on the opinions of Polish respondents. Sustainability 12(4):1466. https://doi.org/10.3390/su12041466

Overtourism and Religious Sites: Two Florentine Examples

Giada Cerri

Abstract Italian art cities have many tourist attractions, including religious sites. Crucial churches and convents charm as many people as well-known museums. For example, about one million people visit the Monumental Complex of Santa Croce in Florence every year, while, in 2023, seven hundred thousand travelers saw the Santa Maria Novella Basilica in Florence. The secular administrative bodies (*fabbricerie*) set up devices to explain the original functioning of the churches. In this context, a painting, fresco, or sculpture is significant beyond being a piece of art. They provide tools, especially to visitors without Catholic backgrounds, to understand church rituals, spaces, and objects. Usually, the implied solutions are similar to those in a museum. The installation of physical apparatuses affects the layout of the space, and the presence of masses of sightseers represents a risk to such fragile environments. In light of these dynamics, the chapter aims to investigate religious sites that are mass tourist destinations and explore the solutions, potentials, and critics of these fragile settings. Using case study methodology, it assesses the museum-like implications that modify the space. In the discussion, it analyzes the different kinds of tourists visiting Santa Croce and Santa Maria Novella in Florence, and considers introducing digital tools to manage the people's flows, attract further visitors, and facilitate their visit. It also evaluates the coexistence of the groups of worshippers and tourists. Such cohabitation requires maintaining prayer and reflection spaces and balancing the desire to visit all spots with the constraints of a purchased ticket.

Keywords Museum-like site · Museum design · Sacred site · Digital solution · Florence

G. Cerri (✉)
Dipartimento di Architettura, Università degli Studi di Firenze, Florence, Italy
e-mail: giada.cerri@unifi.it

V. C. Sharma et al. (eds.), *Architectural Vision of Overtourism*,
SpringerBriefs in Architectural Design and Technology,
https://doi.org/10.1007/978-981-96-5794-0_2

1 Introduction

Religious and sacred sites differ from museums but can offer a similar visitor experience [32]. Secular tourists often view churches primarily for their artistic value, overlooking their spiritual significance [34]. This shift can lead to an understanding of the nature of these places. Along with skilled personnel and effective management, providing suitable spaces and services is essential [3, 21]. Reviews [16] highlight the need for a shared cultural background among visitors to foster cultural exchange and ensure proper interpretation, often requiring behavioral guidance.

The overtourism phenomenon poses a significant threat to the sustainability of cultural sites [31, 33]. Santarossa [30] likens mass tourism to military invasions, as large crowds can damage fragile environments and diminish the quality of visits [20]. While large churches can accommodate crowds, only certain areas are designated for public use, creating challenges in managing groups in spaces meant for smaller gatherings. This surge in tourism has increased maintenance costs, leading to the introduction of ticket fees. Thus, visitors are both a source of damage and revenue.

The chapter aims to interpret the several aspects involved in managing religious sites, inspired by Gathan's question: "what comes first: the touristic aspect or worship?" [15]. In particular, considering overtourism as an endemic aspect of some locations and the risks and impact of mass tourism in fragile settings such as historical religious sites, the essay would like to understand what solutions are regulating and managing the presence of tourists and worshippers and how that impacts the interior spaces.

The chapter illustrates a study of religious sites that are mass tourist destinations. It presents the difficulties and the potentialities of overcrowded churches inside an overtouristic Italian city like Florence. Section two introduces concepts linking religious sites and tourism, presenting a short literature review. Then, it specifies the current meaning of museum-like sites and the origin of the link tourism-churches. Section three briefly presents the chosen methodology and two case studies: the Monumental Complex of Santa Croce and the Monumental Complex of Santa Maria Novella, both in Florence. The essay compares these two sites, which are dealing with mass tourism effects, and presents the respective organization bodies, the services, and the activities. It also reviews the museum-like solutions adopted in both places, the relation with the worshipper's spaces, and their adopted digital solutions. The work is also sensible to several aspects of the United Nations Sustainable Development Goals (SDGs). In particular, it considers the Thematic Indicators for Culture. The chapter ends with an open conclusion on the relationship between overtourism and religious spaces.

2 Context

2.1 Religious Sites and Tourism: A Literature Review

Pilgrimage to sacred sites is one of the oldest forms of tourism, particularly at religious sites [7, 12]. Scholars have studied the connection between tourism and these sites, with O'Gorman [24] identifying religious travel as an early driver of travel behavior. While the Bible forbids the commercialization of religious places [17], souvenirs have been sold at shrines since medieval times [18]. Nolan and Nolan [23] compare devout pilgrims with secular tourists, and Stausberg (2011) notes that the presence of tourists affects the site's artistic value and maintenance revenues [32].

According to Wiltshier and Clarke [40], religious tourism often exists outside the general local touristic network. Understanding the dynamics between religious and sacred site tourism involves considering economic factors [28] and the commodification of experiences [9]. Tourists often see entry fees negatively, while worshippers perceive tourists as disruptors [2]. Urry [38] discusses how consumption of these sites may start before visiting, highlighting that worshippers contribute to the tourist experience.

Managers need to work on balancing the presence of both groups to avoid commodification of the religious sites. Timothy and Olsen [35] note that the latter are prevalent in tourism marketing. Different motivations drive visitors to sacred places, as Leiper [19] categorizes tourists based on expectations. Cohen [10] suggests that differing motives impact tourist behavior, while Shackley [32] argues that visitors to sacred sites often have distinct motives. Finney [13] classifies visitors as seekers, lotus-eaters, pilgrims, or accidental tourists. Nolan and Nolan [23] assert there is no clear divide between pilgrims and tourists, noting overlaps and the potential for coexistence without negative impacts.

2.2 Touristic Churches and Rituals

Religious and touristic rituals are central to monumental churches, particularly within the context of Catholic traditions in Italy. Religious rituals are often limited in time and space, with some churches closing or restricting access during ceremonies. While daily celebrations are observable, private rituals like confessions typically do not attract tourists. However, the cultural aspect of these rituals can fascinate visitors, transforming them into spectators during significant religious events.

Tourism manifests in churches and convents through various visitor behaviors, often resembling worshippers and museum-goers, as noted by Bourdieu and Darbel [4]. Typical activities include walking, observing, and conversing, which can turn churches into museum-like spaces. This analysis is valid for small groups but less for mass tourism, where large, organized groups often disrupt the sacred atmosphere with noise and distractions. Religious sites' sacred and secular meanings are considered part of a community's heritage. Participation in rituals is part of the cultural roots of a place, and the awareness about the behavior during such rituals is a common ground within a community.

2.3 *Museum-Like Sites: Between Worship and Visit*

Museology and museography primarily focus on sacred art, temporary exhibitions, and religious museums [1, 5, 15, 26, 27]. These museums are often managed like conventional spaces, with limited connection to their sacred surroundings. Minucciani's book, *Religions and Museums* [22], examines sacred art display from a curatorial perspective, while research in American museum journals contributes to understanding artifacts but remains within traditional museum contexts. Maia Wellington Gahtan's research on religious spaces in mass tourism culminated in the book *Churches, Temples, Mosques: Places of Worship or Museums* [15]. Based on a thematic forum hosted in Florence this work aims to foster discussion among stakeholders and adopt a cross-disciplinary approach to the complexities of sacred spaces and museums.

Unlike traditional museums, religious spaces express local culture, traditions, and faith. Religious spaces maintain a single peculiarity and powerful intangible messages. Religion and tradition also overlap, determining a solid bond with the local community that interprets the site as part of its roots and heritage. The frequency of monumental churches is linked to local communities and catholic worshippers. Pilgrims are a further kind of tourism. These people are aware of behavior and religious messages. Due to the traditional national context and the shared background (scholastic path), local visitors, including secular people, know the behavioral norms. The rise of visitors without preparations is an issue concerning the supervision of the way of dressing and behaving inside, but also a matter of communication. All the elements inside the religious site are expressions of communication efforts, more or less linked to the catholic doctrine. Such a meaning cannot be clear if people are not previously informed about the catholic religion, its rituals, and traditions. The condition of being a place of worship and a tourist destination weighs upon the possible solution to communicate the inner meaning of a church's several spaces, objects, and elements. A painting or a fresco is not "just" a work of art. Its meaning and reason are linked to a religious message and one historical moment. There is a need for translation of it, but a church cannot be transformed into a forest of signs. Such a situation can be traced in other geographic locations and religious spaces, as highlighted by [15].

3 Methodology and Discussion

3.1 Tourism and Religious Sites: An Old Link

Since the sixteenth century, Italian heritage has drawn visitors and artists. Over time, the length of travel changed significantly—from up to fourteen months during the Grand Tours to four months in Romanticism, influenced by improved transportation, rising bourgeois lifestyles, and declining aristocracy. Travel books began replacing local guides, with Thomas Cook's agency helping clients organize their tours, marking a shift from the Grand Tour to everyday travel.

Cook's 1875 [11] guidebook listed Santa Croce in Florence as a critical sight. The post-war economic boom ushered in general tourism. Frommer [14] guide became a popular resource for American travelers, highlighting significant cities and attractions like Florence's Santa Croce and Santa Maria Novella. Today, various guidebooks reflect the faster paced travel trends and shorter stays have led to increased tourist traffic. This visitor rise contrasts sharply with the dramatic decline in local populations and churchgoers in art cities, resulting in year-round tourist congestion.

3.2 Methodology

The chosen methodology is based on a qualitative approach, and a case study analysis was proposed [39]. After a general introduction about religious sites and tourism (literature review) that allows a determination of the boundaries of the discussion, comparing two case studies allows one to define elements that can help the development of a discussion on the complex link between tourism and religious sites. The case analyses include field notes, specific literature reviews, and site observations.

The cases are the Monumental Complex of Santa Maria Novella and the Monumental Complex of Santa Croce, both in Florence. They are both monastic systems, still hosting a little community of friars; they are part of the mass tourism routes and are outstanding examples of Florentine sacred art and architecture. They share a similar property articulation, and the body in charge of their management are the *fabbricerie*, a secular body whose aim is traced in the following paragraph (Fig. 1).

3.3 Florentine Religious Sites: A Complex Articulation

In recent years, several Italian churches have begun charging admission fees for visitors, a move that has faced internal opposition from the Italian Episcopal Community (CEI—Comunità Episcopale Italiana). Supporters argue that the fees help cover fixed costs and are akin to donations, with entrance generally accessible for worshippers. In Italy, ticketed churches include Santa Croce, Santa Maria Novella, the Cathedrals

Fig. 1 Florence, Italy. Location of the monumental complexes of Santa Maria Novella, close to the rail station, and of Santa Croce, close to the Arno River

of Siena and Pisa, and others in Venice and Ravenna. While Florence's Cathedral, Santa Maria del Fiore, is free, the access to the Baptistery, Giotto's bell tower, and Brunelleschi's Dome requires a ticket.

Similar practices exist in other European locations like Westminster and the Sagrada Familia, primarily to cover maintenance and staffing costs. However, the structure of these sites can blur the lines between secular and religious visits, potentially confusing tourists without a Catholic background. Discussions around ecclesial heritage and the need for educational tools have emerged, notably by the Association of Italian Ecclesial Museums (AMEI) in 2012. Capurro [6] highlights a focus on catechesis rather than a broader religious context, emphasizing the need for better communication strategies for diverse visitors. While interest in artistic heritage is common among tourists, cultural backgrounds vary, and some efforts like *Art et Fides* have worked to prepare church staff better to engage with these visitors.

3.4 The Monumental Complex of Santa Croce in a Nutshell

Santa Croce is one of the most visited sites in Florence, with almost one million people per year (Fig. 2). The structure partially belongs to the State of Italy, specifically the *Fondo edifici di Culto del Ministero degli Interni* and the Municipality of Florence. *Opera di Santa Croce* manages the Complex. It is a *fabbriceria* comprising laypeople in charge of running, preserving, and enhancing the Complex. The convent still hosts a small community of Franciscan friars. The property articulation is also complicated by the National Library, which, starting in 1921, occupies part of the original Complex. Other conventual spaces host the Leather School (*Scuola del Cuoio*). The overlapping interests, owners, and activities complicate the site's management. The Opera is framed also as a museum institution, recognized by the Tuscany Region as a Relevant Regional Museum—*Museo di rilevanza regionale.*

Santa Croce is called the Italian Pantheon, named after Ugo Foscolo's poem *Dei sepolcri*, celebrating the memory of Florentine personality, first, and Italian, later. Here, there are several art masterpieces. Giotto, Coppo di Marcovaldo, Cimabue, Canova, Bronzino, and Donatello are some of the artists who worked there (Fig. 3). Besides, the complex expresses architectural magnificence with the church interiors and construction, Pazzi chapel, and the system of cloisters. The Complex is sadly known for the damages caused by the Arno River flood in 1966. One of the symbols is Cimabue's crucifix, which is now restored and shown in the museum path (vestry).

To enter, one can purchase the ticket on-site, on the left side of the church, or online. The ticket's total price is 8 Euro, the reduced-price ticket is 6 Euro. One can also purchase an audio guide or a guided tour. Florentine inhabitants can enter for free. Worshippers have a reserved entrance and area.

3.5 The Monumental Complex of Santa Maria Novella in a Nutshell

The Dominican convent is located in a hot area of Florence, close to the rail station and at the crossroads of tourist movement. The Complex belongs partly to the Italian State (*Fondo edifici di Culto del Ministero degli Interni*) and the Municipality of Florence. The latter collaborates with the *Opera di Santa Maria Novella* to manage the spaces, while the religious life is entrusted by decree to the Dominican Community (Fig. 4).

The Complex includes artistic artifacts, such as Giotto's and Brunelleschi's Crucifixes and the frescos by Masaccio, Ghirlandaio, and Filippo Lippi. Alberti's façade is a Renaissance masterpiece, and the cloister system is a noteworthy example of conventual architecture. The Complex is under a refurbishment process. Occupied by a military school until 2016, the municipality opted for general cultural conversions of the site. Part of the convent hosts the new museum *Mundi, Museo Nazionale dell'Italiano*, a museum dedicated to the Italian language (expected to open in 2024). Another part will host the art civic storage.

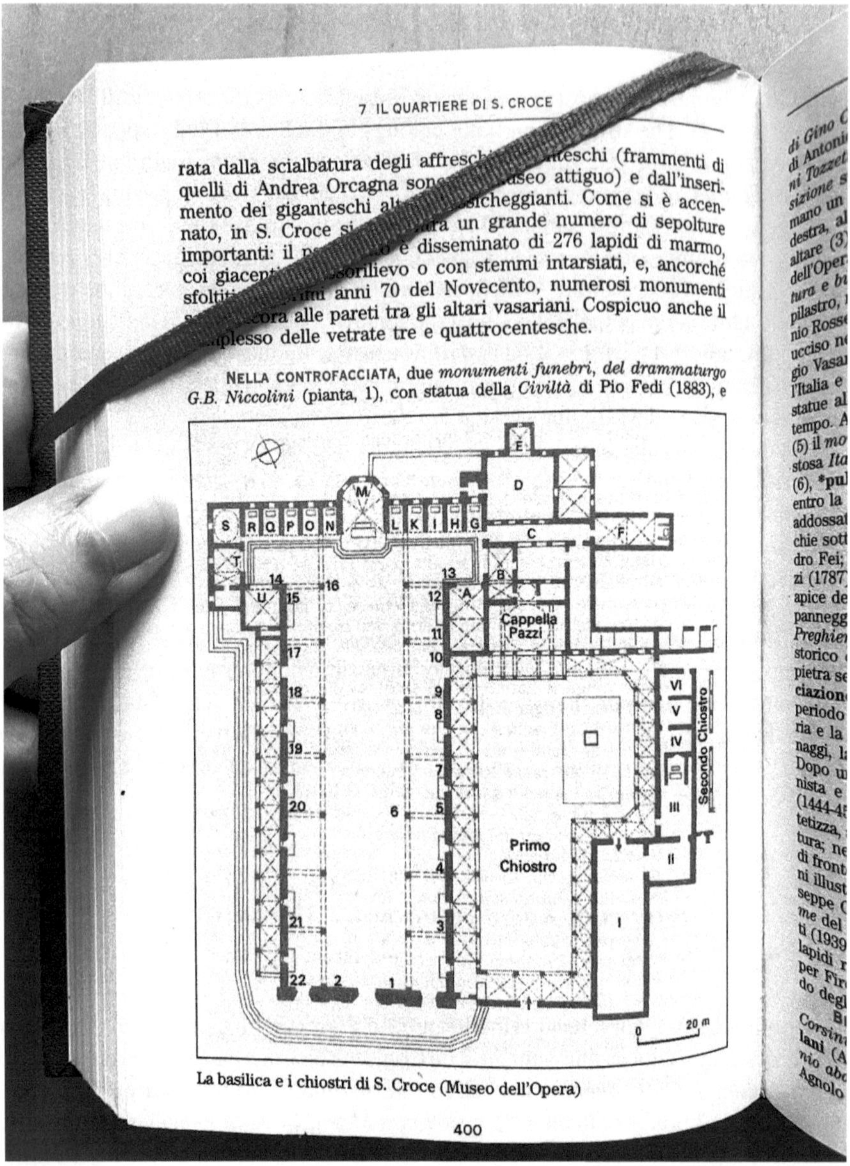

Fig. 2 Pages dedicated to the Monumental Complex of Santa Croce in the "red guides" edited by the Touring Club Italiano [36]

Fig. 3 Some tourists visit the chapels of the church of Santa Croce. (Credit and courtesy: Dimitris Kamaras)

To enter the site, one has to buy a ticket, whose supply is the responsibility of the Florentine civic museum system (MUS.e is the body in charge of this management). The ticket's total price is 7.50 €, and the reduced-price ticket is 5 €.

3.6 Visitors, Digital Tools, Museum-Like Solutions, Online Presence

Tourist Dynamics

Various visitors can be observed in the two Florentine complexes: singles, couples, independent and guided small groups, and accompanied large groups. Generally, single visitors and couples are keenly interested in the content and architecture. They are focused, engaged, and respectful. Some small groups, whether guided or not, display similar behavior. In contrast, large groups tend to follow their guides during the visit and can be pretty noisy and chatty. These larger groups often consist of school trips, day-trippers, cruise passengers, or clients from organized tours. School groups typically visit in the morning, while large groups are present throughout the day.

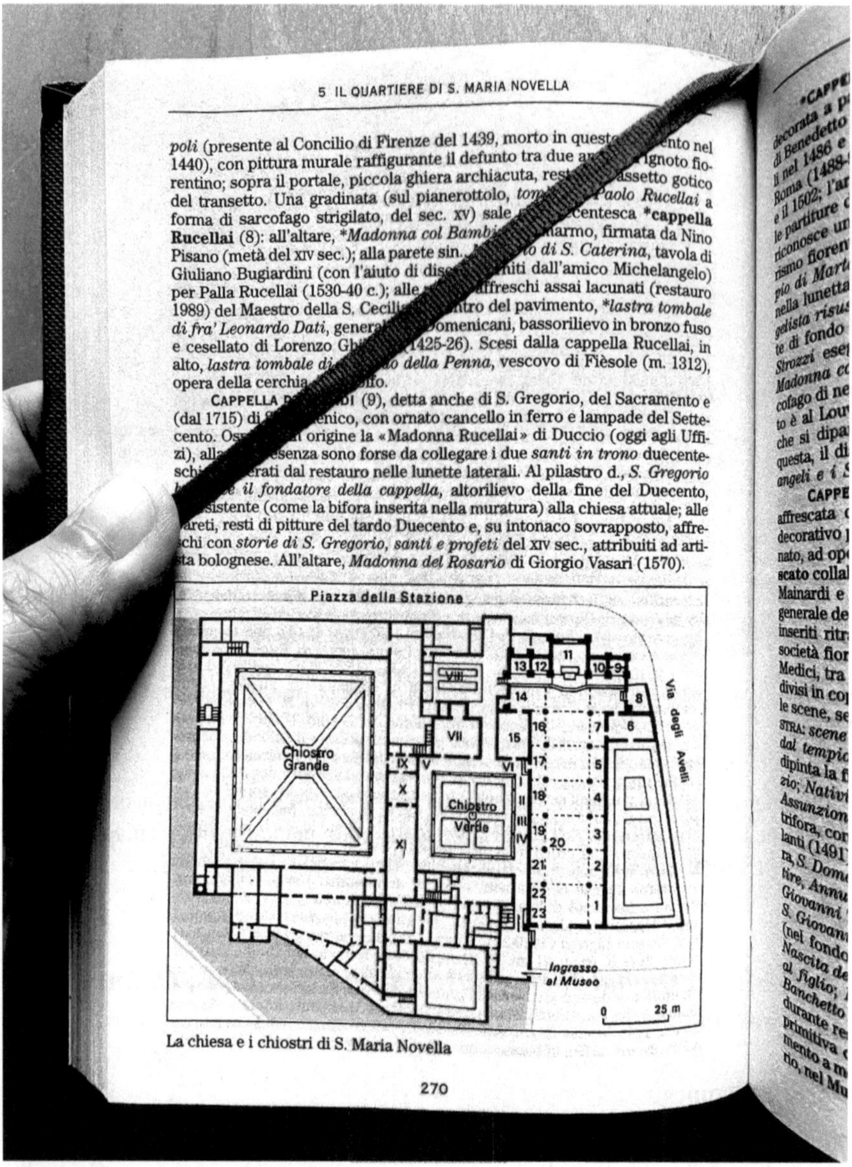

Fig. 4 Pages dedicated to the Monumental Complex of Santa Maria Novella in the so-called "red guides" edited by the Touring Club Italiano [36]

The presence of large and numerous groups weighs on the space organization. In Santa Croce, they have reserved access from the façade instead of the ticket office side, although the purchasing action still happens in the ticket office. Inside, visitors stop in front of the prominent artists' works, in Santa Croce, Giotto's frescos, Cimabues's Crucifix, Pazzi Chapel, and in Santa Maria Novella, Brunelleschi's crucifix, Masaccio's trinity, Spanish chapel, avoiding the less-known areas. The result is a similar dynamic to the urban scene: masses concentrate in the same spots, while others are empty. In order to control the visit fluxes, Santa Croce online ticketing is organized in time slots (every fifteen minutes). Santa Maria Novella, where the number of visitors is lower than in Santa Croce, does not offer that option.

Digital Tools to Audience Development Strategies

The Opera of Santa Croce started testing alternative ways to attract and manage fluxes in the 2000s when it introduced the ticket. In 2020, it launched a digital application for smartphones to sustain the singular visit to the Complex. Such a tool crossed the need to provide alternative ways of visiting cultural sites, booking visits, and responding to sanitarian restrictions due to the COVID-19 pandemic. That app is now disabled. It has been transformed into a native smartphone application called *In Toscana con Dante* (In Tuscany with Dante). It was released during Dante Alighieri's celebrations and was promised as the first chapter of a larger design. Traditional tools are surviving. There is still the possibility of on-demand tours, offering customization tailored to the group's request, besides the physical audioguide service (two-hour lengths). In the 2000s, Proctor and Telli [29] stated that users still considered audioguides central to the visit experience. Twenty years later, traditional audioguides in Santa Croce and Santa Maria Novella suggest the same conclusions. One of the main differences between digital—which needs to run on personal devices—and traditional audioguides—provided by an institution—is the level of customization. The digital content is flexible and has a certain dynamism (it can be easily implemented, upgraded, or changed). It offers an increased level of customization, reaching the taste or expectations of more public targets. Furthermore, some applications allow the user to choose the content to listen to, such as thematic paths in the case of Santa Croce.

Younger targets are involved with a digital service called *Santa Croce in extended reality*. Such experimentation has been developed jointly with the Italian telecommunication company TIM. A digital device accompanies the visitors through two alternative paths inside the Complex, "The Greats of Santa Croce" and "The Franciscan Path." Insight information is provided thanks to augmented reality. All these solutions are intended to serve the visitors. Unfortunately, accessibility is intended only to remove architectural barriers, not as a way to welcome everybody.

Museum-Like Solutions, Way-Finding, Education Tools, and Temporary Installations

The first element linking the complexes to tourist sites is the ticket offices. In the case of Santa Maria Novella, there are two options to access: the entrance from the namesake square (Fig. 5) or the back of the church, the railway square. The latter option complicates the spaces' comprehension and communication since one enters

Fig. 5 Tourists entering Santa Maria Novella's Complex from Santa Maria Novella Square

from an information point and the back of the convent space. In Santa Croce, single visitors and worshippers access the church from the left side, while large groups enter from the square. The group's first sight is the façade of Pazzi Chapel.

Inside the church space, in both cases, the layout mixes museum-like signals, traditional religious elements, votive candlesticks, and bulletin boards dedicated to the faithful community. The visiting space is delimited by ropes, breaking the unity of some parts of the church. Some stanchions integrate caption devices or emergency signages. Scattered in the church space, other communication panels contain additional tourist information. The exhibit elements are generally discreet in the church but are not homogeneous by style, design, shape, materials, and year of installation. Others contain messages to worshippers. The result is a cacophony of signs that, luckily, are lost in the magnificence and power of the architectural spaces.

In Santa Maria Novella, the exhibit elements gain more mass, presence, and coherence in the so-called Museum areas, comprehending the cloisters and relative chapels. The information panels are few and mixed in style and shape [8]. The museum path of Santa Croce develops among the sacristy spaces and the refectory. Here, except the church, other spaces are often used to host temporary exhibitions and art installations.

Locals and the resident community are considered targets to reach. Initiatives and activities, like family kits, contemporary art installations, thematic conferences, or events, are mainly devoted to inhabitants. They would like to attract those who already know the monumental complexes. There is also a willingness to keep the Franciscan and Dominican awareness. There is a continuing calendar of ecclesiastical meetings and a project involving third parties to link civic society and the friar community. In Santa Croce, the initiative "The Seed of the Common Good," for example, tied the local Community of Minor Conventual Friars, the Opera, the Foundation Progetto Valtiberina (located among Tuscany and Umbria Regions), and

Fig. 6 Tourists climbing the scaffolding having erected to the conservation of Masaccio's Holy Trinity in Santa Maria Novella. (*Courtesy* FEC)

Aboca (an Italian healthcare company). The project's goal is to restore the vegetable garden of the Complex. Other initiatives speak to both targets. It is possible to visit the restoration yards at both sites. Santa Croce inaugurated that chance in 2008. During the opening hours, people can climb the scaffolding and see the frescos of the chapel under restoration from an unprecedented point of view. It also gives the possibility to observe the work of the restorers (Fig. 6). Santa Croce also has a space dedicated to temporary exhibitions and hosts cyclically contemporary art installations (such as Felice Limosani video installation dedicated to Dante Alighieri, *Dante. Il Poeta Eterno*, 2021–2022, in Pazzi Chapel). Santa Maria Novella integrates its cloister spaces with other institutions' activities. Since the whole project involving the Complex of Santa Maria Novella is under construction, one has to wait a few years before its completion to evaluate how the religious spirit of the complex balances with contemporary cultural and tourist programs.

Online Presence

The online presence is a further space to investigate. The first visit to a site starts online [25]. The online space is a parallel world where people catch information and visit suggestions. It is fundamental for the official presence of the institutions online to give the visitors or the community the correct information and coherent content to maintain the institution's adherent image.

The Santa Maria Novella website, managed by the Opera, provides information on visiting hours, the history, and the artistic and architectural heritage of the Complex. A section introduces the history of the Dominican order, the monastic life, and the liturgies. Overall, Opera of Santa Maria Novella focuses on the importance of its convent's religious roots, although it uses the same strategies to inform and attract visitors. There is no information about the tickets. The institution in charge of selling tickets is the Civic Museum website. One has to reach the Civici Musei Fiorentini and choose "Museo di Santa Maria Novella" in the drop-down menu. Such a distinction might need to be clarified to visitors. The pastoral life is the least highlighted theme on the Santa Croce website, a topic addressed by an outdated website accessible via an external link. Conversely, the sections dedicated to worshippers and pilgrims are well signaled in situ. The website gives ample space to the visit initiatives, the possible visiting paths, and the Opera itself. It looks like a standard museum website. Here, it is possible to book the museum and event tickets. The management interface forces the visitors to choose the interval hour. Worshippers do not have to book their access, free of timing and charges.

3.7 Thematic Indicators for Culture in the 2030 Agenda

Given the subject's specificity, the aim is to consider the indicator development for culture inside the general framework expressed by UN Sustainable Development Goals (SDGs). The themes concerning the monumental religious complexes are embedded into the four pillars of the Thematic Indicators for Culture (TIC). The latter contribute transversally to SDGs. The TIC are composed by four pillars and related activities. In the case of Florentine Monumental Complexes, the indicator participating in the definition of a sustainable development of these sites has been selected: In the first pillar "Environment and resilience:" "2. Sustainable management of heritage," "4 Cultural facilities," and "5 Open space for culture;" in the second pillar "Prosperity & Livelihoods:" expresses the aspects of "7 cultural employment" and "12 governance of culture;" in the third pillar "Knowledge & Skills:" "14 Cultural knowledge," "15 Multilingual education," "16 Cultural & artistic education," and "17 Cultural training;" in the fourth pillar, "Inclusion & Participation:" "18 Culture for social cohesion," "20 Access to culture," and "21 Cultural participation" [37]. All these aspects are traceable in the mission and the work conducted by the Operas and the other governing bodies. As expressed by the UNESCO document about the cultural indicators, these embed the SDGs and express goals and methodology to achieve them more effectively. In the case of the religious complexes, there is work to be done to achieve some points, still, given the religious and secular mission of the institutions, some aspects (the indicators) are naturally part of their work.

4 Conclusions

The excursus and analysis of two Florentine religious sites present a fragile situation. The institutions organize the fluxes by introducing a ticket and a schedule to staggered entrances. In both cases, the presence of tourists is not a novelty; traditionally, travelers and foreigners visit them. What has changed is the continuous presence of masses of people, especially groups, throughout the year.

Due to the internal organization, there are two different approaches to managing the fluxes and attracting people. Regarding the interior space organization, they are adopting similar museum-like solutions. Aesthetic choices can differentiate the two, but the interior museum-like solutions are similar: ropes, information stanchions, and panels mounted on vertical supports. Santa Croce works more like a traditional museum. Specific interior designs should be adopted to calibrate the needs of tourists, as well as essential sources of revenues to maintain the structures, the space for worshippers, and the maintenance of the link with the locals, which is not necessarily religious.

The online presence analysis highlights a substantial difference between the two sites. Santa Croce works well and is comparable to a well-designed museum website. Santa Maria Novella is fragmented, without a clear overall goal, highlighting the presence of several actors involved in the management and scientific directions. Such fragmentation affects, in general, the whole management. Despite the need for continuity in digital strategy, Santa Croce is attempting to provide digital solutions directed to different targets. The integration with digital tools and implementation of those solutions as instruments to foster public development is relatively recent in Italy. Moreover, many little museums still need the strength and the resources to define a coherent digital development plan, although there is a general effort to achieve some digital targets. The COVID-19 pandemic stressed such a necessity.

It is worth noting that Santa Croce aims to target locals by promoting specific initiatives and activities. The attempts to reach locals and attract them seem contradictory, given the high number of people and the risks of commodifying religious sites. Both Complexes understood that they could not fight against overtourism and its effects. Given the dropping of the inhabitants in the historical center, it is clear that the community of worshippers is limited and that these churches are oversized. However, the presence of residents and the local community is an added value. Traditionally, Santa Croce and Santa Maria Novella are reference points. The two areas gravitated around them, even if they lost their significance in civic society. Their transformation into a sole tourist site is a risk to avoid. That would represent a loss to society. They need to maintain a role and a link with the local community, even if it is not local anymore (most residents move outside the UNESCO skirt, and tourist accommodations characterize both Santa Croce and Santa Maria Novella districts). Initiatives and activities dedicated to specific targets could represent a way to maintain such a link. Santa Croce, in particular, has invested in using digital solutions to attract people (digital natives, international tourists), involve local targets (events and specific paths), and book tickets. All the efforts have more dimensions of trials and beta tests than an articulated strategy. To host different kinds of targets, there are the availability of family kits, of the services to the schools, or public events, as concerts. Such articulations get closer to typical museum services. A way to reconnect with the local community is traced in the thematic openings, the events dedicated to people celebrated in the Basilica. The Franciscan community is involved in the project, and the events telling Francesco d'Assisi's life or the Franciscan doctrine are frequent. The website allows ticket bookings to enter both the Complex and the events. Less updated and fragmented are Santa Maria Novella websites that have no native or

CORRIERE DELLA SERA

CORRIERE FIORENTINO

IN VIA RICASOLI

Overtourism a Firenze, residenti barricati davanti alla Galleria dell'Accademia (con il giallo dei permessi dei nastri per le code)

di Giulio Gori

In via Ricasoli a Firenze, davanti alla Galleria dell'Accademia, il suolo pubblico da anni è occupato dai nastri per dividere le file di turisti: «Certi giorni è quasi impossibile riuscire a entrare in casa». Il Comune: «Nessuna autorizzazione»

Firenze, San Niccolò dice addio a «Spazio Base»: «I fiorentini spariscono, è una città per ricchi»

di Giulio Gori

Affitto raddoppiato per lo spazio artistivo in via San Niccolò da 26 anni. L'ultimo evento aperto al rione. «Il centro è una 'No man's land»

Firenze invasa dagli adesivi «Yankee go home»: un messaggio contro l'overtourism?

di Jacopo Storni

Decine di adesivi gialli nel centro di Firenze dopo le scritte dei mesi scorsi contro i turisti. Impossibile capire chi li abbia messi

Fig. 7 An excerpt from the local section of the National newspaper *Corriere della Sera* highlighting the overtourism's effects and reactions

embedded digital application. As said about the space's visit, the digital and web strategy might be rethought once the rearrangement is completed.

The news highlights the problem of overtourism, and the last inhabitants of the city center see tourists as something to fight (Fig. 7). Tourism industry is essential to Florence. However, a balance between different needs is necessary to avoid its complete commodification [31]. The involvement of the locals in the initiatives can represent a strong signal in inviting people to rediscover an illustrious site of the city and, in a way, take back a shared heritage.

References

1. Ambrose T, Paine C (2012) Museum basics. Routledge
2. Ashworth GJ (2009) Do tourists destroy the heritage they have come to experience? Tour Recreat Res 34(1):79–83
3. Bagnoli L (2006) Manuale di geografia del turismo. Dal Grand Tour ai sistemi turistici. UTET, Torino

4. Bourdieu P, Darbel A (1966) L'amour de l'art: les musées et leur public. Minuit, Paris
5. Capurro R (2015) Dysplaynig the sacred in the museum. Some reflections. In: Gahtan MW (ed) Sacred objects in sacred collections. Edifir, Firenze
6. Capurro R (2010) The Italian diocesan museums: bridging religious and cultural identity. Int J Inclusive Museum 3(2):103–110
7. Carmichael DL, Hubert J, Reeves B (1994) Sacred sites, sacred places. Routledge, Abingdon
8. Cerri G (2024) Sacred architecture and touristic consumption: the role of the interior design strategy. In: Niglio O (ed) Culture of the sacred space. Springer
9. Clarke A (2013) Value creation–creating values: contradictions and constraints in the development of religious tourism. Annon Manage. Rev. 2(1):2013
10. Cohen E (2004) Contemporary tourism, diversity and change. Elsevier Ltd., Oxford
11. Cook T Firm, publishers L (1875) Cook's tourist's handbook for Southern Italy. Cook, son, and Jenkins
12. Cummings PJ (1987. Sacred sites in contemporary cultures. In: US/ICOMOS (ed) Old cultures in new worlds. 8th ICOMOS general assembly and international symposium. Programme report. Washington, D.C., pp 551–557
13. Finney RZ, Orwig RA, Spake DF (2009) Lotus-eaters, pilgrims, seekers, and accidental tourists: how different travelers consume the sacred and the profane. Serv Mark Q 30(2):148–173
14. Frommer A (1957) Europe on 5 dollars a day. Arthur Frommer Incorporated
15. Gahtan MW (2012) Churches, temples, mosques: places of worship or museums? Edifir, Firenze
16. Huang K, Pearce P (2019) Visitors' perceptions of religious tourism destinations. J Destination Market Manage 14
17. Johnston AM (2013) Is the sacred for sale? Tourism and indigenous peoples. Taylor & Francis, London
18. Leightner JE (2013) Stop turning my father's house into a market: secular models and sacred spaces. J Markets Morality 16(2):429–440
19. Leiper N (2004) Tourism management, vol 455. Pearson Education Australia Frenchs Forest
20. Liberatore G, Biagioni P, Ciappei C, Francini C (2023) Dealing with uncertainty, from over-tourism to overcapacity: a decision support model for art cities: the case of UNESCO WHCC of Florence. Curr Issue Tour 26(7):1067–1081. https://doi.org/10.1080/13683500.2022.2046712
21. Lozato-Giotart JP (2008) Geografia del turismo. Hoepli, Milano
22. Minucciani V (2013) Religion and museums. Immaterial and material heritage. Allemandi & Co, Torino
23. Nolan ML, Nolan S (1992) Religious sites as tourism attractions in Europe. Ann Tour Res 19(1):68–78
24. O'Gorman KD (2010) The origins of hospitality and tourism. Goodfellow
25. Orlandi SD, Calandra G, Ferrara V, Marras AV, Radice S (2019) Web strategy museale. Zenodo, Milano. https://doi.org/10.5281/zenodo.3547149
26. Paine C (ed) (2000) Godly things: museums, objects & religion. Leicester University Press, New York
27. Paine C (2013) Religious objects in museums: private lives and public duties. Bloomsbury Publishing
28. Pine BJ, Gilmore JH (1999) The experience economy: work is theatre & every business a stage. Harvard Business School Press, Boston
29. Proctor N, Tellis C (2003) The state of the art in museum handhelds in 2003. In: Museums and the web 2003: selected papers from an international conference (7th, Charlotte, NC, March 19–22, 2003)
30. Santarossa A (2012) Il progetto delle invasioni. Indagine sulla natura militare del turismo di massa. Aracne, Roma
31. Settis S (2014) Se Venezia muore. Einaudi, Torino
32. Shackley M (2005) "Service Delivery" at sacred sites potential contribution of management science. Eur J Sci Theol 1(4):33–40
33. Shirvani DA, De Luca G (2023) A post-pandemic look at short-term tourist accommodations and World Heritage Site management. J Tour Cult Chang. https://doi.org/10.1080/14766825.2022.2164502

34. Stausberg M (2012) Religion and tourism: crossroads, destinations and encounters. Taylor & Francis, London and New York
35. Timothy DJ, Olsen DH (2006) Tourism, religion and spiritual journeys. Routledge, Oxford
36. Touring Club Italiano (2007) Guida d'Italia. Firenze e provincia. Milano: Mondadori
37. UNESCO (2019) Thematic indicators for culture in the 2030 agenda. UNESCO, Paris
38. Urry J (1995) Lo sguardo del turista - il tempo libero e il viaggio nelle società contemporanee. SEAM, Formello
39. Yin RK (2003) Case study research: design and methods, 3rd edn. Sage, Thousand Oaks
40. Wiltshier P, Clarke A (2012) Tourism to religious sites, case studies from Hungary and England: exploring paradoxical views on tourism, commodification and cost-benefits. Int J Tour Policy 4(2):132–145

Managing Overtourism in Historical Villages and Sustainable Strategies: A Comparative Study on Cumalıkızık and Penglipuran

Ahmet Kurnaz

Abstract This paper investigates the growing issue of overtourism in historical villages, focusing on Cumalıkızık in Turkey and Penglipuran in Indonesia, two villages recognized for their cultural and architectural heritage. Overtourism occurs when the number of tourists surpasses a destination's physical, social, and environmental capacity, leading to challenges such as infrastructure degradation, cultural erosion, and diminished quality of life for residents. Cumalıkızık, an Ottoman-era village, has been experiencing a rapid increase in tourist numbers, resulting in the deterioration of its historical structures, overstrained infrastructure, and a loss of cultural authenticity. In contrast, Penglipuran, a traditional Balinese village, has successfully managed its tourist influx through sustainable practices such as controlled visitor management, community involvement, environmental sustainability initiatives, and preservation of cultural heritage. This comparative study examines how Penglipuran's effective strategies can be adapted to address the overtourism challenges faced by Cumalıkızık. The research delves into the importance of integrating local communities into tourism planning, employing visitor control mechanisms, and using technology to enhance tourist experiences. It also discusses the potential barriers to implementing these strategies in Cumalıkızık, such as financial limitations, legal constraints, and cultural differences. By drawing lessons from Penglipuran, this study aims to contribute to developing a sustainable tourism model for Cumalıkızık and other similar heritage sites, ensuring the balance between economic benefits and cultural conservation.

Keywords Overtourism · Sustainable tourism · Cultural heritage conservation

A. Kurnaz (✉)
Istanbul Sabahattin Zaim University, Istanbul, Turkey
e-mail: ahmet.kurnaz@izu.edu.tr

V. C. Sharma et al. (eds.), *Architectural Vision of Overtourism*,
SpringerBriefs in Architectural Design and Technology,
https://doi.org/10.1007/978-981-96-5794-0_3

1 Introduction

Tourism is considered one of the fastest growing and most dynamic sectors of the global economy. According to the World Tourism Organization (UNWTO), international tourists reached 1.5 billion in 2019 [1]. While the economic returns of tourism significantly contribute to the development of countries and regions, overtourism has become increasingly debated in recent years [2].

Overtourism occurs when the demand for tourism in a specific destination exceeds that destination's physical, social, and environmental capacity [3]. This situation lowers the quality of life for the local population, leads to the degradation of natural and cultural heritage, and threatens the sustainability of the tourism destination [4]. Particularly, regions of historical and cultural importance are more vulnerable to the negative impacts of overtourism [5].

Historical and cultural heritage sites stand out as the main attractions in tourism. While these areas provide visitors with unique experiences, they also contribute to the local economy [6]. However, the uncontrolled increase in tourism activities in these regions leads to physical deterioration and cultural erosion [7]. In this context, historical villages, and settlements, with their limited infrastructure and sensitive ecosystems, are more vulnerable to the effects of overtourism [8].

Cumalıkızık Village, located in the Bursa province of Turkey, is included in the UNESCO World Heritage List with its approximately 700-year-old historical fabric dating back to the Ottoman period [9]. The village, with its stone and wooden houses reflecting traditional Ottoman architecture, narrow streets, and authentic rural lifestyle (see Fig. 1), attracts great interest from domestic and foreign tourists [10]. However, the increasing number of visitors in recent years has negatively affected the village's physical and social fabric [11]. Problems such as the deterioration of historical houses, insufficient infrastructure, traffic and parking issues, declining quality of life for the local population, and the threat to cultural values necessitate the development of a sustainable tourism model for Cumalıkızık Village.

Penglipuran Village in Bali, Indonesia, has similar historical and cultural values and is an important destination for preserving traditional Balinese culture and architecture. The village is known for its authentic architecture, orderly streets (see Fig. 2), and strong community ties, and it is considered one of the cleanest villages in the world [12]. By planning tourism activities with the participation of the local population and applying sustainability principles, Penglipuran has successfully managed the negative effects of overtourism [13]. The village controls the number of visitors, preserves its cultural values, and distributes economic benefits equally to the community [14].

This study aims to examine the effects of overtourism on historical and cultural heritage sites by comparing Cumalıkızık and Penglipuran villages and exploring how these effects can be managed. By analyzing the factors underlying the success of Penglipuran Village, the study will discuss how these strategies can be applied in Cumalıkızık. In this context, the role of architecture, urban design, and technology integration in reducing the negative impacts of overtourism will be evaluated.

Fig. 1 Cumalıkızık Village in the Bursa, Turkey (Author 2023)

Fig. 2 Penglipuran Village in Bali, Indonesia (Author 2023)

Although many studies in the literature examine the concept and effects of overtourism [15–17], research focusing specifically on practices in historical villages remains limited. This study offers new perspectives on sustainable tourism management by examining villages in different geographical and cultural contexts.

The significance of the research lies in its contribution to the balance between the preservation of historical and cultural heritage and the conduct of tourism activities. Addressing local participation, community-based tourism approaches, and technology integration will contribute to academic literature and practical applications. This study aims to guide other historical and cultural heritage sites that may face similar challenges in the future by discussing how the negative impacts of overtourism can be reduced and sustainable tourism models can be developed, using the examples of Cumalıkızık and Penglipuran.

2 Literature Review

2.1 Overtourism and Its Impact on Historical Villages

Overtourism is a phenomenon where the number of tourists exceeds the carrying capacity of a destination, leading to negative impacts on the environment, culture, and residents' quality of life [3]. The term has gained prominence in recent years due to the exponential growth of global tourism and the concentration of tourists in popular destinations [2].

Historical villages are particularly vulnerable to overtourism due to their fragile ecosystems, limited infrastructure, and unique cultural heritage [5]. The physical structures in these villages, often centuries old, are not designed to withstand large number of visitors [7]. The influx of tourists can lead to the deterioration of historical buildings, damage to ancient streets, and strain on infrastructure such as water supply and sewage systems [6].

The concept of the "tourist-historic city" suggests that the heritage value of a place can be compromised by the very tourism it attracts [7]. Overcrowding can diminish the authenticity and ambiance that make historical villages appealing in the first place [5]. Moreover, the commercialization of culture can lead to the loss of traditional practices and a shift toward catering to tourist demands rather than preserving local heritage [18].

Social impacts are also significant. Residents may experience a decreased quality of life due to noise pollution, overcrowding, and loss of privacy [4]. The cost of living may increase as local resources become strained, and housing prices may rise due to demand from tourism-related developments [19]. In some cases, tensions can arise between tourists and locals, leading to a phenomenon known as "tourismphobia" [2].

Environmental degradation is another critical issue. Increased waste production, pollution, and overuse of natural resources such as water can harm local ecosystems.

Waste management becomes a significant challenge in destinations experiencing overtourism, leading to littering and pollution of natural habitats [4].

In Turkey and Indonesia, several historical villages have faced these challenges. For example, the village of Şirince in Turkey experienced a surge in tourism due to a media frenzy, resulting in infrastructure strain and environmental issues [20]. In Indonesia, traditional villages like Wae Rebo have struggled to balance preserving their unique culture with the demands of tourism [21].

2.2 *Sustainable Tourism Practices in Historical Villages*

Sustainable tourism is defined as fully considering its current and future economic, social, and environmental impacts, addressing visitors' needs, industry, environment, and host communities [22]. In the context of historical villages, sustainable tourism aims to preserve cultural heritage while fostering economic development and ensuring social equity [6].

Community-based tourism is a central strategy in sustainable tourism, especially in rural and historical settings. Community-based tourism involves the participation of local communities in tourism planning and management, ensuring that they have control over tourism activities and benefit economically. Studies have found that community-based tourism can empower communities by providing income, preserving cultural practices, and enhancing self-reliance [23].

In historical villages, maintaining traditional architecture and cultural practices is essential. Heritage tourism should be managed in a way that conserves cultural assets and provides educational experiences for visitors [6]. This includes using traditional materials and techniques in restoration efforts and supporting local artisans and craftspeople [24].

Environmental conservation is another pillar of sustainable tourism. Implementing effective waste management systems, promoting recycling, and reducing resource consumption are critical [25]. Renewable energy sources and water conservation techniques can mitigate tourism's environmental footprint [26].

Education and awareness are also vital components. Educating both tourists and residents about the importance of preserving cultural and environmental resources can foster responsible behaviors [27]. This includes information on local customs, appropriate conduct, and environmental practices.

Successful examples of sustainable tourism practices can be seen in places like Safranbolu, Turkey, and Penglipuran, Indonesia. Safranbolu has maintained its Ottoman-era architecture and involved residents in tourism planning, helping to preserve its UNESCO World Heritage status [28]. Penglipuran Village is renowned for its cleanliness, community governance, and adherence to traditional values, successfully balancing tourism growth with cultural preservation [13].

2.3 International Policies and Guidelines on Heritage Conservation and Sustainable Tourism

International organizations have developed policies and guidelines to support heritage conservation and promote sustainable tourism. UNESCO's World Heritage Convention, established in 1972, aims to identify, protect, and preserve cultural and natural heritage worldwide [29]. Being listed as a World Heritage Site can boost tourism and oblige countries to ensure the site's conservation and sustainable management [30].

The International Council on Monuments and Sites (ICOMOS) has produced several charters, including the International Cultural Tourism Charter, which provides principles for managing tourism at places of heritage significance [31]. This charter emphasizes balancing conservation and tourism development, advocating for community involvement and sustainable practices.

The World Tourism Organization's Global Code of Ethics for Tourism outlines principles to guide stakeholders in sustainable tourism development [32]. It promotes respect for the cultural heritage of host communities, responsible tourism behaviors, and the equitable distribution of economic benefits [33]. The code encourages governments, tourism operators, and tourists to work together to minimize negative impacts.

The 2030 Agenda for Sustainable Development, adopted by all United Nations Member States in 2015, includes 17 Sustainable Development Goals (SDGs) [34]. SDG 11 focuses on making cities and human settlements inclusive, safe, resilient, and sustainable, with target 11.4 specifically aiming to "strengthen efforts to protect and safeguard the world's cultural and natural heritage." SDG 12 promotes responsible consumption and production patterns, which are directly relevant to managing tourism sustainably.

These international frameworks provide guidelines that can be adapted nationally and locally. They encourage the following.

Community Participation: Engaging local communities in decision-making ensures that tourism development aligns with their needs and values [35].

Conservation of Heritage: Implementing policies that protect cultural and natural heritage sites from overexploitation [31].

Sustainable Economic Development: Promoting tourism that provides economic benefits without compromising environmental and cultural integrity [36].

Education and Awareness: Raising awareness among tourists and residents about preserving heritage and practicing sustainable behaviors [27].

Countries like Turkey and Indonesia have incorporated these guidelines into their national tourism strategies. For example, Turkey has developed regulations to protect heritage sites and promote sustainable tourism practices [37]. Indonesia has implemented community-based tourism models in several villages, aligning with international principles [13].

However, challenges remain in the implementation and enforcement of these policies. Resource constraints, lack of awareness, and conflicting interests can hinder progress. Continuous effort, monitoring, and adaptation are necessary to ensure that international guidelines effectively contribute to sustainable tourism in historical villages.

3 Research Methodology

This study employs a qualitative comparative case study approach to investigate the multifaceted impacts of overtourism on historical villages, drawing insights from Cumalıkızık in Turkey and Penglipuran in Indonesia. These villages, both recognized as UNESCO heritage sites, share commonalities in their traditional architecture and cultural values but diverge significantly in their experiences and management of tourist influx, making them ideal for comparative analysis. Data collection was rigorous and multifaceted, encompassing an extensive literature review of scholarly works and official reports from organizations like UNWTO and UNESCO, providing a foundational understanding. This was complemented by field observations during on-site visits, where the physical and social impacts of tourism were directly observed and documented. Crucially, semi-structured interviews with local residents and business owners offered invaluable qualitative data on community perceptions and experiences. Finally, document analysis of local policies and management plans illuminated the formal frameworks governing tourism and heritage conservation. The collected data underwent thematic analysis to identify recurring patterns and themes related to overtourism impacts and sustainable practices, using systematic coding for organization. A subsequent comparative analysis highlighted the differences and similarities in the villages' approaches, aiming to identify successful strategies from Penglipuran that could potentially be adapted for Cumalıkızık. The selection of these particular case studies is justified by their direct relevance as heritage sites under tourism pressure, their UNESCO status, and the contrasting outcomes in managing overtourism, which offers a nuanced understanding of effective sustainable tourism strategies within diverse cultural and geographical contexts. This systematic methodology, from problem identification to data synthesis and recommendation, aims to contribute significantly to both academic discourse and practical applications in heritage tourism management. The figure below summarizes the research methodology steps of the study concisely (see Fig. 3).

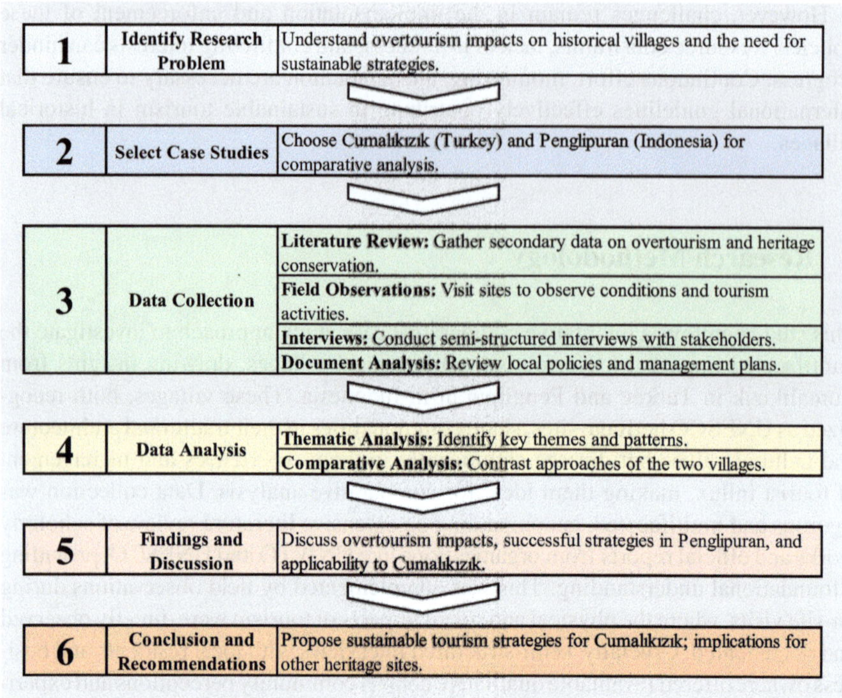

Fig. 3 Research process flowchart (Author 2024)

4 Cumalıkızık: The Struggle of Historical Fabric with Overtourism

Cumalıkızık is one of the rare villages in northwestern Turkey, located in the Yıldırım district of Bursa, that has preserved the early rural settlement fabric of the Ottoman Empire to this day [38]. The village, founded after the Ottomans conquered Bursa in the fourteenth century, is considered one of the most important villages called "kızık" [39]. The term "kızık" derives from the Kızık tribe of the Oghuz Turks, and there are many villages named after this tribe [40].

Cumalıkızık's historical significance stems from its architectural and urban features, which shed light on the founding period of the Ottoman state [10]. The village presents an authentic example of the Ottoman rural settlement model (see Fig. 4), which was included in the UNESCO World Heritage List in 2014 [9]. UNESCO has highlighted that the village is an exemplary representation of Ottoman rural life and early civilian architecture.

Architecturally, Cumalıkızık is characterized by its narrow and cobblestone streets, and bay-windowed and courtyard houses. The houses are typically three stories high, with stone on the lower floors and wooden materials on the upper floors. The streets are narrow and winding due to the proximity of the house facades,

Fig. 4 Historical rural settlement (Author 2023)

giving the village a distinctive atmosphere. The colors and woodwork used on the exterior facades of the houses enhance the village's aesthetic value.

Culturally, Cumalıkızık preserves traces of traditional Turkish village life. The villagers are engaged in traditional agriculture and animal husbandry, and handicrafts and local culinary culture are maintained. Historical structures such as the village mosque, bath, and fountain are important examples of Ottoman period architecture and social life.

4.1 *Development of Tourism in Cumalıkızık*

Until the 1980s, Cumalıkızık remained a secluded, self-sustained village. However, with tourism development in Turkey and the increasing interest in cultural heritage conservation, the village became a tourist destination. Especially after the 2000s, the village gained popularity through media and promotional activities, attracting domestic and foreign tourists [11].

Its inclusion in the UNESCO World Heritage List brought international recognition to the village, significantly increasing tourist numbers. While the number of visitors was around 150,000 in 2010, it reached 500,000 by 2015 [41]. This rapid increase contributed positively to the village's economy and brought about various problems.

4.2 Problems Caused by Overtourism

The rising number of visitors has negatively impacted the village's physical structure. Historical houses and streets are wearing down and becoming damaged due to excessive foot traffic [10]. The crowds in the narrow streets are causing the cobblestone pavements and historical buildings to deteriorate. While some homeowners are restoring their homes to take advantage of the economic opportunities provided by tourism, the use of non-traditional materials and techniques is disrupting the authenticity of the village's original fabric (see Fig. 5).

Additionally, tourists' behaviors that damage historical structures have been observed. Actions such as writing on walls, climbing on structures, and removing parts of buildings as souvenirs are destroying cultural heritage. The inadequacy of restoration and conservation efforts further exacerbates these issues.

Cumalıkızık's existing infrastructure struggles to cope with the increasing number of tourists. Water and sewage systems frequently break down and become inadequate

Fig. 5 The use of non-traditional materials and techniques (Author 2023)

due to heavy use. Especially during the summer and on weekends, the influx of visitors causes overuse of water resources and shortages [11].

There are also serious problems with waste management. The litter left behind by tourists accumulates in the streets and around the village, leading to environmental pollution. Inadequate waste collection services and the absence of recycling systems further aggravate this issue.

Traffic and parking issues represent another aspect of the infrastructure that needs to be improved. The village's narrow streets and limited parking spaces are incapable of handling the vehicle traffic. The uncontrolled entry of tour buses and private cars into the village results in traffic congestion and safety risks. This situation negatively affects the daily lives of the local population and complicates emergency interventions.

Overtourism reduces the quality of life and satisfaction of the local population. Village residents complain about the noise, crowds, and traffic problems caused by tourists. The influx of visitors, particularly on weekends and during holidays, which far exceeds the village population, makes daily life more difficult. Additionally, tourists entering private spaces without permission and violating privacy during photo sessions lead to discomfort among the local population and negative attitudes toward tourists [38, 39]. In interviews with local shopkeepers and residents, demands for controlling and regulating the number of tourists are prominent.

The economic opportunities brought by tourism have led to changes in the village's traditional lifestyle and social structure. Traditional agricultural and livestock activities are decreasing, while commercial activities such as selling tourist souvenirs and running restaurants and cafes are coming to the fore. This situation weakens the village's authentic identity and cultural values [42]. Furthermore, outside investors opening businesses and buying real estate in the village are causing the local population to lose economic control. Traditional handicrafts and products are being replaced by mass-produced tourist items, accelerating cultural erosion, and failing to meet the expectations of tourists seeking authentic experiences.

Overtourism threatens the village's environmental values and natural resources. Increasing waste production and inadequate waste management lead to water and soil pollution. The agricultural lands and forests surrounding the village are being opened for the construction of tourist facilities and parking lots, resulting in a reduction in biodiversity and ecosystem degradation.

In addition, increased energy consumption and more renewable energy sources must be addressed for environmental sustainability. In interviews conducted in the village, it was noted that while the local population is aware of the environmental issues, insufficient support and resources exist to address them [38, 39].

4.3 Current Management Strategies and Their Shortcomings

Although some measures have been taken against the overtourism issues in Cumalıkızık, these efforts are generally insufficient and ineffective. Local governments, non-governmental organizations, and universities have conducted various projects and studies. For instance, initiatives such as closing some streets to vehicle traffic, creating parking areas, and placing information signs have been implemented.

However, these measures have failed to control the number of tourists or resolve the infrastructure problems. A long-term and holistic approach to management strategies has prevented the problems from persisting. Furthermore, the limited involvement of the local population in tourism management and their lack of an active role in decision-making have diminished the effectiveness of the measures taken.

Another significant issue is the need for current legal regulations and conservation plans [39]. The failure to update conservation-oriented zoning plans, weak enforcement mechanisms, and the lack of penalties have led to arbitrariness in construction and restoration processes. This has accelerated the deterioration of the village's historical and cultural fabric.

4.4 The Need for Sustainable Tourism

To preserve and pass on Cumalıkızık's historical and cultural values to future generations, the development of sustainable tourism policies is inevitable. Sustainable tourism is an approach that aims to promote economic development while preserving environmental and cultural values [43]. In this context, local participation, environmental protection, cultural heritage conservation, and visitor management are critically important [22].

Implementing sustainable tourism practices in Cumalıkızık will help mitigate the negative impacts of overtourism and contribute to the long-term development of the village. In this regard, the following steps are crucial:

Increasing Local Participation: Active involvement of the local population in tourism planning and management will make the decisions more effective and acceptable [35].

Developing Visitor Management Strategies: Controlling the number of visitors and implementing entrance fees and reservation systems can help regulate the flow of tourists [44].

Promoting Environmentally Friendly Practices: Waste management, energy efficiency, and renewable energy sources will support environmental sustainability [45].

Conserving Cultural Heritage: Using traditional architectural techniques in restoration works and supporting cultural activities will help preserve the village's authentic identity [6].

Technology Integration: Improving and managing the visitor experience through digital platforms and smart applications is possible [27].

Studying successful examples like Penglipuran Village and adapting applicable strategies will be an important step in helping Cumalıkızık achieve its sustainable tourism goals [13]. This study aims to thoroughly examine Penglipuran Village's experiences and how these strategies can be applied to Cumalıkızık.

5 Penglipuran Village: A Successful Example of Sustainable Tourism and the Keys to Its Success

Penglipuran Village, located in the Bangli region of Bali, Indonesia, is a unique settlement that preserves traditional Balinese culture. The village reflects Bali's authentic characteristics through its architectural structure and cultural life [26]. Despite the pressure of overtourism, Penglipuran has successfully implemented sustainable tourism practices, and preserved its cultural heritage while enhancing the local population's well-being [13].

The history of Penglipuran dates back 700 years. The village's name is derived from the term "Pengeling Pura," which means "remembering a sacred place." Penglipuran is one of the rare settlements that has maintained the traditional Balinese way of life and cultural values. The architectural structure of Penglipuran strictly adheres to traditional Balinese principles, ensuring the village's aesthetic and cultural integrity [46].

The houses are built using local materials, and each house features a traditional gate called "angkul-angkul" at its entrance (see Fig. 6). The streets are flat and clean, with the layout of the houses arranged symmetrically. The village's streets are closed to vehicle traffic, offering visitors a calm and peaceful atmosphere. This layout reflects the social and cultural values of the Balinese community.

The village population continues to live according to traditional beliefs and rituals. Religious ceremonies, festivals, and traditional arts are an integral part of daily life in the village [47]. These cultural practices not only strengthen social solidarity but also ensure the transmission of cultural heritage to future generations.

5.1 Development of Tourism in Penglipuran and Overtourism Issues

Since the 1990s, Penglipuran Village has gained attention as a tourism destination [48]. The village's unique architecture and cultural values have attracted domestic

Fig. 6 Traditional house in Penglipuran Village (Author 2023)

and international tourists. However, the increasing number of visitors began to exert pressure on the village's physical and social fabric [13].

Overtourism has caused various issues in the village. The uncontrolled increase in visitor numbers has led to infrastructure needs to be improved. There have been difficulties related to the excessive use of water resources and waste management. Furthermore, tourists' lack of cultural sensitivity has negatively impacted the daily lives of the local population. For instance, disrespect toward sacred sites, non-compliance with dress codes, and privacy violations have led to social tensions [13].

5.2 Strategies Applied Against Overtourism

Penglipuran Village exemplifies a successful sustainable tourism model through several integrated strategies to counter overtourism. Visitor Management forms a core pillar, with an entrance fee contributing to village conservation and development, alongside checkpoints that enable daily visitor monitoring and limitation [13]. Tourists are restricted to specific visiting hours to respect residents' privacy, regulating flow and preserving tranquility [13]. Guided tours further enhance cultural sensitivity by informing visitors about the village's heritage, preventing unwanted behaviors. Community Participation and Management are central, with the "Desa Adat" traditional community council actively planning and managing tourism policies [49]. The local population receives continuous education on sustainable tourism and environmental protection, complemented by informational materials for tourists to raise cultural awareness. Tourism revenues are transparently reinvested into community needs, infrastructure, and social projects, ensuring equitable economic distribution [49]. Environmental Protection and Sustainability are paramount, with the village lauded for its cleanliness and strict waste management, promoting recycling and composting. Renewable energy and water conservation practices like rainwater harvesting are widely adopted [26] (Fig. 7).

Green areas are meticulously protected, and new constructions adhere to strict supervision, with community efforts dedicated to tree planting. Lastly, Cultural Heritage Preservation is a primary objective. Restoration of new buildings adheres to traditional architectural principles [50], maintaining the village's unique fabric. Religious ceremonies, festivals, and traditional arts are actively maintained and showcased to tourists, providing authentic experiences while preserving cultural values. Traditional handicrafts, weaving, and agricultural products are supported and offered

Fig. 7 Traditional handicrafts, weaving, and agricultural products (Author 2023)

to tourists, generating income and safeguarding cultural heritage. These comprehensive strategies, involving active community participation, integrated environmental approaches, and continuous education, have enabled Penglipuran to mitigate overtourism's negative impacts, preserving its unique identity and ensuring long-term sustainable development [13, 26, 49, 50] (Table 1).

Penglipuran Village offers vital lessons for Cumalıkızık in combating overtourism, given their shared historical and cultural significance. Penglipuran's successful strategies can help Cumalıkızık manage its challenges and foster sustainable tourism. Crucially, active community participation in tourism management, similar to Penglipuran's model, would align decisions with local needs and boost acceptance. Implementing visitor control mechanisms like entrance fees, regulated visiting hours, and guided tours could effectively manage tourist flow, preserving Cumalıkızık's physical and social fabric. Furthermore, adopting Penglipuran's robust environmental practices in waste management, energy conservation, and green space protection is essential for long-term sustainability. Emphasizing the preservation of traditional architecture, local lifestyle, and promoting cultural activities and handicrafts would maintain Cumalıkızık's unique identity. Finally, education programs for both residents and tourists on sustainability and cultural sensitivity are key to preventing undesirable behaviors. By adapting these proven strategies, Cumalıkızık can safeguard its historical and cultural values, transitioning toward a sustainable tourism model.

Table 1 Tourism strategies implemented in Penglipuran (Author 2024) [13, 26, 46–50]

Strategy	Description
Visitor management	Entrance fees, controlled visiting hours, and guided tours to regulate tourist flow
Community participation and governance	Local community council manages tourism activities and decision-making
Environmental sustainability practices	Effective waste management, recycling, composting, renewable energy use, and water conservation
Cultural heritage preservation	Maintaining traditional architecture; supporting cultural events and practices
Education and awareness	Educating locals and tourists on sustainability and cultural sensitivity
Economic benefits distribution	Equitable sharing of tourism revenue among community members
Technology integration	Utilization of technology for visitor information and management

6 Penglipuran's Strategies for Cumalıkızık: Applicability and Challenges

6.1 *Applicability of Penglipuran's Strategies to Cumalıkızık*

Penglipuran Village's successful strategies offer a robust framework for Cumalıkızık to manage overtourism. Visitor Management and Control Mechanisms are paramount; Cumalıkızık could implement an entrance fee system, similar to Penglipuran, with revenue directed toward the village's conservation, restoration, and infrastructure development. This would enable monitoring tourist numbers and imposing necessary restrictions, particularly during peak times [13]. Establishing specific visiting hours would minimize disruption to residents' daily lives, controlling intense crowding on weekends and holidays. Encouraging guided tours would also cultivate tourist awareness of the village's historical and cultural values, preventing undesirable behavior by conveying local sensitivities. Community Participation and Local Governance are crucial, mirroring Penglipuran's model where community input significantly shapes decision-making [48]. Cumalıkızık could establish a community-based management council to determine tourism policies, oversee implementation, and resolve issues, ensuring decisions align with local needs. Education and awareness programs, in collaboration with universities and NGOs, could further empower the local population regarding sustainable tourism and cultural heritage. Fair and transparent distribution of tourism's economic benefits would also bolster community support.

For Environmental Protection and Sustainability Practices, Cumalıkızık can adopt Penglipuran's successful waste management, energy conservation, and water resource utilization strategies [47]. Implementing an effective waste management system with recycling and composting, alongside improving garbage infrastructure, would mitigate pollution. Promoting energy efficiency through renewable sources like solar panels and implementing water-saving methods, including rainwater harvesting, are vital. Strict regulation of new constructions would preserve green spaces and agricultural lands, with reforestation efforts enhancing the village's natural balance. Preservation and Promotion of Cultural Heritage are key. Cumalıkızık should conduct restoration works using traditional architectural principles and local materials to safeguard its historical fabric and unique identity [50]. Supporting traditional handicrafts, local products, and culinary culture would both preserve heritage and generate economic benefits. Organizing cultural events, festivals, and traditional ceremonies would strengthen community bonds and attract tourists. Finally, Education and Technology Integration are essential. Education programs for both locals and visitors on sustainability and cultural sensitivity, delivered through various platforms, coupled with informational signage, would raise awareness [27]. Integrating technology through digital platforms and mobile applications for visitor information, guided services, and reservation systems would improve tourist flow management and overall satisfaction [27].

6.2 Challenges in Implementation

Several challenges could arise while implementing the strategies applied in Penglipuran in Cumalıkızık. Firstly, local resistance to change and economic concerns should not be overlooked. Practices such as entrance fees and visitor limitations could lead to a short-term decline in the income of some local businesses and operators. Therefore, it is important to inform the community about the long-term benefits of these strategies and ensure their active participation in decision-making processes [16]. This would increase social ownership and support for the implementation of these strategies. Another challenge is the need for more financial resources and support. Necessary infrastructure investments, restoration works, and education programs require financial resources. To address this, state support, contributions from local governments, international funds, and private-sector collaborations could be considered [45]. Additionally, the revenue generated from entrance fees could be used to finance these projects. Legal and bureaucratic obstacles also pose a significant challenge to implementation. Conservation-oriented zoning plans need to be updated, legal regulations need to be enacted, and enforcement mechanisms need to be strengthened. In this process, cooperation with local and national governments and the support of non-governmental organizations and academic institutions is important. Strengthening the legal infrastructure will ensure that the practices are effective and sustainable. The lack of technological infrastructure and low levels of digital literacy among the local population could hinder technology integration [27]. Therefore, infrastructure investments should be made to facilitate the implementation of technological solutions, and training should be provided to the local population on the use of technology. Pilot projects and partnerships can ensure the effective use of technology.

Cultural and social differences may also prevent the direct application of strategies. Penglipuran and Cumalıkızık have different cultural, religious, and social structures. Therefore, strategies need to be adapted to the unique conditions and values of Cumalıkızık. Flexible and localized practices should be developed, considering the local population's traditions, beliefs, and needs. Finally, tourist acceptance and satisfaction should also be considered. Limitations on visitor numbers and entrance fees may reduce some tourists' interest in the destination [44]. Therefore, visitor management strategies should be designed to enrich the tourist experience. Guided tours, cultural events, and authentic experiences will enhance tourist satisfaction.

7 Conclusion and Recommendations

This study examined the effects of overtourism on historical and cultural heritage sites through the examples of Cumalıkızık in Turkey and Penglipuran in Indonesia. While Cumalıkızık, with its authentic architecture and cultural values from the Ottoman period, faces physical, environmental, and social challenges brought by overtourism,

Penglipuran Village, which shares similar characteristics, has successfully managed these issues through sustainable tourism practices [13].

In conclusion, for Cumalıkızık to transition to a sustainable tourism model, it is important to benefit from the strategies implemented in Penglipuran Village. However, instead of directly applying these strategies, it is crucial to adapt them to the unique cultural, social, and economic conditions of Cumalıkızık.

Visitor Management: The effective management of visitors is of critical importance in reducing the negative impacts of overtourism in Cumalıkızık. Implementing entrance fees, monitoring visitor numbers, and imposing restrictions, when necessary, will contribute to the preservation of the village's physical and social fabric. The income generated from entrance fees can be used for restoration works, infrastructure development projects, and environmental protection activities.

Encouraging guided tours and regulating visiting hours will help tourists become more aware of the village's historical and cultural values, thus preventing undesirable behavior. These practices will also prevent the local population's daily life from being negatively affected, creating a more positive interaction between tourists and locals.

Community Participation and Local Governance: Ensuring community participation and establishing local governance mechanisms will increase the effectiveness of tourism management decisions. Creating a management council or committee representing the local population will allow strategies to be developed in line with the village's needs and expectations. Education and awareness programs will help raise the consciousness of the local population regarding sustainable tourism, environmental protection, and the importance of cultural heritage [27].

Environmental Protection and Sustainability: Implementing environmental protection and sustainability practices is essential for preserving the ecological balance of Cumalıkızık and preventing environmental issues. Establishing an effective waste management system, promoting recycling and composting practices, and improving energy efficiency will reduce environmental pollution.

Encouraging renewable energy sources, such as solar panels, and implementing water conservation techniques will help minimize energy consumption and carbon emissions. Preserving green spaces and agricultural lands will protect the village's natural fabric and biodiversity.

Cultural Heritage Conservation and Preservation: The preservation and promotion of cultural heritage are critical for maintaining the village's unique identity and enhancing its attractiveness to tourists. Restoration works should adhere to traditional architectural principles and use local materials to preserve the village's historical fabric.

Supporting traditional handicrafts, local products, and culinary culture will provide economic benefits and ensure cultural heritage transmission to future generations. Handcrafted products and local delicacies can be offered to tourists to provide them with authentic experiences. Additionally, organizing cultural events, festivals, and traditional ceremonies will strengthen the cultural bonds of the local community while attracting tourists.

Education and Technology Integration: Education and technology integration will raise awareness among the local population and visitors regarding sustainability and cultural sensitivity [28]. The effective use of technology can improve visitor management and enhance the tourist experience. Digital platforms and mobile applications can provide information about the village, offer guided services, and manage tourist flow more efficiently.

Funding and Resource Allocation: The effective use of funding and resources is crucial for implementing necessary infrastructure investments and programs. State support, international funds, private-sector collaborations, and the income generated from entrance fees can be utilized for this purpose [45]. The transparent and efficient allocation of resources will ensure the sustainability of projects.

Legislation and Policy Strengthening: Strengthening legislation and policies will enhance the effectiveness of the implementations. Updating conservation-oriented zoning plans, enacting legal regulations, and strengthening enforcement mechanisms are necessary steps. Cooperation with local and national governments and support from non-governmental organizations and academic institutions will be essential in establishing the necessary legal infrastructure and ensuring the effectiveness of the implementations.

It is also important to be aware of potential challenges during implementation and take necessary precautions. Local resistance to change and economic concerns should be noticed. Therefore, ensuring the active participation of the local population in decision-making processes and informing them about the long-term benefits of the strategies is crucial for success.

In conclusion, Penglipuran Village's experiences provide valuable guidance for Cumalıkızık's transition to a sustainable tourism model. However, a participatory and flexible approach that considers local conditions and needs is essential for successfully implementing these strategies. Collaborating with academic institutions, non-governmental organizations, and tourism sector stakeholders will enhance the effectiveness of the implementations. Preserving Cumalıkızık's historical and cultural values is a local issue and a matter of national and international importance. Sustainable tourism practices will ensure the village's unique identity is passed on to future generations while balancing economic development with environmental and social responsibility [43]. In this context, adapting Penglipuran's successful strategies to Cumalıkızık will be an important step toward preserving cultural heritage and achieving sustainable development. Future research could delve deeper into the application of sustainable tourism in historical villages like Penglipuran and Cumalıkızık. Case studies and field research on the challenges encountered during implementing these strategies and how they were overcome would be valuable. Ultimately, the preservation of cultural heritage sites and the widespread adoption of sustainable tourism practices will not only enhance the well-being of local communities but also ensure the continuity of a rich heritage that can be passed on to future generations. The steps Cumalıkızık takes in this direction will serve as an example for other historical and cultural heritage sites.

References

1. UNWTO. International tourism highlights. https://doi.org/10.18111/9789284421152. Last accessed 20 Sept 2024
2. Milano C, Novelli M, Cheer JM (2019) Overtourism and tourismphobia: a journey through four decades of tourism development, planning and local concerns. Tour Planning Develop 16(4):353–357. https://doi.org/10.1080/21568316.2019.1599604
3. Goodwin H (2017) The challenge of overtourism. Responsible tourism partnership working paper, vol 4, pp 1–19
4. Koens K, Postma A, Papp B (2018) Is overtourism overused? Understanding the impact of tourism in a city context. Sustainability 10(12):4384. https://doi.org/10.3390/su10124384
5. Russo AP (2002) The "vicious circle" of tourism development in heritage cities. Ann Tour Res 29(1):165–182. https://doi.org/10.1016/S0160-7383(01)00029-9
6. Timothy DJ, Boyd SW (2003) Heritage tourism. Pearson Education, London
7. Ashworth GJ, Tunbridge JE (2000) The tourist-historic city: retrospect and prospect of managing the heritage city. Elsevier, London. https://doi.org/10.4324/9780080519470
8. Su MM, Wall G (2010) Implications of host-guest interactions for tourists' travel behaviour and experiences. Tour Int Interdiscip J 58(1):37–50
9. UNESCO. Bursa and Cumalıkızık: the birth of the Ottoman Empire. https://whc.unesco.org/en/list/1452. Last accessed 8 Nov 2014
10. Bilgin EA, Arslan TV, Durak S (2019) Physical changes in World Heritage Sites under the pressure of tourism: the case of Cumalıkızık village in Bursa. Eur J Sustain Develop 8(2):249–249. https://doi.org/10.14207/ejsd.2019.v8n2p249
11. Kuzuoglu S, Hatipoglu BK (2021) The lack of policy, planning, and governance: the mismanagement of visitor pressure in Cumalıkızık, Bursa—a world heritage site. In: Mediterranean protected areas in the era of overtourism: challenges and solutions. Springer, Cham, pp 241–263. https://doi.org/10.1007/978-3-030-69193-6_12
12. Pradnyantara IKE, Lestari NPNE (2021) Implementation of community based tourism concept in sustainable tourism management in Penglipuran village. Enrichment: J Manage 11(2):491–495. https://doi.org/10.35335/enrichment.v11i2.131
13. Sutawa GK (2012) Issues on Bali tourism development and community empowerment to support sustainable tourism development. Procedia Econ Finance 4:413–422. https://doi.org/10.1016/S2212-5671(12)00356-5
14. Sutrisna M, Bagus I, Saskara N, Ayu I (2020) Prospective analysis of sustainable tourism development in Penglipuran Village as a tourist village role model in Bali. Technium Soc Sci J 13:184–200
15. Dodds R, Butler R (eds) (2019) Overtourism: issues, realities and solutions, vol 1. Walter de Gruyter, Berlin
16. Jamal T, Camargo BA (2014) Sustainable tourism, justice and an ethic of care: toward the just destination. J Sustain Tour 22(1):11–30. https://doi.org/10.1080/09669582.2013.786084
17. Muresan IC, Oroian CF, Harun R, Arion FH, Porutiu A, Chiciudean GO, Todea A, Lile R (2016) Local residents' attitude toward sustainable rural tourism development. Sustainability 8(1):100. https://doi.org/10.3390/su8010100
18. Kvasnová D, Marciš M (2022) Eliminating overtourism in UNESCO destinations: a case study from Slovakia. In: de Lana A, Gravari-Barbas M (eds) Cultural leadership in transition tourism: developing innovative and sustainable models. Springer, Cham, pp 229–247. https://doi.org/10.1007/978-3-031-14121-8_12
19. Gursoy D, Chi CG, Dyer P (2009) An examination of locals' attitudes. Ann Tour Res 36(4):723–726. https://doi.org/10.1016/j.annals.2009.06.003
20. Kilinc-Unlu A (2011) A study of historic towns after "Tourism explosion": The case of Çeşme, Foça and Şirince in Western Turkey. Graduate Program in Historic Preservation. University of Pennsylvania

21. Afifah N, Kurniawan KR (2022) Intangible aspect of revitalization in vernacular heritage architecture in case study of Wae Rebo NTT Village. Appl Eng Technol 1(2):90–100. https://doi.org/10.31763/aet.v1i2.671
22. UNWTO. Sustainable development of tourism: conceptual definition. https://www.unwto.org/sustainable-development. Last accessed 5 Aug 2024
23. Trejos B, Chiang LHN (2009) Local economic linkages to community-based tourism in rural Costa Rica. Singap J Trop Geogr 30(3):373–387. https://doi.org/10.1111/j.1467-9493.2009.00375.x
24. Sebele LS (2010) Community-based tourism ventures, benefits and challenges: Khama Rhino Sanctuary Trust, Central District, Botswana. Tour Manage 31(1):136–146. https://doi.org/10.1016/j.tourman.2009.01.005
25. Magadán-Díaz M, Rivas-García JI (2022) Residents' perception of sustainable tourism in protected mountain areas: the case of Asturias. J Mt Sci 19(12):3597–3614. https://doi.org/10.1007/s11629-021-7182-8
26. Wardhana GD, Valeria DE, Awalina TN, Septemuryantoro SA (2022) Penglipuran Bali Tourism Village (the third cleanest village in the world). In: Proceedings of international seminar on translation, applied linguistics, literature, and cultural studies, vol 1, no 1, pp 58–63. https://doi.org/10.33633/str.v3i1.7293
27. Gretzel U, Sigala M, Xiang Z, Koo C (2015) Smart tourism: foundations and developments. Electron Mark 25(3):179–188. https://doi.org/10.1007/s12525-015-0196-8
28. Ceylan S, Somuncu M (2016) A critical view to tourism diversification in the cultural tourism areas: Safranbolu UNESCO World Heritage Site. Int J Turkic World Tour Stud 1(1):53–64
29. UNESCO. Convention concerning the protection of the world cultural and natural heritage. https://whc.unesco.org/en/conventiontext/. Last accessed 17 Nov 2024
30. Adie BA, Hall CM (2017) Who visits world Heritage? A comparative analysis of three cultural sites. J Herit Tour 12(1):67–80. https://doi.org/10.1080/1743873X.2016.1151429
31. ICOMOS (1999) International cultural tourism charter: managing tourism at places of heritage significance. International council on monuments and sites. https://www.icomos.org/images/DOCUMENTS/Charters/INTERNATIONAL_CULTURAL_TOURISM_CHARTER.pdf. Last accessed 17 Nov 2024
32. UNWTO. Global code of ethics for tourism. World Tourism Organization. https://www.unwto.org/global-code-of-ethics-for-tourism. Last accessed 17 Nov 2024
33. Fennell DA (2006) Tourism ethics. Channel View Publications
34. United Nations. Transforming our world: the 2030 agenda for sustainable development. https://sustainabledevelopment.un.org/content/documents/21252030%20Agenda%20for%20Sustainable%20Development%20web.pdf. Last accessed 17 Nov 2024
35. Jamal TB, Getz D (1995) Collaboration theory and community tourism planning. Ann Tour Res 22(1):186–204. https://doi.org/10.1016/0160-7383(94)00067-3
36. UNWTO. Making tourism more sustainable: a guide for policy makers. https://doi.org/10.18111/9789284408214. Last accessed 18 Nov 2024
37. Tosun C (2001) Challenges of sustainable tourism development in the developing world: the case of Turkey. Tour Manage 22(3):289–303. https://doi.org/10.1016/S0261-5177(00)00060-1
38. Pekerşen Y, Güneş E, Seçuk B (2019) The attitude of local people in the context of conservation and sustainability of cultural heritage tourism values: Cumalıkızık Sample. J Turk Tour Res 3(3):350–368
39. Eren R, Aypek N (2012) Residents' attitudes toward tourism development in rural tourism areas: the case of Cumalıkızık. Int J Soc Econ Sci 2(2):43–47
40. Özen MA (2021) Globalization of the rural and consumption: the case of Bursa Cumalıkızık. Uludag J Econ Soc 40(1):15–35
41. Ministry of Culture and Tourism: Tourism Statistics. https://yigm.ktb.gov.tr/TR-9851/turizm-istatistikleri.html. Last accessed 26 June 2024
42. Sezen B, Boz İ (2021) Visitors satisfaction level of rural tourism activities: the case of Cumalikizik Village. Black Sea J Public Soc Sci 4(1):24–33. https://doi.org/10.52704/bssocialscience.779509

43. Butler RW (1999) Sustainable tourism: a state-of-the-art review. Tour Geogr 1(1):7–25. https://doi.org/10.1080/14616689908721291
44. Mihalic T (2016) Sustainable-responsible tourism discourse: towards 'responsustable' tourism. J Clean Prod 111:461–470. https://doi.org/10.1016/j.jclepro.2014.12.062
45. Bramwell B, Lane B (2011) Critical research on the governance of tourism and sustainability. J Sustain Tour 19(4–5):411–421. https://doi.org/10.1080/09669582.2011.580586
46. Sudarwani MM (2018) The local wisdom form of sustainable architecture in Penglipuran Village. Int J Eng Technol Manage Res 5(3):59–66
47. Astara IWW, Budiartha NP, Wesna PAS, Selamet IK, Wijaya IKM (2023) Penglipuran Tourism Village, Kubu Village, Bangli District, Bali in maintaining Bali local wisdom values. J Legal Cult Anal 2(3):175–190. https://doi.org/10.55927/jlca.v2i3.5804
48. Adi INR, Utama MS, Budhi MKS, Purbadharmaja IBP (2017) The role of government in community-based tourism and sustainable tourism development at Penglipuran Traditional Village-Bali. IOSR J Human Soc Sci 22(6):15–20. https://doi.org/10.9790/0837-2206131520
49. Arismayanti NK, Nogroho S, Sudana IP (2014) Strategi Pengembangan Desa Wisata Berbasis Masyarakat Di Desa Adat Penglipuran Bangli. PARIWISATA 1410:47
50. Rachima I, Maulina Dian P (2015) Traditional architecture local wisdom case study: Penglipuran Village, Bali. In: On dwelling form, pp 59–66

Urban Identity and Resilience: The Impact of Overtourism on Venice

Jose Antonio Lara-Hernandez

Abstract Venice's unique double fractal urban fabric, with its intricate canals and human-scale architecture, forms a crucial part of the city's identity and sense of community. However, overtourism significantly threatens this delicate balance, altering not only the physical environment but also the social and cultural dynamics that make Venice distinct. This chapter explores the interplay between the urban fabric of Venice, and the emergent phenomenon of Temporary Appropriation (TA), which reflects the ways residents adapt public spaces for daily life. Through assemblage theory and agential realism, we analyse how tourism erodes the sense of community by overwhelming the city's public spaces, transforming them from local resources into sites of spectacle and commodification. Drawing from sense of community and spatial data analysis, we uncover the dual role of Venice's architectural design: while it fosters community resilience by encouraging social interaction, it also amplifies the pressures of overtourism. The findings presented here advocates for rethinking tourism management strategies that prioritize both the preservation of Venice's built environment and the urban identity.

Keywords Overtourism · Urban identity · Sense of community · Double fractal design · Venice · Temporary appropriation · Urban resilience

1 Introduction

Venice, Italy, is a very well-known city as a tourist attraction. Its unique canal system, iconic architecture, and rich cultural heritage presents a unique case of urban resilience and identity under pressure. Unlike other historical cities, Venice's urban fabric is distinct due to its double fractal design, a complex spatial pattern that integrates the city's layout at various scales, from *sestiere* (borough) to the *calle* (street).

J. A. Lara-Hernandez (✉)
Auckland University of Technology, Auckland CBD, Auckland, New Zealand
e-mail: antonio.lara-hernandez@aut.ac.nz

V. C. Sharma et al. (eds.), *Architectural Vision of Overtourism*,
SpringerBriefs in Architectural Design and Technology,
https://doi.org/10.1007/978-981-96-5794-0_4

This fractal arrangement is a testament to Venice's organic development, built upon layers of history and architectural exaptation [1], which fosters an unparalleled sense of place and community. However, the city's distinctive urban fabric is increasingly strained by overtourism. The unstoppable influx of tourists transforms public spaces into sites of spectacle, limiting their availability and functionality for residents [2].

Overtourism in Venice not only affects the city's built environment but also undermines its social cohesion. As public spaces become dominated by visitors, the sense of community (SOC) among locals is diminished, eroding the shared cultural practices and interactions that contribute to Venice's urban identity. This phenomenon is exacerbated by the spatial concentration of tourism services, which crowd public areas, disrupt local routines, and transform essential gathering places into commodified landscapes [3]. These dynamics call into question the long-term viability of Venice's urban identity and resilience, raising concerns over whether the city can sustain its cultural and social fabric amidst increasing tourist demands.

This chapter explores the relationship between Venice's urban fabric and its capacity for resilience under the pressures of overtourism. Specifically, it aims to examine how Venice's double fractal design supports and influences SOC and identity. The impact of overtourism on temporary appropriation (TA) understood as the spontaneous use of public spaces that embodies the social life of Venice is analysed. Thus, first we assess Venice's double fractal urban fabric as a facilitator or hinderer of resilience and SOC. Second, we aim to understand in what ways overtourism alters the practice of TA, and how this affects Venice's urban identity.

To do so, this study employs a mix-methodology approach grounded in assemblage theory and agential realism. Assemblage theory, with its focus on the relational and emergent nature of urban systems, provides a framework for analysing the dynamic interactions among Venice's urban fabric, SOC, and tourism pressures [4, 5]. By understanding Venice as an assemblage, this research examines how each of these elements contributes to and reshapes the city's identity and resilience. Additionally, applying agential realism helps to capture the intra-actions between human and non-human agents within Venice's urban landscape. This perspective emphasizes that agency in Venice's public spaces is not confined to individual actors but emerges through the relational interactions among tourists, locals, spatial configurations, and socio-legal frameworks. By integrating these theoretical approaches, the chapter sheds light on the adaptive and resistive capacities of both Venice's urban fabric and urban identity.

2 Literature Review

2.1 *Assemblage Theory and Agential Realism*

To analyse the complex interactions within Venice's urban landscape, this study combines assemblage theory and agential realism. Assemblage theory, as described by scholars like DeLanda [6] and Dovey [7], offers a framework for understanding cities as dynamic, multi-layered systems composed of interconnected parts. According to this theory, urban phenomena emerge from the synergistic relationships between these parts, rather than from any single fixed entity [8, 9]. In the case of Venice, the city can be seen as an assemblage of physical structures, social practices, economic forces, and regulatory frameworks that interact to shape its identity and resilience.

Assemblage theory is particularly useful for examining how overtourism and TA coexist within Venice's urban fabric. Each relationship between elements within the city—tourists, residents, public spaces, and tourism services—affects and is affected by the others. The presence of tourists influences how locals engage with public spaces, which in turn alters the dynamics of TA. Conversely, residents' adaptive use of spaces through TA reshapes the character of these areas, impacting how they are perceived and utilized by tourists. This bidirectional interaction illustrates the downward causality of assemblage theory, where the urban environment imposes constraints and offers opportunities for its individual components.

Agential realism, another post-humanistic theoretical approach employed in this study, emphasizes the role of intra-actions—mutual influences among agents that do not have a fixed identity but gain agency through their relationships with others [10]. This perspective challenges the notion of separate, stable entities within Venice's urban environment, suggesting instead that identities and functions emerge through relational interactions. For example, a square or a street in Venice may function as a community gathering space in the early morning, a playground for children by the afternoon, and a tourist attraction by evening. All of them are examples of TA. Thus, the square's identity is not fixed but is continuously redefined through its interactions with various users at different times. This fluidity underscores the resilience of Venice's urban fabric, as spaces adapt to accommodate different activities and groups.

Through the use of assemblage theory and agential realism, this study offers a nuanced view of Venice's double fractal design as both a product and a process—shaped by interactions among its physical, social, and economic components, not just as a mere container of activities. These frameworks help to explain why TA and overtourism do not merely coexist but actively shape each other within Venice's urban landscape. Understanding these intra-actions provides insights into how Venice's urban identity can be preserved through adaptive strategies that recognize the interdependent nature of the city's assemblage.

2.2 Urban Identity and Place-Identity

Urban identity refers to the distinct character and essence of a city, shaped by both its built environment and the people within it. This identity is a product of various layers, including historical development, architectural style, public spaces, and the collective practices of its inhabitants [11, 12]. Therefore, Venice's identity is tightly bound on the one hand to its double fractal design conformed by an intricated network of canals, and then by its *calli* (streets) and *campi* (squares), both blending in a unique spatial configuration [13]. This design creates an intimate, human-scale environment that fosters a strong sense of place and continuity with Venice's past.

The concept of place-identity as explored by Bolio Arceo [14], and Rios, Vazquez and Miranda [15], emphasizes the role of physical spaces in shaping individual and collective identities within urban settings. They argue that the fragmentation and commercialization of public spaces erode place-identity by disrupting residents' ability to connect meaningfully with their environment. In Venice, overtourism poses a similar threat, as more than 30 millions of tourists yearly [16] transform public spaces into commercial zones, thereby diminishing locals' sense of ownership and attachment to these spaces. This alteration undermines the social and cultural continuity that is essential to maintaining a strong urban identity.

Authors such as Bolio Arceo [17], Lalli [18], and Nylund [19] on place-identity highlight how fragmented or commodified public spaces weaken urban identity—a pattern observed similarly in Venice under the pressures of tourism. Furthermore, when these spaces become overrun by tourists, their function shifts from fostering community to serving as sites of spectacle, thus weakening the intrinsic connection between place and identity for Venetian residents.

2.3 Temporary Appropriation (TA)

Temporary Appropriation (TA) is a concept that refers to the so-called "informal" and spontaneous use of public spaces by residents for activities other than their intended purpose [20]. TA embodies the dynamic relationship between citizens and their urban environment, as individuals and groups reappropriate spaces to meet social, cultural, or immediate needs [21]. Unlike formal, planned urban interventions, TA reflects a grassroots approach to public space use, driven by residents' desires to express and reinforce their connection to the city [22].

In Venice like any other city, TA is a powerful expression of urban identity and resilience, as it allows residents to reclaim parts of the city from the commercialized spaces dominated by tourism. For example, locals might temporarily use a public square to host social gatherings, performances, or community events, creating a shared space for interaction and reinforcing social bonds [23]. However, overtourism limits these opportunities for TA by crowding out locals and saturating public areas with tourist-oriented activities [24]. The commercialization of Venice's public spaces disrupts these organic expressions of community, making it harder for residents to engage in TA and thus eroding their connection to the city.

TA also serves as a mechanism for social resilience by providing a flexible, adaptive means for residents to respond to changes in the urban environment. As assemblage theory suggests, cities are constantly evolving entities shaped by interactions among diverse elements. In Venice, TA acts as a resilient response to the pressures of overtourism, enabling locals to adaptively manage their environment despite external pressures. By understanding TA as an emergent, adaptive practice, urban designers and policymakers can recognize its importance in maintaining a balanced urban ecosystem that serves both residents and visitors.

2.4 *Sense of Community*

The concept of a Sense of Community (SOC) is closely linked to the urban environment and plays a significant role in fostering social resilience. To understand its implications, it is essential to first define the term "community". Gusfield [25] identifies two key interpretations of the term: one that is spatial, referring to a specific geographic area such as a neighbourhood or city, and another that is relational, emphasizing the connections and relationships between individuals. These interpretations are complementary, not mutually exclusive. Expanding on this, McMillan and Chavis [26] describe SOC as the sense of belonging among individuals, the feeling that members matter to one another and to the group, and the belief that collective commitment will meet individual needs.

The built environment plays a pivotal role in shaping SOC. Chavis and Wandersman [27] argue that the physical environment, coupled with interpersonal relationships, can enhance the SOC, boosting residents' satisfaction with their neighbourhoods. Similarly, studies by Talen [28, 29] and Long and Perkins [30] highlight that factors such as residential density, transportation options, and access to public spaces are key to fostering SOC. Human-scale design, which prioritizes the relationship between people and their immediate surroundings, has been shown to enhance social interactions and SOC [31, Long and Ye, 2016]. Research by Melis, Lara-Hernandez, and colleagues (2020, 2022) further indicates that informal and spontaneous uses of space—often overlooked in traditional urban design—play a critical role in community resilience. These unplanned behavioural patterns not only strengthen the sense of place but also contribute to public space resilience. Swyngedouw [32] underscores the complexity of modern urbanization, where socio-technical dynamics often erode natural systems and community ties. Achieving urban resilience and sustainability, therefore, requires innovative socio-technical assemblages, such as TA, and a paradigm shift that embraces nonlinearity, complexity, and uncertainty.

3 Methodology

This research as described by Groat and Wang [33] employs a multi-method approach to investigate the complex dynamics of urban identity, community resilience, and overtourism in Venice. The study draws upon assemblage theory and agential realism to frame Venice as a fluid and interconnected urban assemblage, where identities, uses, and functions emerge from the interactions among residents, tourists, and spatial configurations.

These theoretical frameworks guide the methodological choices, allowing an in-depth examination of both the spatial and social dimensions of Venice's unique urban fabric. Assemblage theory emphasizes the non-hierarchical relationships within cities, where each element contributes to the assemblage's identity while being influenced by its surrounding parts [34, 35]. Agential realism further underscores that these elements gain agency not as isolated entities but through their intra-actions with others. For Venice, this perspective helps capture how public spaces, residents, and tourists continually reshape one another, contributing to a dynamic urban identity that adapts under pressure from overtourism.

These theoretical lenses facilitated an understanding of Venice not as a static urban form but as an evolving interplay of agents, each contributing to the city's overall identity. By mapping these interactions, the study reveals how Venice's built environment both constrains and enables resilience practices, offering a nuanced view of how urban identity persists and adapts within a tourism-saturated context.

This study utilizes Street Network Analysis to assess Venice's urban fabric in relation to TA emergence. This method is particularly useful in studying the relationship between urban form and the social aspect of a city [36, 37]. Regarding SOC, an adapted version of the Brief Sense of Community Scale (BSCS), as detailed by Peterson, Speer, and McMillan (2008), was utilized to assess the Sense of Community (SOC).

3.1 Street Network Analysis

A quantitative analysis of the street network was conducted, following the methodological framework developed by Geoff Boeing [38]. This approach leverages street network analysis to obtain precise metrics of urban morphology and enables a robust comparison of spatial patterns across various cities. By examining differences in street network design, we can better understand how these variations influence the scale and nature of TA. Additionally, quantifying complex network features offers valuable insights into the impact of urban form on social interactions and economic activities, which are essential for promoting urban liveability and sustainability. The purpose of this analysis was to investigate the spatial configuration and connectivity within the urban areas studied. Venice features a large number of blind alleys, which differs significantly from the interconnected lattice structures. Visualizing

Fig. 1 Graph of Venice's street network. Edges represent walkable paths, and nodes represent intersections and path endpoints. *Source* [39]

the betweenness centrality of pathways reveals a higher intensity of commonly used walkways in Venice, creating more secluded low-movement zones or "eddies" (Figs. 1 and 2).

3.2 *Qualitative Observations*

Field observations were conducted in various public spaces across Venice, focusing on resident interactions and the informal use of spaces through TA. Observations documented how residents adapt spaces temporarily for personal or collective activities, such as using squares for social gatherings, performances, or temporary markets. These interactions were analysed to understand how TA serves as a form of spatial resistance to overtourism, allowing locals to reclaim areas temporarily and maintain a connection to their urban environment.

Observations were also extended to high-tourist zones, where overtourism pressures were observed to displace or alter traditional TA practices. This allowed for a comparative analysis of public spaces across the city, distinguishing between areas where TA is actively practiced and those where overtourism limits or transforms these activities. By documenting the diversity of TA practices, the study sheds light on how residents creatively adapt to preserve a sense of community amidst the spatial demands of tourism.

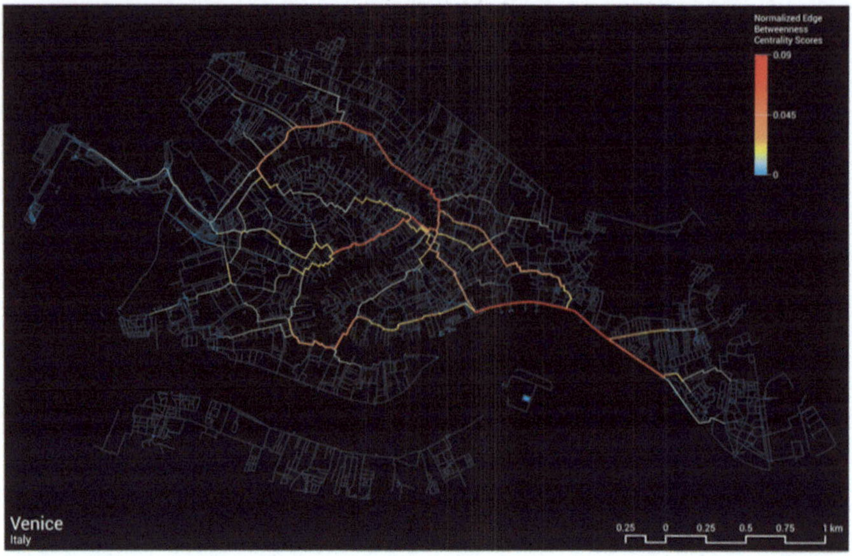

Fig. 2 Edge betweenness centrality of urban pedestrian pathways. Lattice-like structures of Venice. *Source* [40]

3.3 Review of Secondary Sources

The Tourism Services Index (TSI) provides a granular measurement of tourism saturation by evaluating the physical supply of tourist-oriented infrastructure in Venice, specifically focusing on lodging, dining, and shopping services. It integrates spatial datasets from platforms like Inside Airbnb, Tripadvisor, and Google Maps, categorizing them into three main indicators: the total number of Airbnb beds (AB), the Internationality Score (IS) of restaurants (based on the diversity of non-Italian reviews), and the count of tourist shops (TS). These elements are aggregated within a hexagonal grid system overlaid on the historic city of Venice, ensuring spatial consistency and reducing statistical bias. The index reflects the pervasiveness of tourism infrastructure at a micro-scale, offering insights into the spatial dynamics of overtourism without directly measuring tourist numbers. Through this structure, TSI identifies areas highly dedicated to tourism services, making it a critical tool for urban and tourism management strategies.

The TSI methodology utilizes advanced geospatial analyses, including hexagonal tessellation to normalize spatial data and control for inconsistencies like irregular boundaries. Each hexagon (with an area of 0.5 hectares) aggregates the three indicators into a standardized composite index using the Mazziotta-Pareto Index method. The result is a detailed, normalized score that links TSI values to urban form parameters, such as proximity to landmarks, street widths, and bridge density. Regression analyses reveal strong correlations between TSI and urban morphology, highlighting how wider streets, more intersections, and closer proximity to iconic features amplify tourism saturation. By operationalizing these connections, the TSI becomes a valuable framework for formulating spatially tailored policies to manage overtourism and its impact on Venice's socio-spatial fabric.

3.4 Sense of Community

In this study, an adapted version of the Brief Sense of Community Scale (BSCS) by Peterson, Speer, and McMillan [41] was utilized to evaluate the Sense of Community (SOC). The BSCS, which is rooted in McMillan and Chavis's [42] model of SOC dimensions—needs fulfilment, group membership, influence, and emotional connection—was modified to incorporate a new dimension: TA. This adjustment ensured that the questionnaire captured both the traditional SOC dimensions and the nuances introduced by TA practices. Consistent with the methodology of Peterson, Hughey, and Hughey [43], only positively worded items were included to maintain consistency and clarity. A five-point Likert scale ranging from "strongly agree" to "strongly disagree" was used for responses. The study recorded a Cronbach's alpha of 0.80 ($M = 4$, $SD = 0.74$) for SOC, reflecting good internal consistency. The specific items included in the BSCS are outlined in Appendix 1.

While the sample size might appear statistically limited if Venice were analysed as a conventional city, agential realism provides a different perspective. Venice's unique urban morphology, characterized by its walkability and compact design, enables most residents to live, work, and perform daily tasks within a 20-min radius of their homes. This distinct spatial configuration supports the smaller sample size by offering context-specific insights into SOC and TA dynamics. Agential realism critiques conventional epistemologies, emphasizing the value of alternative frameworks for understanding urban phenomena. This study's integration of street network analysis offers a richer understanding of how spatial configurations shape TA practices, highlighting how urban design fosters resilience and diversity in public spaces. By combining qualitative and spatial methodologies, the study provides a nuanced perspective on the relationship between urban form and community practices.

Limitations

Several limitations were acknowledged in the research methodology:

Spatial Specificity: The spatial analysis is specific to Venice's unique urban fabric, and results may not be directly applicable to other cities without similar urban forms.

In terms of temporal constraints, observations were conducted over a limited time frame, which may not capture the full variability of TA practices across seasons or in response to fluctuating tourism levels. Qualitative observations were limited to selecting public spaces within Venice, which could limit the generalizability of findings on TA practices across the entire city.

Despite these limitations, the methodology provides a comprehensive approach to exploring the complex dynamics of urban identity and resilience in Venice, offering insights that can inform sustainable tourism strategies and urban planning.

4 Results

4.1 Double Fractal Design and Community Resilience in Venice

Venice's **double fractal urban fabric** is characterized by a hierarchical network of spaces ranging from narrow alleyways to larger public squares, forming a spatial structure that accommodates both intimate community interactions and larger, collective activities. This unique layout, with its interlocking scales, supports a distinctive rhythm of life that is both flexible and resilient. The design enables a balance between private, enclosed areas and open, shared spaces, fostering a SOC and continuity for residents.

The spatial analysis revealed that Venice's fractal structure not only contributes to a coherent urban identity but also enhances community resilience by facilitating TA, specifically:

Small-Scale Interactions: Narrow alleyways and smaller piazzas are conducive to informal gatherings and everyday social interactions. These spaces are often appropriated by locals for brief, spontaneous uses—such as socializing, resting, and displaying personal artworks—reinforcing a sense of belonging and ownership.

Larger Public Squares: Major squares, such as Piazza San Marco, have historically served as focal points for community events, celebrations, and public gatherings. Although these areas are frequently dominated by tourists, they still offer potential for TA during quieter periods or off-peak hours, allowing residents to reclaim these spaces temporarily.

The fractal design, therefore, acts as a structural enabler of community-driven practices by providing a diverse range of adaptable spaces. This flexibility allows Venetians to navigate and repurpose their environment in response to the shifting pressures of tourism. The human-scale architecture of Venice, which aligns closely with residents' needs and social patterns, supports TA by making it easy for individuals and groups to informally adapt spaces for short-term use without extensive planning or modification. As such, Venice's urban form emerges as a resilient assemblage, where spatial configurations actively contribute to the city's social resilience.

4.2 Tourism Services Index (TSI)

A recent study by Ignaccolo, Zheng and Williams [24] introduces the Tourism Services Index (TSI), a highly granular metric designed to quantify the presence of tourism services in Venice's historic city at the city-block scale. The TSI combines three indicators: the capacity of Airbnb listings, the internationality of restaurants based on review diversity, and the density of tourist-oriented shops such as souvenir and designer stores. Using a hexagonal grid of 1642 cells covering Venice, each 90 m in diameter, the index maps the spatial distribution of these services (see Fig. 3). This

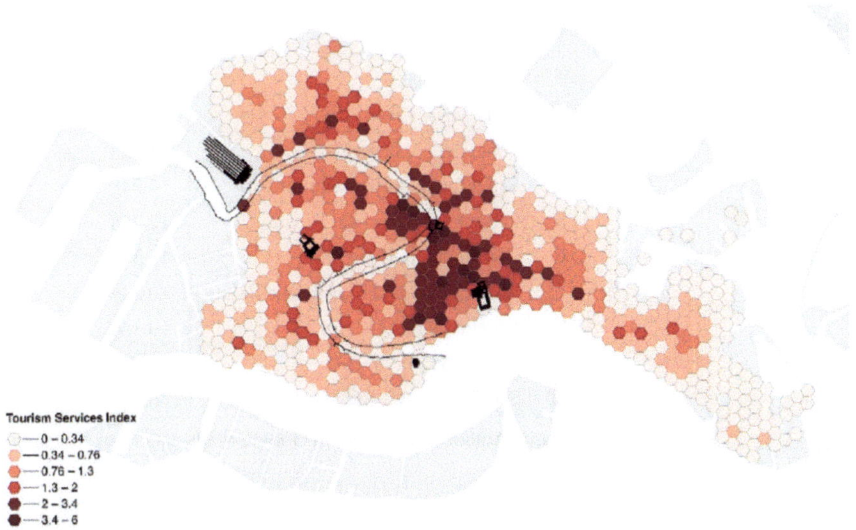

Fig. 3 Tourism services index (TSI) scores (in a 0–5 continuous scale) within the historic city of Venice. *Source* [24]

approach provides a detailed snapshot of tourism saturation in specific areas. They found a strong association between high TSI values and proximity to landmarks and bridges, which act as focal points for tourism services due to their structural and symbolic importance [24].

The TSI also reveals that areas with higher values are associated with an increase in unoccupied housing and a decrease in permanent residents, signalling a correlation between tourism intensity and demographic shifts. From a built-environment perspective, street intersections and widths show minimal association with TSI values compared to landmarks and bridges, underlining the unique urban dynamics of Venice. Thus, suggesting that urban form, particularly the distribution of physical infrastructure like bridges and tourist landmarks, significantly shapes the spatial patterns of TA and tourism services (Table 1).

4.3 Impact of Overtourism on Temporary Appropriation (TA) Practices

Displacement of TA in High-Impact Tourism Zones

Overtourism disrupts the spatial dynamics of Venice by transforming public spaces into sites primarily intended for visitors, which has a displacement effect on TA practices. The TSI highlighted specific areas where tourism infrastructure is heavily concentrated, including high-profile destinations like Piazza San Marco and the

Table 1 Ten most reviewed landmarks of Venice

#	Landmark	Number of reviews
1	Rialto Bridge	173,643
2	Saint Mark's Basilica	22,910
3	Ponte dell'Accademia	21,612
4	Santa Lucia railway station	14,348
5	Palazzo Contarini del Bovolo	12,758
6	Giardini della Biennale	11,584
7	Basilica di Santa Maria della Salute	10,546
8	Venetian Arsenal	10,364
9	Campo San Polo	6,747
10	San Giorgio Maggiore	6,617

Source Author elaboration on Google Maps data from November 2024

Rialto Bridge. In these zones, the spatial and social dynamics are predominantly shaped by tourist flows, reducing opportunities for residents to engage in TA.

Field observations confirmed that in high-tourist areas, local appropriations of space are often curtailed or entirely displaced by commercial activities catering to visitors. For instance, some of the campo (squares) with high-tourist density, are often dominated by organized tours, and seating arrangements set up for cafes and restaurants. These arrangements physically occupy the space, leaving little room for residents to appropriate these areas for social gatherings or other informal uses. Similarly, canals and waterfronts that were traditionally used by locals for leisure or social activities are increasingly commercialized, with gondola services and sightseeing boats monopolizing these spaces. This commercialization reduces the flexibility of these spaces, making it difficult for residents to practice TA in these iconic areas.

The displacement effect not only limits the physical space available for TA but also alters the perception and accessibility of these spaces. Residents reported feeling alienated from certain areas of Venice, particularly during peak tourist seasons, as these spaces no longer support local social practices. The transformation of public spaces into tourist-centric zones erodes the sense of ownership and belonging that is central to Venice's urban identity.

4.4 *Relocation of TA*

While overtourism has limited TA opportunities in central areas, it has led to the emergence of TA in secondary or less frequented spaces. Residents have adapted by relocating their social and cultural practices to peripheral zones or underutilized areas of the city. These spaces, often less accessible or attractive to tourists, provide a quieter setting where residents can engage in TA with fewer interruptions.

For instance, in less touristy neighbourhoods, smaller squares and alleyways have become focal points for community activities. Observations documented locals using these spaces for social gatherings, impromptu performances, and children's play areas, practices that were once common in the city's main squares. These peripheral locations have thus taken on a greater role in preserving community interactions. Another example are the church courtyards and private courtyards: Some residents have also resorted to using semi-private spaces, such as church courtyards or the inner courtyards of residential buildings, as alternatives to public squares. These areas allow residents to carry out community rituals, celebrations, and other forms of TA without interference from tourism activities. The shift to these alternative spaces highlights the adaptability of TA as a resilience practice in response to overtourism.

The transformation of TA practices to secondary spaces illustrates the resilience of Venetian communities in reclaiming their social practices despite the spatial pressures of overtourism. However, this relocation also underscores a form of spatial segregation, where residents are gradually pushed out of central, historically significant areas. Here spatial segregation refers to "the extent to which individuals of different groups occupy or experience different social environments" [44]. This shift has implications for the socio-spatial fabric of Venice, as the separation of locals and tourists disrupts the organic mix of social interactions that traditionally characterized the city.

4.5 The Role of Assemblage in Venice's Adaptive Identity

Venice's response to overtourism can be understood through the lens of assemblage theory, which views the city as an evolving, interconnected system where each element contributes to the overall identity. In Venice, TA acts as a form of assemblage adaptation, allowing residents to reclaim spaces and maintain a SOC despite external pressures. The interactions among residents, tourists, public spaces, and regulatory frameworks continuously reshape the city's urban identity, illustrating how identity and resilience are co-constitutive processes within Venice's urban assemblage.

Assemblage theory also explains the fluidity of Venice's urban identity under overtourism. Rather than a static entity, Venice's identity is shaped by ongoing intra-actions between its architectural form, social practices, and economic dynamics. These intra-actions result in a city that is simultaneously fragile—susceptible to the overwhelming impact of tourism—and resilient, capable of adaptive responses that preserve its character. The relocation of TA to secondary spaces, for instance, reflects an assemblage adaptation that allows the city to retain its community spirit even as its central areas are redefined by tourism.

5 Findings

The findings reveal a nuanced picture of Venice as a resilient, adaptive urban assemblage under the pressures of overtourism. First, this study identified the double fractal design as a resilience enabler. Venice's unique urban morphology supports community resilience by providing adaptable spaces for TA, fostering social interactions, and reinforcing a sense of belonging. Second is displacement and transformation of TA. In other words, overtourism has led to a displacement of TA practices from central areas to secondary spaces, illustrating both the adaptability of Venetian communities and the spatial segregation imposed by tourism. Third, Venice has a dynamic urban identity through assemblage. Venice's identity is an emergent property of interactions within the city's assemblage. The practice of TA exemplifies how identity is continually negotiated and adapted in response to changing social and economic conditions.

These results underscore the importance of Venice's double fractal design in supporting TA and community resilience, while highlighting the complex ways in which overtourism reshapes the city's social and spatial dynamics. The implications of these findings are explored further in the Conclusion section, where strategies for sustainable tourism and preservation of Venice's urban identity are proposed. This data-driven approach provides a robust framework for addressing overtourism while preserving the socio-economic fabric and cultural heritage of Venice.

6 Conclusion

Venice exemplifies the intricate interplay between urban morphology, social dynamics, and the profound impacts of overtourism. This chapter has illustrated how the city's distinctive double fractal design, characterized by its interwoven canals, streets, and public squares, simultaneously fosters community resilience and attracts global tourism. While this unique urban fabric enhances social interaction and a sense of belonging, the pressures of overtourism have transformed public spaces into commodified zones, challenging Venice's social and cultural fabric.

Temporary Appropriation (TA) emerged as a critical mechanism for residents to navigate these pressures. TA reflects the adaptability of Venetians in reclaiming urban spaces, from major piazzas to peripheral zones, underscoring their resilience and commitment to preserving their social identity. However, the spatial segregation between locals and tourists, driven by overtourism, risks eroding the organic interactions that define Venice's communal essence. This shift not only disrupts the socio-spatial equilibrium but also raises urgent questions about the sustainability of urban identity under the weight of global tourism.

Through the lens of assemblage theory and agential realism, this chapter highlights Venice as a dynamic, evolving entity. The city's urban identity is not static but emerges from the continuous interactions between its physical structures, social practices, and economic forces. Recognizing the co-constitutive relationship between resilience and identity is essential for designing sustainable tourism strategies that prioritize the preservation of both the built environment and the socio-cultural life of Venice. Future policies must embrace these complexities, ensuring that Venice remains not just a spectacle for the world but a living, thriving community for its residents.

References

1. Melis A, Pievani T, Lara-Hernandez JA (2024) Architectural exaptation: when function follows form. Routledge, London, UK. https://doi.org/10.4324/9781003347118.
2. Lara-Hernandez JA (2024) Living in Venice: a local's perspective in a tourist's paradise'. In: D'Alessandro J, Medas B, Melis A, Menichini D (eds) We are out of time. Pacini Editore, Pisa, Italy, pp 156–160
3. Hayes M, Zaban H (2020) Transnational gentrification: the crossroads of transnational mobility and urban research. Urban Stud 57(15):3009–3024. https://doi.org/10.1177/0042098020945247
4. DeLanda M (2006) A new phylosophy of society. Continuum, London, UK
5. Stevens Q, Daly J, Dovey K (2024) Designing for possibility in public space: affordance, assemblage, and ANT. Urban Des Int 29(2):94–104. https://doi.org/10.1057/s41289-023-00235-y
6. DeLanda M (2016) Assemblage theory. In: Speculative realism. Edinburgh University Press, Edinburgh, UK
7. Dovey K (2012) Informal urbanism and complex adaptive assemblage. Int Dev Plan Rev 34(4):349–368. https://doi.org/10.3828/idpr.2012.23
8. Lara-Hernandez JA (2022) Resilience, architectural exaptation, and temporary appropriation. In: Carta M, Perbellini MR, Lara-Hernandez JA (eds) Resilient communities and the Peccioli charter: towards the possibility of an Italian charter for resilient communities. Springer International Publishing, Cham, Switzerland, pp 257–264. https://doi.org/10.1007/978-3-030-85847-6_27.
9. Melis A, Pievani T, Lara-Hernandez JA (2024) Architectural exaptation: when function follows form. Architecture Now. Wellington, NZ. https://doi.org/10.4324/9781003347118
10. Barad K (2007) Meeting the universe halfway: quantum physics and the entanglement of matter and meaning. Kindle ver Duke University Press. https://doi.org/10.2307/j.ctv12101zq
11. Eliot TS (1949) Notes towards the definition of culture
12. Boaz F (2017) Anthropology and modern life. W.W. Norton & Company Inc., New york
13. Brown F, Crompton A (2007) The double fractal structure of Venice. In: Proceedings, 6th international space syntax symposium, İstanbul, 2007, no. January 2007. pp 11.1–11.12
14. Bolio Arceo E (2012) Urban transformations and place-identity: the case of Merida. Oxford Brookes University, UK
15. Rios M, Vazquez L, Miranda L (2012) Place as space, action, and identity. Taylor & Francis Group, Routledge, London, UK
16. Hosany S An entry fee may not be enough to save Venice from 20 million tourists. The Conversation
17. Ujang N (2012) Place attachment and continuity of urban place identity. Procedia Soc Behav Sci 49:156–167. https://doi.org/10.1016/j.sbspro.2012.07.014
18. Lalli M (1992) Urban-related identity: theory, measurement, and empirical findings. J Environ Psychol 12(4):285–303. https://doi.org/10.1016/S0272-4944(05)80078-7
19. Nylund K (2000) 'Place and cultural identity in the segregated city', *Yhteiskuntasuunnittelu.* Finn J Urban Stud 38(August):8–27

20. Lara-Hernandez JA, Melis A (2018) Understanding the temporary appropriation in relationship to social sustainability. Sustain Cities Soc 39C:1–14. https://doi.org/10.1016/j.scs.2018.03.004
21. Lara-Hernandez JA (2019) Temporary appropriation: theory and practice of the street. University of Portsmouth, United Kingdom
22. Lara-Hernandez JA, Melis A, Coulter CM (2020) Temporary appropriation and informality: a nuanced relationship. In: Lara-hernandez JA, Melis A, Thompson J (eds) Temporary appropriation in cities: human spatialisation in public spaces and community resilience. Springer Nature Limited, London, UK. Ch. 4
23. Mews G, Lara-Hernandez JA, Sbarra M (2024) Radical imaginaries and new places for children in cities. In: 60th ISOCARP world planning congress. ISOCARP, New Clark City, Phillipines
24. Ignaccolo C, Zheng Y, Williams S (2023) Tourism morphometrics in Venice: constructing a tourism services index (TSI) to unmask the spatial interplay between tourism and urban form. Cities 140(June):104369. https://doi.org/10.1016/j.cities.2023.104369
25. Gusfield JR (1975) Community: a critical response
26. McMillan DW, Chavis DM (1986) Sense of community: a definition and theory. J Community Psychol 14(1):6–23. https://doi.org/10.1002/1520-6629(198601)14:1%3c6::AID-JCOP2290140103%3e3.0.CO;2-I
27. Chavis DM, Wandersman A (1990) Sense of community in the urban environment: a catalyst for participation and community development. Am J Community Psychol 18(1):55–81. https://doi.org/10.1007/BF00922689
28. Talen E (2010) The spatial logic of parks. J Urban Des (Abingdon) 15(4):473–491. https://doi.org/10.1080/13574809.2010.502335
29. Talen E (2000) Measuring the public realm: a preliminary assesment of the link between public space and sense of community. J Archit Plann Res 17(4):344–360
30. Long DA, Perkins DD (2007) Community social and place predictors of sense of community: A multilevel and longitudinal analysis. J Commun Psychol 35(5):563–581. https://doi.org/10.1002/jcop.20165
31. Kim J, Kaplan R (2004) Physical and psychological factors in sense of community: new urbanist Kentlands and nearby orchard village. Environ Behav 36(3):313–340. https://doi.org/10.1177/0013916503260236
32. Swyngedouw E (2018) More-than-human constellations as immuno-biopolitical fantasy in the urbicene. New Geographies 9:18–23
33. Groat L, Wang D (2002) Architectural research methods. John Wiley & Sons Inc., New York, USA
34. Hayles NK (2016) Cognitive assemblages: technical agency and human interactions. Crit Inq 43(1):32–55
35. Nail T (2017) What is an assemblage ? SubStance 46(1):21–37
36. Hillier B (2005) Between social physics and phenomenology: explorations towards an urban synthesis?. In: Proceedings 5th international space syntax symposium. pp 1–21. [Online]. http://eprints.ucl.ac.uk/1679/1/hillier05-socialphysics.pdf
37. Qi J, Mazumdar S, Vasconcelos AC (2024) Understanding the relationship between urban public space and social cohesion: a systematic review, vol 7, no 2. Springer International Publishing. https://doi.org/10.1007/s42413-024-00204-5
38. Boeing G (2021) Street in network models and indicators for every urban area the world. Geogr Anal 54(3):519–535. https://doi.org/10.1111/gean.12281
39. Lara-Hernandez JA, Walker C, Burgess A (2024) Venice's assemblage: infrastructural resilience and sense of community in times of crisis. In: International conference of smart and sustainable built environment (SASBE 2024). Auckland, New Zealand
40. Lara-Hernandez JA, Walker C, Burgess A (2024) Venice's assemblage: infrastructural resilience and sense of community in times of crisis. In: International conference of smart and sustainable built environment (SASBE 2024). Emerald Group Publishing Limited, Auckland, New Zealand
41. Peterson NA, Speer PW, McMillan DW (2008) Validation of a brief sense of community scale: confirmation of the principal theory of sense of community. J Community Psychol 38(1):61−73. 10./jcop

42. McMillan DW (1996) Sense of community. J Community Psychol 24(4):315–325
43. Peterson NA, Hughey PWS, Hughey J (2006) Measuring sense of community: a methodological interpretation of thr factor structure debate. J Community Psychol 34(4):453–469. https://doi.org/10.1002/jcop.20109
44. Reardon SF, Sullivan DO (2004) Measure of spatial segregation. Sociol Methodol 34(650):121–162. https://doi.org/10.1111/j.0081-1750.2004.00150.x

Smart Cities Versus Overtourism: Balancing Urban Innovation with Sustainable Tourism

Bassam Samir Al-Romeedy and Pankaj Kumar Tyagi

Abstract This chapter explores the intersection of smart city concepts and overtourism, highlighting how urban innovations can support sustainable tourism in cities facing challenges from rising urban populations and global tourism. It discusses how smart cities, through technologies like IoT, big data, and AI, can address overtourism by implementing smart infrastructure, real-time monitoring, and AI-driven decision-making to manage tourist flows, improve visitor experiences, and reduce environmental and social impacts. The roles of stakeholders such as governments, businesses, and local communities in these efforts are also examined. Additionally, the chapter looks at future trends, including the adoption of new technologies and climate resilience, while addressing ethical issues like privacy and the digital divide. It concludes with policy recommendations for urban planners, emphasizing the need to align smart city development with sustainable tourism goals for balanced and resilient urban environments.

Keywords Smart cities · Overtourism · Sustainable tourism · Urban innovation · Artificial intelligence · Sustainable urban development

1 Introduction

Smart cities use advanced technologies like IoT, AI, big data analytics, and cloud computing to improve urban living, optimize services, and promote sustainability by enabling real-time monitoring and data-driven decision-making in areas such as transportation and energy efficiency. However, challenges like privacy concerns,

---

B. S. Al-Romeedy (✉)
Tourism Studies Department, Faculty of Tourism and Hotels, University of Sadat City, Sadat City, Egypt
e-mail: bassam.samir@fth.usc.edu.eg

P. K. Tyagi
University Institute of Tourism and Hospitality Management, Chandigarh University, Mohali 140413, Punjab, India

V. C. Sharma et al. (eds.), *Architectural Vision of Overtourism*,
SpringerBriefs in Architectural Design and Technology,
https://doi.org/10.1007/978-981-96-5794-0_5

data security, and the digital divide must be addressed. Overtourism, driven by factors like affordable travel, social media, and higher disposable income can lead to overcrowding and environmental damage. Smart city technologies can help manage overtourism by regulating tourist flows and implementing strategies like dynamic pricing or visitor caps. Additionally, smart transportation systems alleviate congestion, and digital platforms guide tourists to less crowded destinations, encouraging responsible tourism. Urban innovations such as intelligent infrastructure, adaptive transportation, and e-governance enhance city efficiency and livability. To ensure sustainable tourism, cities must adopt eco-friendly practices and regulatory measures such as visitor caps, zoning laws, and tourism taxes to manage overtourism and support conservation efforts while aligning with community needs [1–4].

Smart city technologies offer various solutions to manage tourist flows and address overtourism. Digital platforms provide real-time updates on crowd levels, allowing tourists to plan visits during off-peak times, while geofencing controls access to sensitive areas. Mobile apps help guide tourists to less crowded sites, easing congestion. AI and machine learning predict tourist behavior and demand, enabling proactive strategies like dynamic pricing or temporary closures to avoid overcrowding. When combined with sustainable tourism practices, these technologies contribute to a more balanced tourism ecosystem [5, 6].

Governments play a key role in advancing smart city initiatives and tourism management by setting policies, providing funding, and fostering public–private partnerships. The private sector, including tourism and technology companies, supports sustainable urban development with eco-friendly practices and innovative solutions. Community engagement ensures that tourism respects local values, preserves cultural heritage, and promotes inclusive growth. Emerging technologies such as 5G, blockchain, and augmented reality (AR) are expected to enhance smart city functions and tourism management by improving connectivity, data security, and tourist experiences [7–9]. Additionally, smart cities must address climate change risks, integrating strategies like climate-resilient infrastructure and low-carbon tourism to safeguard communities and preserve their status as sustainable tourist destinations [10, 11].

The adoption of smart city technologies raises ethical concerns, particularly regarding privacy, data security, and the digital divide. To address these challenges, cities must develop strong data governance frameworks, implement clear regulations, and ensure transparent communication to protect privacy and manage data responsibly. It is also essential to make smart technologies inclusive, ensuring that vulnerable populations can benefit from these innovations. Integrating smart technologies with sustainable tourism requires a holistic approach, including policies that support sustainability goals, collaboration among stakeholders, fostering innovation, and establishing regulations to mitigate the negative impacts of tourism. Ongoing monitoring, evaluation, and the sharing of best practices are crucial for refining strategies and improving results [12, 13].

This study closely aligns with the Sustainable Development Goals (SDGs) by examining the relationship between technological innovation, urban sustainability, and tourism management. Smart cities utilize advanced technologies like IoT, AI, and data analytics to enhance urban infrastructure, improve resource efficiency, and elevate the quality of life for residents, contributing directly to SDG 11, which focuses on creating inclusive, safe, resilient, and sustainable cities. However, uncontrolled growth of urban tourism can lead to overtourism, intensifying issues such as overcrowding, pressure on local resources, environmental damage, and cultural loss. This ties into the critical need to address SDG 12, which advocates for responsible consumption and production, and SDG 13, which calls for urgent action on climate change. Smart city strategies can be designed to balance the economic benefits of tourism (SDG 8: Decent Work and Economic Growth) with the social and environmental well-being of urban areas. For example, smart mobility and dynamic crowd management systems can alleviate congestion in tourist hotspots, preserving cultural heritage and ensuring fair access for both residents and tourists (SDG 16: Peace, Justice, and Strong Institutions). By incorporating sustainable tourism practices into smart city models, this research highlights the potential to reconcile urban innovation with sustainability, supporting the broader goal of achieving a balanced, inclusive future as outlined by the SDGs [14].

In this vein, the combination of smart city innovations with sustainable tourism practices presents a promising strategy for addressing the challenges of overtourism while improving urban quality of life. By leveraging technology, data-driven decision-making, and collaborative efforts, cities can create environments that are sustainable, resilient, and inclusive. The active participation of governments, the private sector, and local communities is crucial for achieving a balance between urban development and tourism, ensuring that future generations can enjoy both vibrant urban spaces and thriving tourism destinations.

2 Definition and Concepts

2.1 Smart Cities

Smart cities utilize advanced technologies and data-driven strategies to enhance urban life, optimize services, and promote sustainability. Central to the smart city concept are information and communication technologies (ICT), the Internet of Things (IoT), and big data analytics, which enable continuous monitoring and management of urban infrastructure, fostering a responsive and adaptable governance system. These cities aim to address challenges like rapid population growth, resource scarcity, and environmental concerns by promoting livability and resilience. Key components include intelligent transportation systems, energy-efficient infrastructure, smart health care, and enhanced public safety, which together create a cohesive urban environment [15–20]. The integration of technology is crucial for developing smart cities, ensuring that digital innovations are smoothly incorporated into urban operations. This involves deploying ICT infrastructure such as high-speed Internet, wireless networks, and cloud-based platforms, which are essential for coordinating various urban systems. For instance, smart grids optimize energy management, intelligent transportation systems improve traffic flow, and e-governance platforms streamline public services. This interconnected framework enables cities to operate more efficiently, reducing costs while improving service quality. Additionally, the continuous advancement of technology ensures that smart cities remain adaptable, evolving to meet new challenges and capitalize on emerging opportunities [21, 22]. The Internet of Things (IoT) is a key component in smart city infrastructure, enabling the continuous collection and exchange of data through interconnected devices and sensors embedded across urban areas. These IoT devices are integrated into various city systems, such as streetlights, traffic signals, waste management, and public transportation, collecting data on environmental conditions, energy consumption, and traffic patterns. This data is transmitted to central control systems for analysis, which helps inform decision-making and automate urban operations. For example, IoT-based traffic management systems can adjust signal timings in real time to reduce congestion and improve travel efficiency. Additionally, IoT plays a vital role in enhancing public safety, monitoring environmental factors, and improving energy efficiency, making cities smarter and better equipped to meet the needs of residents [23–26].

Cities across the globe are adopting innovative Intelligent Transportation Systems (ITS) to address urban mobility issues and promote sustainability. In Singapore, the Smart Mobility 2030 initiative utilizes real-time traffic monitoring and predictive analytics to improve road efficiency and reduce congestion. The city's Electronic Road Pricing (ERP) system employs GPS technology to adjust vehicle charges based on traffic conditions, encouraging drivers to avoid peak times and opt for public transport. In Barcelona, Spain, the IoT-enabled smart parking system directs drivers to available spaces, minimizing time and fuel spent searching for parking and lowering emissions. Stockholm, Sweden, integrates ITS with its congestion pricing system, using traffic data to adjust toll rates based on demand, resulting in a 20% reduction in congestion and a 14% decrease in air pollution. In the United States, Los Angeles utilizes AI-powered traffic signals through the Automated Traffic Surveillance and Control (ATSAC) program to manage traffic flow in real time, reducing delays and fuel consumption. Copenhagen, Denmark, has introduced a smart bike system that prioritizes cyclists at traffic lights, promoting sustainable commuting and decreasing car dependency. Curitiba, Brazil, a leader in Bus Rapid Transit (BRT), employs ITS to coordinate bus schedules, ensuring efficient public transit services that offer a compelling alternative to private vehicles. These examples demonstrate how cities are adopting ITS to improve urban mobility, minimize environmental impacts, and foster sustainable urban development.

Data-driven decision-making is essential for smart cities, relying on technologies like AI, machine learning, and big data analytics to guide urban planning and management [27, 28]. By using insights from IoT devices, social media, and digital platforms, smart cities can optimize resource allocation and policy development, such as predicting traffic congestion or improving energy efficiency. Sustainable urban planning includes the use of renewable energy, energy-efficient buildings, and eco-friendly transportation like electric buses and bike-sharing programs, alongside efforts in green space creation, waste management, and water conservation to balance environmental, economic, and social goals [29, 30]. Additionally, citizen engagement is promoted through digital platforms and mobile apps, improving communication with officials, promoting transparency, and encouraging participatory budgeting. This approach enhances civic engagement and a sense of community ownership, ultimately improving residents' quality of life and fostering long-term sustainability [31, 32].

2.2 *Smart Cities in Egypt*

Egypt has significantly advanced its urban development strategy over the past forty years, embarking on ambitious projects to establish a new generation of cities designed with contemporary technological advancements in mind. These fourth-generation cities represent an evolved model incorporating sophisticated network systems, activities supporting the knowledge economy, and advanced research centers. Their planning integrates principles of green and smart cities, along with competitive economic functions such as global tourism (e.g., New Alamein City and Ras El Hekma) and international trade and business service hubs (e.g., East Port Said City). The New Administrative Capital, a pivotal economic zone for Egypt [33], serves as a new center for state activities and a magnet for global institutions. Egypt officially entered the smart city era in 2017, allocating a special budget for the establishment and implementation of 13 cities adhering to global technological standards. The government's push for smart cities stems from their potential to drive significant growth in communications and real estate. Efforts are ongoing to imbue all new cities with a smart city system, aiming for a digital society that streamlines services for both institutions and individuals. The importance of smart cities extends beyond development and innovation; they are crucial for managing burgeoning urban populations and their immense resource and management demands [33, 34]. Investing in robust basic and technological infrastructure is seen as a key driver of economic and social benefits, impacting the population directly and indirectly, including productive activities. This has prompted the Egyptian government to incentivize infrastructure investors, promoting projects with design and operational standards exceeding national averages. New cities are strategically divided into areas with varying investment incentives, prioritizing low- and medium-income housing zones and areas requiring rapid development [33–35]. The benefits derived from these new cities are not solely economic and social, but also encompass cultural, urban, and environmental returns. The Egyptian government has developed a comprehensive set of indicators to assess urban policy effectiveness across both new and existing urban areas. The New Urban Communities Authority, under the Ministry of Housing, has finalized plans for 44 new cities as part of Egypt's broader National Strategic Plan for Construction and Urban Development and the State's Sustainable Development Plan 2030 [33–35].

2.3 Overtourism

Overtourism occurs when tourist numbers exceed a destination's capacity, overwhelming local infrastructure, the environment, and communities. This leads to the degradation of the destination's appeal, diminishing the visitor experience and residents' quality of life. Initially affecting iconic cities and landmarks, overtourism has become a global issue. It results in resource depletion, daily life disruptions, and the erosion of the cultural and natural features that attract tourists [1, 2]. Contributing factors include the affordability of travel, social media-driven tourism trends, and the uneven distribution of visitors to popular attractions. Poor urban planning and inadequate tourism management further exacerbate overcrowding and strain local resources [2–4].

Overtourism has significant negative effects on both urban areas and local communities. In cities, overcrowding strains public infrastructure like transportation, waste management, and health care, while historic sites and public spaces deteriorate. For residents, increased noise, pollution, and congestion lower quality of life, and rising demand for accommodations drives up prices, making housing and goods unaffordable. Overtourism also risks the commercialization of local culture, distorting traditions to cater to tourist expectations. In natural areas, it causes ecological damage, including ecosystem degradation, litter, and increased carbon emissions, threatening long-term sustainability [4, 9, 36, 37].

To address these impacts, smart city technologies can help manage tourist flows and promote sustainability. By using sensors, GPS tracking, and mobile apps, cities can gather real-time data on tourist behavior, enabling better management of tourism patterns. Strategies like dynamic pricing and directing tourists to less crowded attractions through mobile apps can ease pressure on popular sites and support more balanced tourism distribution [15, 38–42].

Smart city technologies are essential in promoting sustainable tourism by improving infrastructure and resource management. For instance, smart transportation systems reduce congestion in tourist areas by optimizing traffic flow and adjusting schedules during peak seasons, minimizing environmental impact. Smart waste management systems handle the increased waste from tourism, maintaining cleanliness in public spaces [6, 7, 10]. Additionally, AI-powered apps personalize tourist experiences, guiding visitors to less crowded sites and offering virtual tours through AR/VR, alleviating overcrowding at popular attractions [7, 8]. These technologies also encourage community engagement, allowing residents to participate in tourism management and promoting responsible tourist behaviors. By integrating renewable energy and energy-efficient practices into tourism infrastructure, smart cities can ensure that tourism benefits are equitably distributed between locals and visitors [18, 19].

2.4 Urban Innovation in Smart Cities

Technological Solutions in Smart Cities

Urban innovation in smart cities focuses on using advanced technologies to improve efficiency, sustainability, and comfort. Key innovations include smart infrastructure, such as sensors and IoT devices to manage utilities like water, energy, and waste. Smart grids optimize energy distribution, and smart water systems monitor consumption and detect leaks. Smart transportation enhances traffic flow, reduces congestion, and improves public transit through intelligent systems. Autonomous vehicles and shared mobility platforms also reduce reliance on private cars and lower emissions. Smart parking helps with congestion, and smart governance uses digital platforms and blockchain technology to improve government services, reduce corruption, and foster transparency [20, 43–45].

Data analytics is essential in smart cities, supporting informed decision-making and enhancing urban living. By integrating data from IoT devices, sensors, and digital platforms, cities use big data, AI, and real-time analytics to optimize functions like energy management and traffic control. Big data analytics processes large amounts of information from sources such as social media and environmental monitors to uncover patterns that guide urban planning. For example, predictive traffic analytics help adjust signals to reduce congestion, while energy data promotes efficient distribution and renewable energy use. AI improves operational efficiency by automating tasks and offering predictive insights, such as forecasting traffic jams and optimizing emergency service routes. It also monitors environmental factors like air quality and noise levels, providing real-time alerts to mitigate health risks [11, 26, 27]. Real-time analytics is crucial for improving both resident and visitor experiences in smart cities. By processing data as it is collected, cities can offer adaptive services, such as directing tourists to less crowded sites and adjusting public transport schedules based on demand. Real-time data also enhances public safety and emergency response by monitoring incidents and organizing evacuation routes during crises like natural disasters [13, 15–17].

3 Balancing Urban Innovation with Sustainable Tourism

Achieving a balance between urban innovation and sustainable tourism requires practices that protect the environmental, cultural, and community integrity of tourist destinations. Responsible travel, which encourages eco-friendly choices and respect for local cultures, plays a key role. This includes supporting local businesses, staying at sustainable accommodations, and visiting off-peak destinations to ease pressure on popular sites. The concept of slow tourism, where travelers spend more time in fewer locations, further reduces environmental impact and promotes deeper cultural engagement. Additionally, safeguarding cultural heritage sites through controlled access and educational programs helps preserve these assets. Revenues from tourism can fund site restoration, and involving local communities in planning ensures that tourism benefits are shared equitably while preserving cultural heritage and fostering community pride [22, 23, 25, 39]. To address the negative impacts of overtourism, governments and local authorities must implement regulations that balance tourism growth with sustainability. One effective strategy is setting visitor caps, which limit the number of tourists allowed at specific sites during peak times to prevent overcrowding, reduce pressure on infrastructure, and protect fragile environments. Many popular destinations, such as national parks and UNESCO World Heritage sites, already enforce daily or seasonal visitor limits to preserve natural and cultural resources. Zoning regulations are another approach, helping manage tourist flow within urban areas by designating specific zones for tourism activities, thus preserving local communities and their character. Tourism taxes, such as those on hotel stays or entry fees, can generate funds for conservation, infrastructure improvements, and community development, ensuring that tourism benefits are reinvested in the destination [10, 15, 28, 46].

Smart city technologies provide innovative solutions to tackle the challenges of overtourism by improving the management of tourist flows, enhancing visitor experiences, and mitigating the adverse effects on urban environments. Real-time data analytics play a crucial role in tracking tourist movements and behaviors, enabling cities to identify peak times and overcrowded areas for implementing dynamic crowd control measures. Mobile apps can provide real-time crowd density information, guiding tourists to alternative attractions or times, thus promoting a more even distribution of visitors and reducing congestion. Smart transportation systems optimize public transit routes and schedules based on real-time demand, ensuring efficient movement while minimizing congestion and environmental impact. Additionally, digital engagement platforms promote sustainable tourism by fostering community involvement and educating tourists about the importance of preserving cultural and natural resources. AR and VR technologies can provide immersive educational experiences, reducing the need for physical visits to sensitive locations, while platforms facilitate feedback between tourists and local authorities, supporting continuous improvements in tourism services [29, 30, 40, 47].

3.1 Role of Stakeholders in Smart Cities and Sustainable Tourism

Local and national governments play a vital role in developing smart city initiatives and promoting sustainable tourism. At the national level, governments set policies and frameworks that guide smart city development, addressing issues such as data privacy, cybersecurity, and system interoperability, while providing financial support and incentives to encourage the adoption of smart technologies and sustainable practices. Locally, governments are responsible for implementing smart city projects, including smart grids, public transportation, waste management, and water supply networks. In tourism management, local authorities regulate tourist flows, enforce zoning laws, and establish visitor limits to prevent overtourism and protect cultural and natural resources. Additionally, they facilitate public–private partnerships to support collaboration on smart city and sustainable tourism initiatives, creating balanced environments that benefit both residents and tourists [31, 32, 38, 41].

The private sector, particularly in tourism and technology, plays a key role in sustainable urban development and smart city solutions. Tourism businesses, such as hotels and tour operators, adopt sustainable practices like green certifications, energy efficiency, and promoting responsible travel. Technology companies provide essential tools such as sensors, IoT devices, and AI systems to optimize urban functions like traffic, energy, and public safety. By collaborating with local governments, these companies help address urban challenges, build digital infrastructure, and fund smart city projects, ensuring economic viability and environmental sustainability through public–private partnerships [1, 6, 7, 39].

Community engagement is crucial for the success of smart city initiatives and sustainable tourism. Involving local residents in decision-making ensures that urban projects benefit the entire community and meet their needs. This fosters support, inclusivity, and a sense of ownership. In tourism, community participation helps preserve cultural and natural resources, balance economic growth with environmental protection, and develop strategies to address tourism impacts. Community-based tourism initiatives allow locals to share their traditions with tourists, benefiting both the community and tourism development. Engaging the community also helps mitigate the negative effects of overtourism, such as displacement, cultural erosion, and environmental harm [7, 18, 20, 21].

4 Future Trends and Challenges

As smart cities evolve, emerging technologies such as 5G networks are poised to transform urban living and sustainable tourism. 5G promises faster speeds and enhanced connectivity, supporting a greater number of connected devices and enabling real-time data exchange across IoT systems. This advancement will improve key smart city functions like autonomous vehicles, intelligent energy grids, and advanced traffic management, making cities more efficient and adaptable. In tourism, 5G will enhance visitor experiences by enabling seamless AR and VR applications, providing interactive guides, real-time language translation, and immersive cultural experiences, all without overburdening physical infrastructure [9, 37, 43, 44].

Blockchain technology is emerging as a key innovation for enhancing data security and transparency in smart cities by providing a secure, tamper-resistant platform for transactions such as real estate transfers, energy usage, and tourist bookings. This can increase trust in urban operations and improve efficiency, while in tourism, it helps reduce fraud in booking and payment processes. AI also plays a critical role by offering predictive analytics to tackle urban challenges and personalize tourist experiences. Smart cities promote sustainability through urban planning that includes green infrastructure, such as parks and green roofs, to improve air quality and biodiversity. Advanced water management systems help cities optimize water use, reduce flood risks, and adapt to droughts. In the energy sector, smart grids and energy storage solutions support the transition to renewable energy by enhancing solar, wind, and other green sources. Smart transportation systems reduce carbon footprints by encouraging public transit, cycling, and walking, while sustainable tourism practices protect ecosystems and cultural sites, ensuring their long-term appeal and sustainability [5, 8, 10, 24–26].

As smart city technologies become more embedded in urban life, addressing ethical issues related to privacy, data security, and digital equity is essential. Privacy concerns arise from the extensive use of sensors, cameras, and data analytics, leading to potential misuse of personal information. To mitigate these concerns, smart cities must adopt robust privacy protections, such as data anonymization, encryption, and strict regulations on data collection, while ensuring transparency and offering residents the option to opt out. Data security is equally critical, as smart cities face increasing vulnerability to cyber threats. To protect infrastructure and services, comprehensive cybersecurity measures like advanced firewalls, intrusion detection systems, and regular security audits are necessary. Additionally, fostering a cybersecurity-conscious culture among residents and city employees is crucial to safeguarding sensitive data and maintaining public trust [16, 17, 19, 33].

Lastly, the digital divide presents a significant challenge to equitable access to smart city benefits. Not all residents have equal access to the Internet, digital devices, or the skills necessary to navigate a technology-driven environment. This disparity can exacerbate social inequalities and limit who benefits from smart city advancements. To address this issue, smart cities should focus on digital inclusion by providing affordable Internet access, offering digital literacy programs, and designing accessible technologies for all, including the elderly and individuals with disabilities. Ensuring that the advantages of smart city innovations are distributed fairly is essential for creating an equitable and inclusive urban future [20–22, 44, 45].

Recommendations and Conclusion

To integrate smart technologies with sustainable tourism, policymakers must develop strategies that balance technological advancements with environmental and community welfare. Key actions include creating a digital infrastructure to collect and analyze real-time data on tourist movements, environmental impacts, and resource usage. This infrastructure can help manage tourism more efficiently, preventing issues like overcrowding and excessive consumption. Policies should also promote the use of renewable energy and energy-efficient technologies in tourism facilities to reduce carbon footprints. Digital tools, such as mobile apps, can guide tourists toward eco-friendly accommodations and responsible travel practices, while raising awareness about the importance of preserving cultural and natural sites. Public–private partnerships can drive innovation, ensuring that smart technologies benefit the tourism sector. Urban planners should prioritize sustainability by designing walkable, bike-friendly cities with efficient public transport and involving local communities in planning to protect cultural assets. Adopting a phased approach, starting with high-impact areas like traffic and energy management, is recommended for cities integrating smart technologies. Cross-sector collaboration, continuous monitoring, and sharing best practices will help refine strategies and foster global cooperation for sustainable urban development and tourism. These efforts ensure that smart cities remain resilient and that tourism remains sustainable for future generations.

In summary, integrating smart city innovations provides a transformative solution to managing the challenges of overtourism, urban development, and sustainability. Key insights reveal that advanced technologies like IoT, big data analytics, and AI enable cities to better control tourist flows, enhance resource use, and improve visitor experiences while reducing overcrowding and environmental damage. Policies promoting sustainable tourism practices, along with active community engagement, help ensure that tourism benefits are fairly distributed and that cultural and natural assets are preserved for future generations. Ultimately, balancing urban development with sustainable tourism requires a comprehensive and inclusive approach, using smart city technologies not only for innovation but as tools to build resilient, vibrant, and sustainable urban environments that meet the needs of both residents and visitors.

References

1. Séraphin H, Gladkikh T, Vo Thanh T (2020) Overtourism, vol 448. Springer International Publishing
2. Dodds R, Butler R (2019) Overtourism: issues, realities and solutions, vol 1. Walter de Gruyter GmbH & Co KG
3. Pechlaner H, Innerhofer E, Erschbamer G (2019). Overtourism: tourism management and solutions. Routledge
4. Ribeiro de Almeida C, Quintano A, Simancas M, Huete R, Breda Z (2020) Handbook of research on the impacts, challenges, and policy responses to overtourism. IGI Global
5. Jover J, Díaz-Parra I (2023) Who is the city for? overtourism, lifestyle migration and social sustainability. In: Migration, tourism and social sustainability. Routledge, pp 9–32
6. Pham K, Vogt C, Andereck K (2024) Overtourism: a potential outcome in contemporary tourism—concepts, indicators, and impacts. Sustainable development and resilience of tourism: wellbeing and quality of life perspectives. Springer International Publishing, Cham, pp 165–186
7. Al-Romeedy B, Tyagi P, Singh A (2024) Peaks and valleys: navigating over-tourism in mountain destinations. In: Balancing mountain tourism, cultural heritage, and environmental stability. IGI
8. Blázquez-Salom M, Cladera M, Sard M (2023) Identifying the sustainability indicators of overtourism and undertourism in Majorca. J Sustain Tour 31(7):1694–1718
9. Sharma A, Hassan A (2021) Overtourism as destination risk: impacts and solutions. Emerald Publishing Limited
10. Al-Romeedy B (2024) Sustainable tourism innovations-a green entrepreneurship perspective. In: Sustainability in travel and tourism. Bharti Publications, pp 54–63
11. Khang A, Gupta S, Rani S, Karras D (2023) Smart cities: IoT technologies, big data solutions, cloud platforms, and cybersecurity techniques. CRC Press
12. Jonek-Kowalska I, Wolniak R (2023) Smart cities in Poland: towards sustainability and a better quality of life? Taylor & Francis

13. Mishra P, Singh G (2023) Sustainable smart cities. Springer International Publishing
14. United Nations (2025) Sustainable development. Department of Economic and Social Affairs
15. Gracias J, Parnell G, Specking E, Pohl E, Buchanan R (2023) Smart cities—a structured literature review. Smart Cities 6(4):1719–1743
16. Halegoua G (2020) Smart cities. MIT press
17. Willis K, Aurigi A (2017) Digital and smart cities. Routledge
18. Song H, Srinivasan R, Sookoor T, Jeschke S (2017) Smart cities: foundations, principles, and applications. John Wiley & Sons
19. Al-Romeedy B, Singh A (2024) From data to destination: how digital traces can be transformed into targeted marketing campaigns for tourists. In: Decoding tourist behavior in the digital era: insights for effective marketing. IGI
20. Mahmood Z, Wheeler M (2018) Smart cities. Springer
21. Peris-Ortiz M, Bennett D, Yábar D (2017) Sustainable smart cities. In: Innovation, technology, and knowledge management. Springer International Publishing Switzerland, Cham
22. Sun H, Wang C, Ahmad B (2017) From internet of things to smart cities: enabling technologies. CRC Press
23. Khang A, Rani S, Sivaraman A (2022) AI-centric smart city ecosystems: technologies. CRC Press, Design and Implementation
24. Eiza M, Cao Y, Xu L (2020) Toward sustainable and economic smart mobility: shaping the future of smart cities. World Scientific
25. Powell A (2021) Undoing optimization: civic action in smart cities. Yale University Press
26. Ahmad K, Maabreh M, Ghaly M, Khan K, Qadir J, Al-Fuqaha A (2022) Developing future human-centered smart cities: critical analysis of smart city security, data management, and ethical challenges. Comput Sci Rev 43:100452
27. Al-Romeedy B (2024b) Breaking chains, building futures: blockchain's impact on sustainable tourism. In: El Archi Y, Benbba B, Dávid L, Cardoso L (eds) Promoting responsible tourism with digital platforms. IGI Global, pp 90–107. https://doi.org/10.4018/979-8-3693-3286-3.ch006
28. Van Twist A, Ruijer E, Meijer A (2023) Smart cities & citizen discontent: a systematic review of the literature. Gov Inf Q 40(2):101799
29. Huda N, Ahmed I, Adnan M, Ali M, Naeem F (2024) Experts and intelligent systems for smart homes' transformation to sustainable smart cities: a comprehensive review. Expert Syst Appl 238:122380
30. Al-Romeedy B, Emam M (2024) Exploring waste reduction potential in rural tourism: what solutions can be implemented?. In: Solid waste management for rural regions. IGI
31. Pandiyan P, Saravanan S, Usha K, Kannadasan R, Alsharif M, Kim M (2023) Technological advancements toward smart energy management in smart cities. Energy Rep 10:648–677
32. Al-Romeedy B, Singh A (2024a) From search to stay: mapping the digital journey of the modern traveler. In: Decoding tourist behavior in the digital era: insights for effective marketing. IGI
33. National Institute of Managament (2020) Egypt and the eleventh goal of the sustainable development goals (sustainable cities and communities). Arab Republic of Egypt, Governance Center
34. Ministry of Housing, Utilities and Urban Communities (2020) Strategic vision for urbanization in egypt, developing youth leadership capabilities and skills. Arab Republic of Egypt
35. State Information Service (2020) Egypt enters the era of smart cities. https://sis.gov.eg/Story/169475?lang=ar. Accessed 12 Dec 2020
36. Honey M, Frenkiel K (2021) Overtourism: lessons for a better future. Island Press
37. Milano C, Cheer J, Novelli M (2019) Overtourism: excesses, discontents and measures in travel and tourism. Cabi
38. Fang B, Yu J, Chen Z, Osman A, Farghali M, Ihara I, Yap P (2023) Artificial intelligence for waste management in smart cities: a review. Environ Chem Lett 21(4):1959–1989
39. Alshamaila Y, Papagiannidis S, Alsawalqah H, Aljarah I (2023) Effective use of smart cities in crisis cases: a systematic review of the literature. Int J Disaster Risk Reduct 85:103521

40. Hui C, Dan G, Alamri S, Toghraie D (2023) Greening smart cities: an investigation of the integration of urban natural resources and smart city technologies for promoting environmental sustainability. Sustain Cities Soc 99:104985
41. Javed A, Ahmed W, Pandya S, Maddikunta P, Alazab M, Gadekallu T (2023) A survey of explainable artificial intelligence for smart cities. Electron 12(4):1020
42. Al-Romeedy B (2024a) From waste to value: a circular economy approach to sustainable rural tourism. In: Solid waste management for rural regions. IGI
43. Santos-Rojo C, Llopis-Amorós M, García-García J (2023) Overtourism and sustainability: a bibliometric study (2018–2021). Technol Forecast Soc Chang 188:122285
44. Séraphin H, Chaney D (2023) A research agenda for the sustainability of the tourism industry: a childism perspective on overtourism. J Clean Prod 414:137556
45. Yrigoy I, Horrach P, Escudero L, Mulet C (2024) Co-opting overtourism: tourism stakeholders' use of the perceptions of overtourism in their power struggles. J Sustain Tour 32(4):818–834
46. Cai M, Kassens-Noor E, Zhao Z, Colbry D (2023) Are smart cities more sustainable? an exploratory study of 103 US cities. J Clean Prod 416:137986
47. Spicer Z, Goodman N, Wolfe D (2023) How 'smart' are smart cities? Resident attitudes towards smart city design. Cities 141:104442

Technology and Overtourism

Part-II Technology and Overtourism, explores the dual nature of technology in tourism management. While acknowledging technology's potential contribution to overtourism, this part also investigates how digital solutions, artificial intelligence, and smart infrastructure can be harnessed to mitigate its negative impacts. The chapters in this part present we delve into the complex relationship between technology and overtourism, exploring both its promises and pitfalls in shaping our architectural and urban landscapes. The chapters in this part present a critical examination of how technological solutions, particularly artificial intelligence and smart infrastructure, are being deployed to address the challenges of excessive tourism. From the dark side of technology's influence on tourism trends to the practical application of AI in managing tourist flows in Nainital's lake town, these contributions reveal the multifaceted nature of technological interventions. Through case studies spanning from Florence's Historic Centre to Slovakia's High Tatras, our authors examine how smart tourism infrastructure and digital innovations are reshaping destination management strategies. However, this part goes beyond mere technological optimism. It critically evaluates whether technology truly offers solutions to overtourism or potentially exacerbates the problem. The authors present both the opportunities and limitations of digital tools in preserving architectural heritage, managing tourist flows, and maintaining community balance. Their insights suggest that while technology can be a powerful ally in addressing overtourism, it must be implemented thoughtfully and in conjunction with sound urban planning principles.

The Dark Side of Technology When Addressing Overtourism: A Critical Overview

Marco Martins, Ricardo Jorge da Costa Guerra, Lara Santos, Luísa Lopes, and Ana Rita Conde

Abstract This chapter aims to provide an overview on the relationship between overtourism and technology, shedding light on the perverse effects of technology when addressing overtourism. An exploratory study was adopted because one should move away from the formal testing of hypotheses and attempt to analyse the key issues around the core concepts we are concerned with, i.e., the dark side of technology when addressing overtourism and its links to theory. The aim of this chapter is, therefore, to take a broader look at this thematic, and thus broaden the discussion around this topic. Results point out the fact that, although technology can be used to better manage overtourism, it is also linked to fashion trends, and several tourism destinations find themselves promptly overbooked in result. The originality of this chapter lies in the fact that unlike other studies, this research focuses on uncovering the dark side of technology, which has generally been promoted as the ideal tool to mitigate the impacts of overtourism. This critical overview might help both scholars and practitioners to reflect on and/or rethink how technology is really helping the destinations to overcome the challenges that come with overtourism.

M. Martins (✉)
CiTUR, Coimbra, Portugal
e-mail: marco.mpm@gmail.com

R. J. da C. Guerra
Polytechnic University of Guarda, Seia, Portugal

Present Address:
L. Santos
Polytechnic University of Cávado and Ave, Barcelos, Portugal
e-mail: laramgsantos@gmail.com

L. Lopes
Instituto Politécnico de Bragança, Mirandela, Portugal
e-mail: luisa@ipb.pt

A. R. Conde
University Lusófona, Porto, Portugal
e-mail: rita.conde@ulusofona.pt

V. C. Sharma et al. (eds.), *Architectural Vision of Overtourism*,
SpringerBriefs in Architectural Design and Technology,
https://doi.org/10.1007/978-981-96-5794-0_6

Keywords Overtourism · Dark side of technology · Destination's management · Tourism industry · Tourism destinations

1 Introduction

In the present times, the global tourism industry needs formulating revised strategies to better manage the crescent influx of tourists, and with that intent, many destinations are now putting almost all their bets in technological tools. According to Martins [1, p. 138], "tourism, as an activity, promotes the interaction of visitors with destination places and with their communities resulting in a set of effects for the local community, the physical space and for the tourists themselves. It is in the current context of contesting the massification of tourism that research on the impacts becomes so relevant". Milano et al. [2] addressing this issue express that since the 1970s, the neoliberalism planning approach to tourism development has been widely prioritised, and aiming for profit they have exploited places and their resources without any regards for anything or anyone.

Overtourism refers to the phenomenon of a destination having an excessive number of visitors, resulting in the overload of its infrastructure, disruption of local quality of life and also environmental and social degradation. Popular tourism destinations such as Venice, Barcelona and Machu Picchu have been struggling with overtourism for years. Because overtourism is considered as a threat to destinations' sustainable tourism development, the issue has been and is being addressed through structural and policy level initiatives besides resorting to technology [3, 4]. At the level of technology, destinations are gradually adopting the concept of smart development. Various technologies have been created in the last years to manage diverse facets of overtourism [5], and these interventions happened mainly on urban environments where the concept of smart city has been adopted seeking to tackle or even override the overtourism issue (see [6–8]). Implementing technological measures without careful consideration can lead to negative consequences. It is important to strike a balance between technological interventions with human-centred approaches to ensure that tourism remains a positive force for both visitors and local residents. However, basing ourselves on what has just been said, we may wonder if the usage of technology in tourism development only has a bright side or does it have a dark side as well. In order to answer this question, this chapter was divided into six sections. It begins with the introduction, which sets out the main reasons for choosing this thematic and the question that has guided our research. Afterwards, the chosen methodology is then presented, followed by a section addressing why, when and how overtourism rose to be a contemporary problem. Moving forward, one shows why and how technologies are being used to tackle overtourism, and further on we ask ourselves if in the context of overtourism, technologies do not have also a dark side. Finally, we discuss the results, wrapping it up with a conclusion in which we discuss nor only the study in itself, but also possible future avenues for research in this field.

2 Methodology

This chapter was developed basing in a critical overview of the literature and on the researcher's own perspective. In a critical review, one addresses purposefully and critically a given issue through a literature synthesis [9, 10] that in turn, will allow us to "evaluate what is 'valuable' in the available body of work" [11, p. 93].

To choose the criteria to guide our literature exploration process and posterior research selection, one followed Lambert et al. [12] steps, i.e., it is encompassing both inclusion and exclusion criteria. The inclusion criteria established consisted of articles that had to be (a) theoretical, empirical and peer-reviewed; (b) published in English; (c) an exploration of the dark side/effects of technology on enhancing overtourism; and (d) published between 2016 and 2024, inclusively. This period was defined because, from 2016 onwards, digital tourism platforms exponentially increased, resulting from technological innovations and the massive use of social networks and smartphones. The research was carried out in the following databases: Web of Science, Scopus and Google Scholar. However, like Newman [13] one also did not intend this critical overview to be neither systematic or comprehensive, but instead to approach those methodological or conceptual gaps that can be revealed when doing an extensive reading in this field of research. Basing ourselves on Wall et al. [14, p. 262], one sought that our critical overview method could identify "taken-for-granted assumptions in a research community to broaden the community's perspectives" and to "identify the taken-for-granted ideas and assumptions that dominate a body of literature and identify alternative ideas and assumptions that are equally feasible and worth considering in open debate".

Nevertheless, it is important to refer to the limitations inherent to all literature reviews, including critical reviews. These limitations consist of the fact that literature reviews depend on the authors' unique perspectives, and, consequently, different authors will likely yield different insights and interpretations besides being generally non-systematic [15].

3 The Rise of Overtourism

It is possible to say that "tourism may be regarded as a fantasy selling industry striving upon people's dreams, hopes, beliefs, expectations and imageries, which are in turn transformed into consumable commodities or experiences" [16, p. 2]. However, the commodification of places has brought with it many opportunities but also challenges, among which is the need to address the excessive touristification of places also known as overtourism. In itself, overtourism, although a recently adopted term, is not a new problem, particularly in well-known destinations, such as Barcelona, Venice or the Niagara Falls. In truth, one of the first academic studies addressing the excessive number of tourists in a destination was the one of Doxey's [17] that showed the Niagara Falls nearby resident's negative reactions to what today we call overtourism. Overtourism is a contemporary issue, one that is evolving constantly and characterised by the more than evident over-visitation to some tourist destinations [18], which leads to negatively impacting the natural, economic and social environments of a destination [19]. In that sense, the United Nations World Tourism Organization [20] defined overtourism as the set of impacts that tourism has on a destination, or parts thereof, that to a great degree influences the residents' quality of life and/or the visitors' experience quality in a negative way. Others like Cheung and Li [21, p. 1] observed that 'overtourism' started being defined as a "massive and uncoordinated influx of tourists to popular destinations", what according to Goodwin [22] and Milano et al. [23] leads hosts or guests, locals or visitors to feel that there exist too many visitors in a given area what results in a loss of life/experience quality and leading to all considering that such a deterioration is unacceptable.

According to Dodds and Butler [24, p. 520], "overtourism has resulted from a combination of a number of factors, which together have created a 'perfect storm' of visitors to specific sites. Although many forces have existed for years, many other influences have increased the issue to a tipping point and this phenomenon has spread to many more destinations apart from solely cities".

Thus, in our literature review one identified several factors, summarised in Fig. 1, that can explain the appearance of this 'perfect storm', among which are lack of a scenario analysis methodology that can be used to predict future developments, and therefore future issues (*see* [25]); lack of strategic planning (*see* [26]); selling destinations solely as commodities (*see* [18, 23]); unwillingness to address and tackle the problem (*see* [24]) and/or an inexistence of a coherent vision for destination tourism development (*see* [27]); social media platforms and other technologies' influence (*see* [28]); absence or inadequate regulations (*see* [29]); unwillingness of the generality of tourism stakeholders to adopt conveniently the instruments available to measure overtourism (*see* [30]).

One must however mention that although these can be pointed out as some of the main reasons why and how visitors' numbers are rising, others may be identified, especially because any destination is unique in itself. Furthermore, there is still no precise method to define what is having an excessive number of visitors [31]. Going forward, one can say that as a concept, overtourism can be seen as the end result of both the tourism's demand and supply acceleration and growth, the overuse of the tourist destinations' natural assets (see Fig. 2), the damaging of its cultural attractions, and the constatation that the tourism excessive growth comes with great impacts in the social and economic environments besides the psychological impacts [19, 32].

Among these impacts, one detaches the increase in the prices related with hospitality, tourist attractions and housing; gentrification; severe alterations in the local business fabric; increased insecurity and the disruption of citizens' lifestyle including labour instability and precariousness; the overuse of resources and an increase in waste generation [33, 34]. As a result of this, the attitudes of the communities towards tourism development have changed and change so drastically that in many cases it led to the development of tourismphobia [2, 26, 35, 36]. In what regards to the concept of tourismphobia, after showing that in Barcelona demonstrations against tourists began to be noticed as far as 2004, Martins [26, p. 5] defined it as "(...) a dislike or

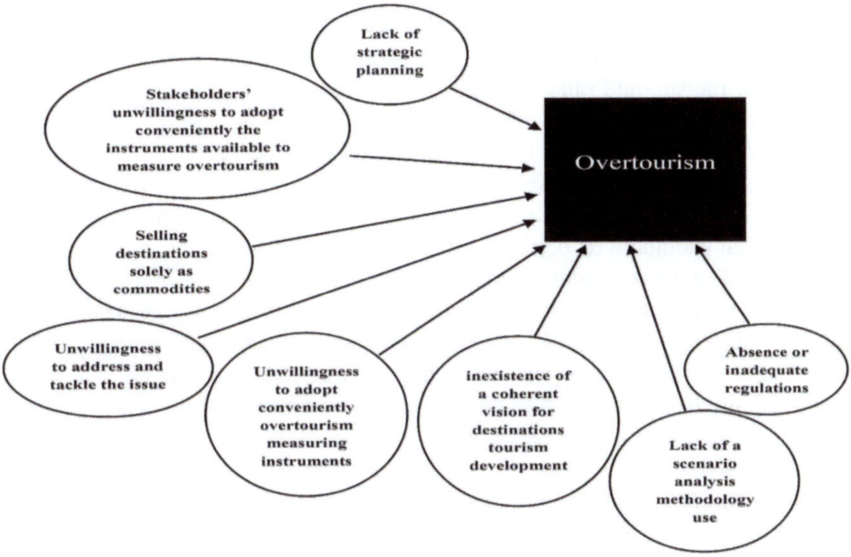

Fig. 1 Main factors leading towards overtourism. *Source* Developed by the authors based on the literature review

Fig. 2 Visitors and/or tourists trying to watch and photograph the Mona Lisa (Louvre). *Source* Author and date unknown

hatred against tourists independently of their ethnic or social origin, race, religion, gender, sexual orientation or other sort of discrimination. However, paradoxically, xenophobia, racism, and other radicalism forms can trigger and/or be the trigger to the increase of hostile and violent actions against tourists. These actions aim to humiliate denigrate and/or hurt the tourists and those associated to them".

Although regarded by some scholars as overused words [34], both overtourism and tourismphobia have launched an interesting discussion on the boundaries between the rights that should be recognised to tourists and the rights that should be recognised to resident communities. This debate also has shown so far that opinions are highly polarised, while some view tourism as the perfect example of what neoliberal is capable of, i.e., strongly embedded in power relations characterised by its exclusionary and resource-degrading agency, which needs to be tamed. For others, tourism should be reckon as still being genuinely coherent and still seeking to be a 'force' that can do good for all [2].

Nevertheless, when adopting a critical stance, Milano et al. [2, p. 2/3] went further stating that "in fact, a glance at the societal debate, channelled through the media and academic scholarship in the field, shows that a much larger number of places around the world and a wider range of destinations, already subject of previous discussions on the 'excesses of tourism', have become entrenched with strategic prioritisation for capital accumulation through mass tourism promotion the ensuing strain of wealth extraction and redistribution."

4 Resorting to Technologies to Tackle Overtourism

Overtourism is widely recognised as a threat when destinations seek to attain sustainability, and it is being spoken for resorting to structural and policy initiatives besides the usage of technology [3, 4]. As global tourism continues to increase, especially post pandemic, innovative technologies are increasingly being explored as solutions to cope with the pressure. In that sense, and in a positive perspective Buhalis [37, p. 267] says that "technology innovations bring the entire range of stakeholders together in tourism service ecosystems" and that "technology-empowered tourism experiences increasingly support travellers to co-create value throughout all stages of travel". Several scholars see technology as crucial to overcome the negative effects of overtourism by expressing that these can help the tourism sector to provide personalised and sustainable customer's experiences while at the same time helping the sector to be more resilient and sustainable and in consequence being able to mitigate the burden that comes along with overtourism [38–41]. From AI-driven systems to virtual experiences and mobile apps, these technologies are enabling destinations to manage visitor numbers, improve sustainability and enhance the overall tourism experience. However, for long-term success, these technology-driven solutions must be implemented alongside effective policy measures and community engagement.

Based on similar ideas, several destinations have in recent years adopted the concept of smart tourism development (see [42]) and of smart tourism city in particular to address the overtourism problem (see [43]). Generally speaking, and as Vargas-Sánchez [44], p. 26] very well puts it, "smartness (or intelligence) represents the ability for a better understanding and solving problems (such as those related to overtourism) using knowledge, that is, based on data and information". Already [42], p. 3] developing on the smart tourism destinations say that these "can be defined as a tourism system that takes advantage of smart technology in creating, managing and delivering intelligent touristic services/experiences and is characterized by intensive information sharing and value co-creation". Hence, the ultimate objective of smart tourism destinations is to increase competitiveness and enhance the life quality of all stakeholders, including local communities and tourists [45].

However, as Koens et al. [27] very well put it, turning tourism in any destination into a transition arena is not an easy task. Tensions between the perspectives and interests of the panoply of stakeholders involved in local tourism development (local communities, tourists, policymakers and decision-makers, tourism industry—local or otherwise) are key to really comprehend issues like overtourism and even the impact of the sharing economy. Therefore, if one really wants to tackle overtourism, then, tourism governance needs to embrace reconceptualising the tourism role within societies and clearly recognising and addressing the existing tensions. Thus, more than relying on just one variable to solve problems, destinations must implement and rely on an integrated system of governance.

5 Technologies and Overtourism: Looking for the Dark Side

Consumers and potential consumers can now constantly share their experiences before, during and after experiencing destinations and their tourist services and products through the creation of digital content characterised for transmitting extremely emotional and personal opinions. Today, the tourism industry is experiencing a transformation with the emergence of new players, including online travel agencies (OTAs), meta-search engines and travel service platforms, which are reshaping the entire tourism value chain. In most developed countries and in many developing ones, the travel and tourism sector heavily rely on major platforms that provide travel information searches, reservations, bookings, accommodations, transportation and financial services. As a result, a powerful elite of private enterprises in software, Internet services, hardware, finance, logistics and infrastructure have gained significant control over destination marketing, operations and management. This is largely facilitated by dominant travel service platforms such as Google for travel information searches, Booking.com and Expedia as online travel agencies (OTAs), TripAdvisor for travel reviews, and Uber, Grab and Airbnb in urban transportation and lodging, respectively [46]. Hence, within the tourism sector, the appearance and the endurance of digital platforms supplying different services have reinforced and continue to reinforce the landscape of conflicts happening on and around tourism destinations, with a special emphasis on cities [1, 32].

On the other hand, positive comments on social media can prove to have a deep impact on the wish to visit some destination. Hence, social media can have what scholars usually denominate as the 'called effect' over destinations, consequently contributing to overcrowding and overtourism [28].

But it is not only the excessive number of tourists that cause problems in destinations; tourist's behaviour is also a variable that has great influence in moulding the opinions of residents. In truth, one of the great issues linked with overtourism is what became known as 'Millennials Instagrammability' that consists in taking photos and selfies to put them on Instagram; the problem is that particularly Generations Y and Z want to take and post the 'best picture ever' on social media without thinking or looking into the consequences [47]. Focusing on this issue particularly, Justin Francis, prominent member of Responsible Travel, declared that "years ago, tourism was about experience seeking. Now it's about using photography and social media to build a personal brand. In a sense, for a lot of people, the photos you take on a trip become more important than the experience" [48, p. 69]. Thus, and although some continue to defend the potential technology and social media platforms as means to mitigate the effects of overtourism [48], the reality shows that these are now, and instead, one of the greatest catalysts of overtourism [49]. It is our belief regarding this issue that one does not need to be a genius to guess where to put all our bet in this fight, especially because it is one between non-influencers versus influencers.

In what concerns to the issue of smart destinations, the literature identifies several problems in its creation design. Consequently, scholars questioned its techno-centrism, because it is one that looks for the privatisation of public services, and therefore there is a greater concern with the private corporations' interests than with the greater good. Furthermore, it seeks a greater dependency on technologies while at the same time seeking to depoliticise the destinations management; besides, it is also a major concern with the possibility of these interventions tending to invade the citizens' privacy [50–52]. In that sense, overtourism would perfectly serve as the perfect escape goat to create smart destinations as initially intended, i.e., less politicised, less democratic and less concerned with the rights of tourists and local communities alike.

6 Conclusion

The novelty of this study resides in its incursion into an area that has been majorly overlooked by research, the dark side of technology when considering overtourism. In fact, as Chaney and Séraphin [53] demonstrated through a systematic literature review and a lexicometric analysis based on 236 articles, studies conducted on overtourism have been excessively concentrated and articulated around management, experiences and around the context of overtourism and its negative impacts. Furthermore, Séraphin et al. [54] also point out that existing studies focus mainly on the consequences of overtourism and not in what it is upstream. Thus, seeking to uncovering the dark side of technology, which in general has been publicised as the ideal tool for mitigating the impacts of overtourism, and even to fight against it, this critical overview might help both scholars and practitioners to think and/or rethink how technology is really helping destinations in overcoming the challenges that come with overtourism.

The results of this study point out the fact that, although technology can be used to better manage overtourism, it is also linked to fashion trends, and many tourist destinations find themselves fastly overbooked because of it. Besides, one also suggests, as Séraphin [55] before us, that it is possible to overcome the dark side of technologies if the adoption of new technologies (incremental strategies) is thought to contribute to tackling overtourism, but successes will only arise if destinations seek profound and meaningful alterations (radical changes).

Along this text is possible to image numerous possibilities for future avenues, among which, to what extent is the visitor experience impacted by technology? What happens when visitor's land in an overcrowded destination following good reviews on technological platforms? What occurs when after co-creating an experience through technological tools visitors discover its non-replicability due to the onsite excess of tourists? These are some of the questionings that one considers that researchers could try to address in future studies.

In the end and basing ourselves on what has been outlined in this chapter, it is our belief that technology will not solve all problems; in truth, if the initiatives to tackle overtourism are isolated and uncoordinated then technology will certainly have an adverse effect. It will probably contribute to amplify the problem instead of making it shorter—it will show how dark it can be. The result should effectively be the opposite, so tourism destination managers have to be aware of the need for careful planning, the involvement of the local community and also the commitment to preserve the unique characteristics that make each destination special. The public sector has here an important role and should articulate with the private businesses and other institutions, the more efficient strategy. Nowadays, after the post-pandemic period, tourism is rapidly growing and brings a wide range of challenges to tourism destinations.

References

1. Martins M (2020) The impact of touristification in city neighbourhoods–the case of Lisbon. In: Oskam J (eds) The overtourism debate. Emerald Publishing Limited, UK, pp 137–150. https://doi.org/10.1108/978-1-83867-487-820201010
2. Milano C, Novelli M, Russo AP (2024) Anti-tourism activism and the inconvenient truths about mass tourism, touristification and overtourism. Tour Geogr 1–25. https://doi.org/10.1080/14616688.2024.2391388

3. Gowreesunkar V, Seraphin H (2019) Introduction: what smart and sustainable strategies could be used to reduce the impact of overtourism? Worldw Hosp Tour Themes 11(5):484–491. https://doi.org/10.1108/WHATT-06-2019-0044
4. Mohanty P, Nair N, Sharma AK (2022) Overcoming overtourism through technology: the case of Asian cities. In: Hassan A (eds) Technology application in tourism in Asia. Springer, Singapore, pp 395–405. https://doi.org/10.1007/978-981-16-5461-9_24
5. Bourliataux-Lajoinie S, Dosquet F, del Olmo Arriaga JL (2019) The dark side of digital technology to overtourism: the case of Barcelona. Worldw Hosp Tour Themes 11(5):582–593. https://doi.org/10.1108/WHATT-06-2019-0041
6. Camatti N, Bertocchi D, Carić H, Van der Borg J (2020) A digital response system to mitigate overtourism. The case of Dubrovnik. J Travel & Tour Mark 37(8–9):887–901. https://doi.org/10.1080/10548408.2020.1828230
7. Fontanari M, Traskevich A (2023) Smart-solutions for handling overtourism and developing destination resilience for the post-COVID-19 era. Tour Plan & Dev 20(1):86–107. https://doi.org/10.1080/21568316.2022.2056234
8. Pasquinelli C, Trunfo M (2020) Reframing urban overtourism through the smart-city lens. Cities 102:102729. https://doi.org/10.1016/j.cities.2020.102729
9. Sutton A, Clowes M, Preston L, Booth A (2019) Meeting the review family: exploring review types and associated information retrieval requirements. Health Info Libr J 36(3):202–222. https://doi.org/10.1111/hir.12276
10. Tai J, Ajjawi R, Umarova A (2024) How do students experience inclusive assessment? a critical review of contemporary literature. Int J Incl Educ 28(9):1936–1953. https://doi.org/10.1080/13603116.2021.2011441
11. Grant MJ, Booth A (2009) A typology of reviews: an analysis of 14 review types and associated methodologies. Health Info Libr J 26(2):91–108. https://doi.org/10.1111/j.14711842.2009.00848.x
12. Lambert SI, Madi M, Sopka S, Lenes A, Stange H, Buszello C-P, Stephan A (2023) An integrative review on the acceptance of artificial intelligence among healthcare professionals in hospitals. Npj Digit Med 6:111. https://doi.org/10.1038/s41746-023-00852-5
13. Newman J (2024) Promoting interdisciplinary research collaboration: a systematic review, a critical literature review, and a pathway forward. Soc Epistemol 38(2):135–151. https://doi.org/10.1080/02691728.2023.2172694
14. Wall JD, Stahl BC, Salam AF (2015) Critical discourse analysis as a review methodology: an empirical example. Commun Assoc Inf Syst 37:257–285. https://doi.org/10.17705/1CAIS.03711
15. Greenhalgh T, Thorne S, Malterud K (2018) Time to challenge the spurious hierarchy of systematic over narrative reviews? Eur J Clin Invest 48(6):e12931. https://doi.org/10.1111/eci.12931
16. Martins M (2015) The tourist imagery, the destination image and the brand image. J Tour Hosp Manag 3(2):1–22. https://doi.org/10.15640/jthm.v3n2a1
17. Doxey GV (1975) A causation theory of visitor-resident irritants, methodology and research inferences:the impact of tourism. In: The Sixth annual conference proceedings of the travel research association. San Diego, CA, pp 195–198
18. Milano C, Novelli M, Cheer JM (2019) Overtourism and tourismphobia: a journey through four decades of tourism development, planning and local concerns. Tour Plan Dev 16(4):353–357. https://doi.org/10.1080/21568316.2019.1599604
19. Mihalic T (2020) Conceptualising overtourism: a sustainability approach. Ann Tour Res 84:103025. https://doi.org/10.1016/j.annals.2020.103025
20. UNWTO (2018) Overtourism'? understanding and managing urban tourism growth beyond perceptions: executive summary. https://doi.org/10.18111/9789284420070
21. Cheung KS, Li L (2019) Understanding visitor-resident relations in overtourism: developing resilience for sustainable tourism. J Sustain Tour 27(8):1197–1216. https://doi.org/10.1080/09669582.2019.1606815

22. Goodwin H (2017) The challenge of overtourism (Responsible Tourism Partnership, Working Paper 4. p 1. https://haroldgoodwin.info/wp-content/uploads/2020/08/rtpwp4overtourism012017.pdf. Accessed 24 September 2024
23. Milano C, Novelli M, Cheer JM (2019) Overtourism and degrowth: a social movements perspective. J Sustain Tour 27(12):1857–1875. https://doi.org/10.1080/09669582.2019.1650054
24. Dodds R, Butler R (2019) The phenomena of overtourism: a review. Int J Tour Cities 5(4):519–528. https://doi.org/10.1108/IJTC-06-2019-0090
25. Martins M, Guerra R, Santos L (2022) Tourism development policy and the use of scenario analysis: a synthesis model. PASOS Rev Tur Y Património Cult 20(5):1103–1112. https://doi.org/10.25145/j.pasos.2022.20.074
26. Martins M (2018) Tourism planning and tourismphobia: an analysis of the strategic tourism plan of Barcelona 2010–2015. J Tour, Herit & Serv Mark 4(1):3–7. https://doi.org/10.5281/zenodo.1247519
27. Koens K, Melissen F, Mayer I, Aall C (2021) The smart city hospitality framework: creating a foundation for collaborative reflections on overtourism that support destination design. J Destin Mark Manag 19:100376. https://doi.org/10.1016/j.jdmm.2019.100376
28. Alonso-Almeida MD, Borrajo-Millán F, Yi L (2019) Are social media data pushing overtourism? the case of Barcelona and Chinese tourists. Sustain 11(12):33–56. https://doi.org/10.3390/su11123356
29. De Giorgi I (2024) Big cruise ships going feral: an ecocritical reading of overtourism in Venice. Lagoonscapes 4(1):139–162. https://doi.org/10.30687/LGSP/2785-2709/2024/01/007
30. Eckert C, Zacher D, Pechlaner H, Namberger P, Schmude J (2019) Strategies and measures directed towards overtourism: a perspective of European DMOs. Int J Tour Cities 5(4):639–655. https://doi.org/10.1108/IJTC-12-2018-0102
31. Leahy K (2023) What's the problem with overtourism? National Geographic. https://www.nationalgeographic.com/travel/article/what-is-overtourism. Accessed 15 September 2024
32. Colomb C, Novy J (2016) Protest and resistance in the tourist city. Routledge, Abingdon, UK. https://doi.org/10.4324/9781315719306
33. Cocola-Gant A, Gago A, Jover J (2020) Tourism, gentrification and neighbourhood change: an analytical framework–reflections from southern European cities. In: Oskam JA (eds) The overtourism debate. Emerald Publishing Limited, Leeds, pp 121–135. https://doi.org/10.1108/978-1-83867-487-820201009
34. Koens K, Postma A, Papp B (2018) Is overtourism overused? understanding the impact of tourism in a city context. Sustain 10:4384. https://doi.org/10.3390/su10124384
35. Cheer JM, Milano C, Novelli M (2019) Afterword: over overtourism or just the beginning? In: Milano C, Cheer JM, Novelli M (eds) Overtourism: excesses, discontents and measures in travel and tourism. CAB International, pp 227–232. https://doi.org/10.1079/9781786399823.0227
36. Milano C (2018) Overtourism, malestar social y turismofobia. Un debate controvertido. Pasos-Revista de Turismo y Patrimonio Cultural 16(3):551–564. https://doi.org/10.25145/j.pasos.2018.16.041
37. Buhalis D (2019) Technology in tourism-from information communication technologies to eTourism and smart tourism towards ambient intelligence tourism: a perspective article. Tour Rev 75(1):267–272. https://doi.org/10.1108/TR-06-2019-0258
38. Buhalis D, Lin MS, Leung D (2023) Metaverse as a driver for customer experience and value co-creation: implications for hospitality and tourism management and marketing. Int J Contemp Hosp Manag 35(2):701–716. https://doi.org/10.1108/IJCHM-05-2022-0631
39. Capocchi A, Vallone C, Pierotti M, Amaduzzi A (2019) Overtourism: a literature review to assess implications and future perspectives. Sustain 11(12):3303. https://doi.org/10.3390/su11123303
40. Kouroupi N, Metaxas T (2023) Can the metaverse and its associated digital tools and technologies provide an opportunity for destinations to address the vulnerability of overtourism? Tour Hosp 4(2):355–373. https://doi.org/10.3390/tourhosp4020022

41. Prados-Castillo JF, Guaita Martínez JM, Zielińska A, Gorgues Comas DA (2023) Review of blockchain technology adoption in the tourism industry from a sustainability perspective. J Theor Appl Electron Commer Res 18:814–830. https://doi.org/10.3390/jtaer18020042
42. Gretzel U, Sigala M, Xiang Z, Koo C (2015) Smart tourism: foundations and developments. Electron Markets 25:179–188. https://doi.org/10.1007/s12525-015-0196-8
43. Boes K, Buhalis D, Inversini A (2016) Smart tourism destinations: ecosystems for tourism destination competitiveness. Int J Tour Cities 2(2):108–124. https://doi.org/10.1108/IJTC-12-2015-0032
44. Vargas-Sánchez A (2019) Overtourism a challenge for smart cities. In: Mateo Sanguino TJ, Lozano Domínguez JM, Fernández de Viana González IJ, Rodríguez Román MA, Redondo González MJ (eds) Actas de las II Jornadas ScienCity 2019: Fomento de la Cultura Científica, Tecnológica y de Innovación en Ciudades Inteligentes, 26–29, Universidad de Huelva. https://rabida.uhu.es/dspace/handle/10272/19330
45. Caragliu A, Del Bo C, Nijkamp P (2011) Smart cities in Europe. J Urban Technol 18(2):65–82. https://doi.org/10.1080/10630732.2011.601117
46. Abbasian Fereidouni M, Kawa A (2019).Dark side of digital transformation in tourism. In: Nguyen N, Gaol F, Hong TP, Trawiński B (eds) Intelligent information and database systems, pp 510–518. ACIIDS 2019. Lect Notes Comput Sci 11432. Springer, Cham. https://doi.org/10.1007/978-3-030-14802-7_44
47. Unger S, Grassl W (2020) Insta-holidays and instagrammability. J Tour Leis Hosp 2(2):92–103. https://dergipark.org.tr/en/download/article-file/1189819
48. Gretzel U (2019) The role of social media in creating and addressing overtourism. In: Dodds R, Butler RW (eds) Overtourism. Issues, realities and solutions. De Gruyter Oldenburg, Berlin, pp 62–75. https://doi.org/10.1515/9783110607369-005
49. Song Q, Wondirad A (2023) Demystifying the nexus between social media usage and overtourism: evidence from Hangzhou, China. Asia Pac J Tour Res 28(4):364–385. https://doi.org/10.1080/10941665.2023.2230313
50. Fabrègue BFG, Bogoni A (2023) Privacy and security concerns in the smart city. Smart Cities 6(1):586–613. https://doi.org/10.3390/smartcities6010027
51. March H, Ribera-Fumaz R (2016) Smart contradictions: the politics of making Barcelona a self-sufficient city. Eur Urban Reg Stud 23(4):816–830. https://doi.org/10.1177/0969776414554488
52. Vanolo A (2014) Smartmentality: the smart city as disciplinary strategy. Urban Studies 51(5):883–898. https://doi.org/10.1177/0042098013494427
53. Chaney D, Séraphin H (2023) A systematic literature review and lexicometric analysis on overtourism: towards an ambidextrous perspective. J Environ Manage 347:119123. https://doi.org/10.1016/j.jenvman.2023.119123
54. Seraphin H, Zaman M, Olver S, Bourliataux-Lajoinie S, Dosquet F (2019) Destination branding and overtourism. J Hosp Tour Manag 38(1):1–4. https://doi.org/10.1016/j.jhtm.2018.11.003
55. Séraphin H (2021) Theorizing new technologies as potential incremental management tools to tackle overtourism. In: Hassan A, Sharma A (eds) Overtourism, technology solutions and decimated destinations. Springer, Singapore. https://doi.org/10.1007/978-981-16-2474-2_1

Digital Technology and AI as Factors that Contribute to the Architectural Vision of Over-Tourism: A Critical Review

Maximiliano E. Korstanje and Shem Wambugu Maingi

Abstract Doubtless, specialists and experts of all pundits have eloquently agreed that over-tourism is a major problem for the future of the tourism industry globally. Recent reports have alerted us to the issues of overcrowding in established international destinations. Particularly, this has ignited an anti-tourist sentiment mainly marked by a harsh hostility against foreign tourists. Although over-tourism—as a social malady—seems not to be new, because it is circumscribed to the outset of tourism research, it has accelerated a deep social decomposition. Anthropologically speaking, the anti-tourist discourse rests on a paradoxical situation. On the one hand, locals appeal to international tourism as a form of poverty relief as well as development, but on the other, its effects are undesired. To resolve this paradox some studies have emphasized the use of AI and digital technologies to organize not only the infrastructure but also its relationship with the demand. Having said this, the present book chapter interrogates the weaknesses and strengths of digital technologies applied to resolve the problems left by mass tourism. Exegetes and detractors of technology offer their arguments for the configuration of a new sustainable tourism agenda for the next decades. What this book chapter looks to answer is: is technology part of the solution for over-tourism?

Keywords Mass tourism · Sustainability · Over-tourism · Technology · Smart tourism · Critical review

M. E. Korstanje (✉)
University of Palermo, Buenos Aires, Argentina
e-mail: mkorst@palermo.edu

S. W. Maingi
Kenyatta University, Nairobi, Kenya
e-mail: maingi.shem@ku.ac.ke

V. C. Sharma et al. (eds.), *Architectural Vision of Overtourism*,
SpringerBriefs in Architectural Design and Technology,
https://doi.org/10.1007/978-981-96-5794-0_7

1 Introduction

Even though over-tourism is not new, it has become a buzzword just after the turn of the twentieth century [1]. Most certainly, off-the-beaten-track destinations are facing the negative effects of mass tourism on their skin as never before [2]. Some experts agree that over-tourism seems to be conditioned by an anti-tourist discourse dubbed tourist-phobia. Others, rather, are alert to the problems of infrastructure to receive mass tours. These shortcomings are plausibly aggravated by the excess demand derived from a combination of factors such as the recent technological breakthrough that makes faster transport means, a reduction of working hours in Westernized economies, and the proliferation of digital platforms and applications that offer informal accommodation, as well as the lack of planning in the public sector to organize the visitors in the ground (only to name a few) [3–5]. Here one must distinguish the phobia from over-tourism. Whereas the former signals a deeper sentiment of apathy for foreigners, most closely linked to chauvinism, the latter is part of serious shortcomings in the infrastructure that affects the destination´s carrying capacity [6, 7]. In this context, some voices have emphasized the urgency to adopt new creative methods to alleviate the saturation generated by mass tours. One of these solutions behooves us in digital technology and artificial intelligence (AI). Cities globally have adopted these creative innovations to provide more integrated services for a more sustainable footprint: For instance, Amsterdam's smart city initiative set to improve traffic management as well as the Amsterdam's beautiful noise project by analyzing social media posts and "ambient geo-social data" produced by tourists as well as local communities and sending alerts on queues and delays in public transit. AI has also been used to advise tourists in real time on congested areas or when nonpeak hours exist by mapping visitor behavior patterns therefore helping cities spread visitor load while maximizing facility yield.

Although Generative AI solutions and dynamic data analytics solutions offer a fertile ground for resolving the dilemma of over-tourism in tourism cities, some authors have claimed that the problem lies in the culture of the industry. To put this bluntly, the tourism industry rests on the need to amass profits. With this in mind, the industry—sooner or later—will go toward a stage of saturation and inevitable decline. Technology does not resolve the trouble rather it reduces and delays the environmental consequences. Under some conditions, digital applications—unregulated by the local authorities—often hasten devastating effects on destinations. What is important to discuss? First, world destinations want (mass) tourism because of its multiplier effect on the local economy while avoiding its resulting costs [8].

As the previous argument is given, the present book chapter focuses on two primordial aims. On the one hand, it discusses critically the anti-tourist discourse that originated in the dissociation between the gains and costs of tourism. On the other hand, it reviews the exegetes and detractors of smart technology as a solution to over-tourism. Lastly, the chapter provides a mixed-balanced position to expand the current understanding of AI and digital technology within the architectural vision of over-tourism.

1.1 Origins of Mass and Over-Tourism

The discipline has been concerned by the fast (untamed) growth of the tourism industry. The industry has been successfully expanded worldwide over the recent two decades [9, 10]. This commercial expansion has been accompanied by negative effects or challenges policymakers failed to resolve. Mass and over-tourism are part of the negative consequences of market growth [11, 12]. The effects of over-tourism have been more evident in cities such as Venice, Dubrovnik, and Amsterdam with high tourist density levels (tourist numbers per inhabitant). Over-tourism presents a complex paradox where, while the industry relies on profits and foreign investment for survival, it must also adopt more sustainable consumption practices without compromising its financial performance. The impacts of over-tourism exceed the physical, ecological, social, psychological, and economic thresholds [13]. Additionally, many developing destinations remain hesitant to embrace proximity or small-scale tourism as sustainable alternatives, as these practices often undermine their primary economic gains.

Nevertheless, the origins of over-tourism are deeply rooted in the cultural foundations since the tourism industry's inception [14, 15]. Although over-tourism has gained traction as a buzzword for these decades, the concept has two important conceptual caveats. On the one hand, it is very hard to measure or even share international standards on how to measure tourism in different countries and cultural backgrounds [16]. Since over-tourism seems to be a question of collective perception, it varies from country and culture. The levels of visual contamination are mainly marked by symbolic borders collectively negotiated by locals. What seems to be acceptable (in number of visitors) in a destination is simply rejected in others. At least empirically, one must confess that *over-tourism measure* is a task very hard to perform. On the other hand, the lack of international legislation makes it almost impossible to adopt shared planning for mitigating its negative effects on the terrain [17]. Further, we must discern between over-tourism and tourism phobia. While the former refers to the problems of capacity or infrastructure to receive mass tourism, the latter signals a much deeper sociological force oriented to demonize the foreign visitor [18]. At a closer look, both concepts operate anyway in a grey zone simply because some anti-tourist discourses are apparently shaped by social discontent, which is generated by overcrowding, but behind this sentiment there lies xenophobia. This means that the host–guest´s encounters are not only mainly marked by history but also by social forces like racism and chauvinism. Under some conditions of class subordination, which is close to the host–guest relationships in the tourism industry, racism remains covert [19]. Echoing this, some voices have alerted on the multilayered aspects of over-tourism as well as its uncontrollable consequences today [20]. As Rachel Dodds and Richard Butler put it, over-tourism seems to be a neologism for an old problem that has whipped the tourism industry from its outset. It derives from a combination of factors which include the ever-growing world population, cheaper travel costs, faster transport means, and better working conditions, as well and the lack of efficient regulation in some developing destinations. Besides, there is a clear disconnection between the supply and the demand sides. While developed nations have encouraged their citizens to travel abroad, developing ones lack the necessary infrastructure to receive mass organized tours. One of the main difficulties in the implementation of mitigation-related policies rests on the economic factor, the destination depends upon. Reducing the demand may very well threaten some local economies [21]. In this context, over-tourism still remains a slippery matter for policymakers and scholars of all pundits. Harold Goodwin argues convincingly that over-tourism resulted not only from the lack of effective planning but also from the ignorance of what some local stakeholders really need. In fact, the discomfort of locals with tourism can be explained by a combination of factors which can be summarized in two poles. At a closer look, tourists consume free resources they ultimately do not pay for. This happens, as Goodwin alludes, because the public realm is free in Westernized societies. Beyond the inaction of local governments, the second factor is associated with the urgency to create jobs. Although tourism (as well as its multiplying effects) favors the creation of jobs, they are low-paid or workers are subject to labor exploitation [22]. In addition, the economic growth achieved in Western nations is detrimental to the mass tourism in developing ones. Millions of foreign visitors coming from the US and Europe saturate destinations in developing economies. The grim future of the tourism industry has led some scholars to applaud the application of artificial intelligence (AI) and digital technologies to placate the negative effects of over-tourism [23, 24]. Smart tourism is a promising toolkit to resolve the pending questions (problems) generated by mass tourism globally. Technology not only instrumentalizes effective solutions to the dilemma posed by tourism but also accommodates tours according to more rational conditions of development [25].

2 Over Versus Smart Tourism

The concept of *smart tourism* refers to the integration of Information and Communication Technologies (ICT) to facilitate access to destinations through innovative digital tools, improving operational efficiency, enhancing competitiveness, and minimizing the negative environmental and social impacts of tourism. In recent years, scholars have highlighted the potential of smart tourism ecosystems to promote sustainable tourism practices by leveraging technologies such as ubiquitous computing, ambient intelligence, machine learning, social media, and context-aware recommender systems. These digital ecosystems offer significant promise to mitigate the negative over-tourism impacts and to a greater extent address mass tourism hazards in cities as well as simultaneously managing the flow of visitors in urban settings [26–28]. As indicated in Table 1, cities like Seoul, South Korea, and Las Vegas, USA, have implemented smart armbands that track visitor behavior in real time, allowing for itinerary adjustments to prevent overcrowding (IoT-based tracking), supporting SDG 11 (Sustainable Cities and Communities) and SDG 9 (Industry, Innovation and Infrastructure). Las Vegas also utilizes IoT initiatives such as the smart parks program to enhance safety and optimize law enforcement and city resources. Similarly, Amsterdam, Netherlands, employs smart mapping applications to manage crowds, while Barcelona, Spain, integrates embedded sensors in urban locations to monitor visitor density and air quality, showcasing the city's leadership in smart city solutions and contributing to the same SDGs.

Moreover, Manchester, UK, utilizes AI-controlled smart streetlights to guide tourists during peak times, and, in 2020, rolled out AI-managed smart traffic junctions to alleviate congestion. This initiative has contributed to a greener and more efficient urban environment. Cities like Venice, Italy, and Kyoto, Japan, use machine learning technologies to forecast tourist demand and predict peak visitation periods, enabling proactive measures to manage visitor flow and supporting SDG 11 and SDG 13. Additionally, Singapore and Dubrovnik, Croatia, apply crowd simulation models to optimize resource allocation and distribute visitors more evenly. Social media analytics in New York City, USA, and Bali, Indonesia, help identify emerging over-tourism hotspots, while Reykjavik, Iceland, and Bangkok, Thailand, use geotagging data to direct visitors to less crowded attractions, contributing to sustainable tourism management and equitable access in line with SDG 10 (Reduced Inequalities). Tokyo, Japan, and Zurich, Switzerland, offer context-aware recommender systems that suggest alternative, off-peak activities, while Sydney, Australia, and Cape Town, South Africa, issue contextual alerts to provide real-time alternatives to visitors during periods of overcrowding. These technologies promote more responsible and sustainable visitation patterns by recommending off-peak times or alternative attractions, supporting SDG 11 and SDG 9.

In Kyoto, Japan, and Copenhagen, Denmark, smart sensors monitor the environmental impact of tourism, such as air quality and noise levels, aligning with SDG 13 and SDG 15 (Life on Land) by minimizing harmful ecological effects. While Machu Picchu, Peru, and Yellowstone National Park, USA, address SDG 11, SDG 12 by deploying visitor counting systems to enforce capacity limits and prevent frugal overuse of fragile sites, London, UK, and Dubai, UAE, adopt digital twin technologies to simulate visitor flow and test management strategies, while Paris, France, and Berlin, Germany, use 3D models to visualize crowd densities and optimize site layouts. Blockchain applications in Venice, Italy, and Siem Reap, Cambodia, manage visitor quotas to ensure equitable access, while Singapore and Los Angeles, USA, implement smart contracts to improve the transparency of resource allocation. Immersive technologies, such as virtual reality tours in Rome, Italy, and Cairo, Egypt, divert visitors from congested locations, while Edinburgh, UK, and Seoul, South Korea, use augmented reality to enhance experiences at less visited sites. Hong Kong and Amsterdam use geo-fencing to issue real-time crowd alerts and encourage movement toward quieter locations, while Santorini, Greece, and Barcelona implement dynamic entry restrictions to regulate visitor numbers.

Cities like New York City, Kyoto, Florence, Italy, and Yellowstone have adopted smart ticketing systems and timed-entry reservation platforms to manage visitor quotas and ensure a more even distribution of tourism throughout the day. Singapore and Berlin use chatbots and virtual assistants to provide personalized itinerary suggestions, while Dubai and London ensure accessibility through multilingual

Table 1 Smart tourism technologies to address over-tourism in tourism cities

Smart tourism technologies	Examples	AI uses to address over-tourism hazards	Cities adopting the technology	Relevant UN SDGs
AI-induced devices	Smart arm bands monitoring visitor behavior	Real-time visitor behavior tracking and itinerary adjustments	Seoul, South Korea; Las Vegas, USA	SDG 11 SDG 9
Smart phones	Smart mapping apps for crowd control	Managing crowd movement and visitor dispersion	Amsterdam, Netherlands	SDG 11, SDG 9
Ambient intelligence technologies	Sensors embedded in urban tourism sites	Monitoring visitor density and environmental conditions	Barcelona, Spain	SDG 13, SDG 11
Smart streetlights	Adaptive lighting to guide tourists during peak times	Guiding tourists during peak periods	Manchester, UK	SDG 11, SDG 9
Machine learning technologies	Predictive analytics for visitor forecasting	Anticipating peak tourist times and suggesting alternative plans	Venice, Italy; Kyoto, Japan	SDG 11, SDG 13
	Crowd simulation models	Optimizing visitor dispersal and resource allocation	Singapore; Dubrovnik, Croatia	SDG 11, SDG 9, SDG 12
Social media	User-generated content analysis	Monitoring sentiment and detecting over-tourism hotspots	New York City, USA; Bali, Indonesia	SDG 11, SDG 16
	Geotagging data	Identifying overcrowded locations and promoting lesser known sites	Reykjavik, Iceland; Bangkok, Thailand	SDG 11, SDG 12
Context-aware recommender systems	Personalized trip planners	Recommending off-peak hours or alternative attractions	Tokyo, Japan; Zurich, Switzerland	SDG 11, SDG 9, SDG 12
	Contextual alerts	Warning tourists about overcrowding and providing real-time alternatives	Sydney, Australia; Cape Town, South Africa	SDG 11, SDG 13
Smart sensors and IoT	Environmental monitoring devices	Detecting ecological impacts of over-tourism (e.g., noise, littering)	Kyoto, Japan; Copenhagen, Denmark	SDG 13, SDG 15
	Visitor counting systems	Managing carrying capacity limits at attractions	Machu Picchu, Peru; Yellowstone National Park, USA	SDG 11, SDG 12

(continued)

Table 1 (continued)

Smart tourism technologies	Examples	AI uses to address over-tourism hazards	Cities adopting the technology	Relevant UN SDGs
Digital twin technology	Virtual replicas of destinations	Simulating visitor flow and testing management strategies	London, UK; Dubai, UAE	SDG 9, SDG 11
	Interactive 3D models of landmarks	Visualizing crowd density and optimizing site layouts	Paris, France; Berlin, Germany	SDG 9, SDG 11
Blockchain applications	Visitor tracking through digital tokens	Ensuring equitable access and managing site quotas	Venice, Italy; Siem Reap, Cambodia	SDG 10, SDG 16
	Smart contracts for tourism services	Enhancing transparency in resource allocation	Singapore; Los Angeles, USA	SDG 16
AR and VR technologies	Virtual reality tours	Diverting tourists to virtual experiences of crowded destinations	Rome, Italy; Cairo, Egypt	SDG 11, SDG 12
	Augmented reality guides	Enhancing off-the-beaten-path experiences	Edinburgh, UK; Seoul, South Korea	SDG 11, SDG 9
Geo-fencing and location services	Real-time crowd alerts via apps	Encouraging dispersion to less crowded areas	Hong Kong; Amsterdam, Netherlands	SDG 11, SDG 9
	Dynamic entry restrictions at sites	Automating visitor limits based on real-time data	Santorini, Greece; Barcelona, Spain	SDG 11, SDG 12
Digital reservation systems	Smart ticketing systems	Managing entry quotas and reducing on-site congestion	New York City, USA; Kyoto, Japan	SDG 11, SDG 12
	Timed-entry reservation platforms	Encouraging spread of visitation throughout the day	Florence, Italy; Yellowstone National Park, USA	SDG 11, SDG 12
Chatbots and virtual assistants	AI-powered tourist guides	Providing real-time alternative itineraries during peak periods	Singapore; Berlin, Germany	SDG 11, SDG 9
	Multilingual conversational agents	Enhancing accessibility for international tourists	Dubai, UAE; London, UK	SDG 10,

(continued)

Table 1 (continued)

Smart tourism technologies	Examples	AI uses to address over-tourism hazards	Cities adopting the technology	Relevant UN SDGs
				SDG 11
Big data analytics platforms	Destination management dashboards	Aggregating visitor data for informed decision-making	Paris, France; Sydney, Australia	SDG 9, SDG 11
	Predictive models for peak period forecasting	Identifying and mitigating potential crowding in advance	Dubrovnik, Croatia; Venice, Italy	SDG 11, SDG 13
Drones and aerial monitoring	Overhead crowd monitoring	Providing real-time data for visitor distribution and safety management	Rio de Janeiro, Brazil; Santorini, Greece	SDG 11, SDG 13
	Drones for ecological impact assessments	Reducing human disturbance in fragile environments	Galápagos Islands, Ecuador; Banff, Canada	SDG 15, SDG 13
Gamification applications	Reward-based apps encouraging sustainable behavior	Incentivizing visits to lesser known destinations	Reykjavik, Iceland; Barcelona, Spain	SDG 12, SDG 13
	Challenges for eco-friendly practices	Promoting responsible tourism habits	Sydney, Australia; Cape Town, South Africa	SDG 12, SDG 13
Multimodal mobility platforms	Integrated transport apps	Reducing traffic congestion by optimizing public and private transport	Tokyo, Japan; Amsterdam, Netherlands	SDG 11, SDG 9
	AI-based route planners	Suggesting alternative travel routes to avoid crowded areas	Singapore; Hong Kong	SDG 11, SDG 9

Source By Authors

agents. Big data platforms in Paris and Sydney aggregate visitor information to support informed decision-making, while predictive models in Dubrovnik and Venice help forecast peak periods and mitigate overcrowding. Drones in Rio de Janeiro, Brazil, and Santorini offer real-time crowd monitoring and environmental assessments in sensitive areas like the Galápagos Islands, Ecuador, and Banff, Canada. Additionally, Reykjavik and Barcelona employ gamification applications to incentivize visits to lesser known destinations, and Sydney and Cape Town encourage eco-friendly tourism behaviors. Tokyo and Amsterdam use multimodal mobility platforms to reduce congestion, while Singapore and Hong Kong employ AI-based route planners to avoid crowded areas, promoting sustainable mobility and reducing the environmental footprint of tourism.

Despite the benefits of smart tourism, scholars such as Seraphin et al. have cautioned on its potential limitations, coining the term *Trexit* (tourism and exit) to describe the strategy of some managers and official authorities for constraining the tourist demand. Trexit centers on restrictive measures oriented to balk at the mass tours. These measures include tourist infrastructure for a smaller niche with high-purchasing power, or simply the imposition of visitor quotas to deter mass tourism [20]. The concept of smart tourism, still evolving, was initially tied to smart cities, where operational data is analyzed through algorithms to optimize decision-making processes. In tourism, the focus has shifted to managing mass data to reorganize destinations and tourism flows, enhancing visitor experiences through technological innovation.

However, some scholars such as Ulrike Gretzel et al. remind us that beyond the popularity of smart tourism, scholars must recognize the consequences of ICT applied to mass data compilations are far from being predictable in our world. Of course, smart tourism still is an ill-defined concept—even today. Smart tourism rests on the business logic waking up new global (unknown) trends that need to be discussed. The term was originally associated with smart city conceptualizing smart as exploiting operational data, generated through complex algorithms with the end of optimizing the operational decision-making process. In the constellations of tourism, the concept is inscribed to its capacity to handle mass data to reorganize the destinations and the tourist fluxes. The axiom assumes that tech opens the doors to more competence and innovation enriching the tourist experience. Most plausibly, smart tourism is changing the morphological borders of tourism research passing from classic to e-tourism (virtual tourism), a new paradoxical tendency where tourism does not involve physical displacement. The concept is doubtlessly a starting point for a transformational momentum (stage) where the classic definitions of tourism are withered away [29]. In this vein, Andres Coca-Stefaniak holds that over- and smart tourism are interlinked as two different sides of the same coin, urban tourism. To some extent, over-tourism is based on a negative sentiment against visitors. This sentiment is molded by the failure of branding strategies to give solutions to local urban morphologies. Palpable or not, the negative effects of tourism on the local morphology are certainly lived as a threat to the autonomy of the city. This sentiment varies on time and culture but can be rechanneled to a tolerable condition with tech-based solutions. Innovative thinking has a potential role in the configuration of smart destinations [30]. In a seminal book, which is entitled *Over Tourism, Technology Solutions, and Decimated Destination,* Azizul Hassan & Anukrati Sharma offer an interesting diagnosis of the future of the tourism industry through tech solutions. The authors see technological solutions help policymakers to revise the current strategies toward more resilient destinations. Technology not only facilitates faster and more efficient decision-making processes but also alleviates the carrying capacity in saturated destinations—by the re-distributions of mass tours in the territory [28]. The concept is often associated with smart cities (Pasquinelli and Trunfio 2020). In this respect, Ko Koens et al. call attention to the importance of urban planning to cope with the negative effects of over-tourism and city tourism. As the authors eloquently note, tourism has historically contributed to the economic growth of urban and rural destinations, but recently, its effects have been placed into the tapestry. For some reason, tourism is affecting the local quality of life negatively. This has been aggravated by the changes in demography in developed economies as well as the change in travel behavior cultivated by new digital platforms such as couch surfing and Airbnb. Local authorities of established destinations such as Barcelona, Venice, and Amsterdam are questioning mass tourism and the experts´ current management strategies. The authors go on to write: "*Engagement between tourism and the multiplicity of public and commercial organizations with varying levels of involvement with tourism in urban areas" is required to help develop a coherent vision of tourism that relates to and integrates with wider city development.... This means re-conceptualising the role of tourism within urban societies, rather than limiting the discussion to an unbalanced development of tourism and the way tourism policy creates "tourist bubbles*" [28]. Smart tourism should be circumscribed in the fields of collaborative planning and community-based destination governance. This means that digital technologies should be regulated to respect (or meet) the interests of all involved stakeholders. As the authors alert, since stakeholders optimize their interests while preferring effect-oriented solutions over collective resilience, technology should accompany society in changing its current cultural values. Technology should be used to implement radical transformations toward a more sustainable society. Of course, smart tourism has exegetes and detractors. In the next section, we shall review the argument of detractors who consider smart tourism—far from solving—aggravates the problem of over-tourism.

3 Detractors of Smart Tourism

Critical studies or detractors of smart tourism come in part of critical sociology. Unlike other disciplines, sociology has developed a critical position revolving around the role played by technology and smart tourism to reverse the environmental degradation produced by the tourism industry [31, 32]. Rodanthi Tzanelli gives a critical snapshot that expands the current understanding of the environmental crisis. The recent COVID-19 pandemic has covered a set of unaccepted risks such as human trafficking, inter-class inequalities, and an unparalleled ecological crisis. Since the capitalist machine never stops, the world has entered a new grand mobility characterized by the presence (hegemony) of viruses on the earth. The virocene has come to stay displacing the Anthropocene to a peripheral situation. To wit, the virocene enacts not only knowledge monopolization through the medical discourse but also places a whole portion of humankind in a state of immobility. As micro-organisms, viruses escape to the nation-state control affecting travel behavior worldwide. The circulation of COVID-19 engages with long-dormant racist discourses ideologically forged in the colonial period. The role played by bio-technology gravitates in two different dynamics. On the one hand, it domesticates consumers while legitimating a sense of immobility. The question is whether the pre-COVID-19 world was marked by a class structure where only a privileged elite was encouraged to move, in the new normal, higher classes stayed at home. This sense of immobility hotly ignites a sentiment of recrimination directed against foreigners and local authorities. As Tzanelli notes, far from reverting mass tourism, COVID-19 calls for the application of new technological dispositifs designed to keep consumers safe at home. Technology is today associated with a bio-capital culture oriented to domesticate the body. Tzanelli sheds light on the quandaries of technification in a new world where the "Other" is simply negated or subordinated to technophilia [8]. Other studies have objected to some privacy concerns regarding smart tourism and big data management. The ramifications of smart tourism may very well engender some unseen risks associated with the misuse of data or the consumers' unfamiliarity with future data management practices. The consumers´ privacy, as well as their personal information, can be manipulated by AI or big data management shortly [33–35]. What is more important? Produced knowledge or information can be mechanized undermining the critical thinking of consumers. Whenever consumers have incorporated a strong dependence on smart technology, their critical reasoning as well as autonomy applied in the decision process declines. Under some circumstances, ICT and smart tourism can mold stereotyped forms of consumption that lead to visual saturation [36–38].

In a trailblazing paper, Stephane Bourliataux-Lajoinie, Frederic Dosquet & Josep del Olmo Arriaga hold that there is an (under-evaluated) dark side in technology when it is applied to resolve over-tourism. The authors define over-tourism as a negative feeling—directed against tourism—because of the overload in a place. For some voices, over-tourism is linked but not limited to urban tourism. This means that some cases of over tourism are found in rural areas. In time, these sentiments evolved from resentment (or apathy) to explicit demonstrations or bans such as *tourists you are the real terrorists! And the well-known slogan Tourists go Home!* These expressions exhibit a manifest refusal to offer hospitality to foreign tourists, as well as selling other products. The authors exert a radical criticism of smart tourism—and the adoption of smart technology—to reverse the devastating consequences of mass tourism. Per their stance, technology does not suffice to changing the current visitor travel behavior. Smart tourism depends on deep analysis of tourism consumption and its patterns—through big data optimization. As they eloquently adhere, Barcelona is the best example that explains why smart tourism does not work. Although new technologies have an important impact on our daily lives, even in the evolution of tourist destinations, no less true seems to be that digital media recreates a condition of engagement between the tourist demand and the site, which invariably leads to over-tourism. Digital tech recreates an emotional strong line between the visitor and the place, which is unsustainable at the time. As they overtly assert: "*What began as an opportunity, sharing experiences on the web is now a problem in the fight against over-tourism" (p 588).* In this line of inquiry, the excess of technology insufflates the demand avoiding the costs for the affected destination. The specialized literature has lamented how the rise of digital applications (i.e., E-commerce, data economy, and tour blogs) escapes to the government´s regulations. In fact, smart tourism is mainly marked by a clear dissociation in what experts named *the hyper-reality of consumption,* which is recreated when the consumers believe they know the product before going there. The hyper-reality of consumption offers a fanaticized image of the site appealing to the emotional drives of each consumer. Over-tourism should be understood as a standardized source to meet the mechanized expectancies of a global (fabricated) demand. Smart tourism seems not to be part of the solution but from the problem, as the authors conclude [39]. Last but not the least, Rodolfo

Baggio, Roberto Micera and Giacomo del Chiappa examine some of the contradictions of smart tourism regarding its interaction with the physical world. Their study is designed to scrutinize to what extent the digital and physical components coexist while achieving higher levels of efficiency in the tourist system. With bases on three study cases (Elba Island, Gallura, and Livigno), the obtained results show that *smart tourism* has higher levels of efficiency only when shared information and knowledge flows are efficient in the physical scenario too. This invariably means that the digital component (as well as AI) seems not to be sufficient to embrace a smart destination successfully [40].

3.1 A Mix-Balanced Approach is Possible?

Critical anecdotes or reviews on technology in modern cinema abound. Films and series like Minority Report, Terminator, Matrix, or Battlestar Galactica speak to us of a dystopian world controlled by AI and machines. In this grim world, the autonomy of humans is determined by the imposition of impersonal ties among humans. Robots have been adapted to satisfy humans´ needs even in the tourism and hospitality industries. The introduction of AI and robots in tourism woke up a hot debate that is open to date. All this leads us to question to what extent we must trust in technology [41–43].

Some points alert on the fact that human reasoning cannot be replaced by AI. Although AI keeps more capacity of analysis on the basis of the data generation, AI has some failures to discern because it lacks critical reasoning. The opposite is true, humans have critical reasoning—with the autonomy of imagine things that not happen in reality, but are particularly subject to emotional distortions that may very well affect the final decision-making process [41]. Most certainly, there are underlying sentiments of hostility from humans directed to robots and AI. This rivalry is mainly marked by the lack of trust in robots or fear of losing the jobs. The literature speaks readers of the fact of reducing the workers' conflict but says nothing how these troubles should be efficiently managed [44]. Last but not the least, a mix-balanced approach in the application of digital technologies should be certainly subject to ethical regulation. Having said this, smart tourism reconnects with the possibility to improve data analysis to reach better decisions or at the least the most feasible ones. Beyond the used technique and its dichotomies, which are employed to get visitor-oriented products, no less true seems to be that digital tech as well as AI should be scrutinized by stronger ethical codes that prioritize the human capital from other economic-centered criteria [45].

4 Conclusion

International destinations are coping today with the problem of over-tourism. Specialists of all pundits have alerted people about the complexity of over-tourism globally. Particularly, over-tourism derives from a combination of factors but is intertwined with the rise of faster transport means, improving the current working conditions and leisure time in developing economies, and the urgency of developing nations to adopt tourism as their main economic activity. In many already classic tourist destinations, a new anti-tourist sentiment sparks. As we have debated, this sentiment of discomfort rests on a paradoxical situation. Locals appeal to international tourism because they amassed profits but they want to avoid their direct effects. To resolve this, some voices have emphasized the incorporation of AI and digital tech to organize the visits in the territory. In an ever-changing world, smart tourism offers an innovative solution to the problems of over-tourism. Needless to say, some detractors have cautioned on the risks of smart tourism. Based on a mix-balance discourse, the present chapter gives readers a conceptual snapshot of the strengths and weaknesses resulting from smart tourism. While tech solutions appear to be significant progress in resolving some open points in the crisis the industry is facing today, it may generate dependency in some circles. Technology, unless regulated by humans, leads normally to standardized models that accelerate the destination decline.

References

1. Boissevain J (Ed) (1996) Coping with tourists: European reactions to mass tourism (Vol 1). Berghahn Books, Oxford
2. Koh E (2020) The end of over-tourism? opportunities in a post-Covid-19 world. Int J Tour Cities 6(4):1015–1023
3. Dodds R, Butler R (2019) The phenomena of overtourism: a review. Int J Tour Cities 5(4):519–528
4. Becker E (2016) Overbooked: the exploding business of travel and tourism. Simon and Schuster, New York
5. Koh E, Fakfare P (2020) Overcoming, "over-tourism": the closure of Maya Bay. Int J Tour Cities 6(2):279–296
6. González-Reverté F (2022) The perception of overtourism in urban destinations. Empirical evidence based on residents' emotional response. Tour Plan & Dev 19(5):451–477
7. Gürsoy İT (2021) Beauty and the beast: a fairy tale of tourismphobia. In: Travel and tourism in the age of overtourism, pp 82–99. Routledge, Abingdon
8. Tzanelli R (2021) Cultural (im) mobilities and the Virocene: Mutating the crisis. Edward Elgar Publishing, Cheltenham
9. Pulido-Fernández JI, Cárdenas-García PJ, Espinosa-Pulido JA (2019) Does environmental sustainability contribute to tourism growth? An analysis at the country level. J Clean Prod 213(1):309–319
10. Simo-Kengne BD (2022) Tourism growth and environmental sustainability: trade-off or convergence? Environ Dev Sustain 24(6):8115–8144
11. Cheer JM, Milano C, Novelli M (2019) Tourism and community resilience in the Anthropocene: accentuating temporal overtourism. J Sustain Tour 27(4):554–572

12. Seraphin H, Sheeran P, Pilato M (2018) Over-tourism and the fall of Venice as a destination. J Destin Mark Manag 9:374–376
13. Maingi SW (2019) Sustainable tourism certification, local governance and management in dealing with overtourism in East Africa. Worldw Hosp Tour Themes 11(5):532–551. https://doi.org/10.1108/WHATT-06-2019-0034
14. Korstanje ME, George BP (2020) Education as a strategy to tackle over tourism for overtourism and inclusive sustainability in the twenty-first century. In: Seraphin H, Gladkikh, Thanh V (eds) Overtourism: Causes, implications and solutions, 341–359. Springer Nature. Cham
15. Milano C, Cheer JM, Novelli M (eds) Overtourism: excesses, discontents and measures in travel and tourism. Cabi, Wallingford (2019)
16. Koens K, Postma A, Papp B (2018) Is overtourism overused? Understanding the impact of tourism in a city context. Sustainability 10(12):4384–4395
17. Blázquez-Salom M, Cladera M, Sard M (2023) Identifying the sustainability indicators of overtourism and undertourism in Majorca. J Sustain Tour 31(7):1694–1718
18. Veríssimo M, Moraes M, Breda Z, Guizi A, Costa C Overtourism and tourismphobia: a systematic literature review. Tour: Int Interdiscip J 68(2):156–169
19. Korstanje ME (2011) Influence of history in the encounter of guests and hosts. Anatolia 22(2):282–285
20. Seraphin H, Gowreesunkar V, Zaman M, Lorey T (2019) Limitations of Trexit (tourism exit) as a solution to overtourism. Worldw Hosp Tour Themes 11(5):566–581
21. Dodds R, Butler R (eds) (2019) Overtourism: Issues, realities and solutions (Vol 1). Walter de Gruyter GmbH & Co KG. Berlin
22. Goodwin H (2021) City destinations, overtourism and governance. Int J Tour Cities 7(4):916–921
23. İştin AE, Turpcu E (2021) Technology application to manage overtourism in Turkey. Overtourism, Technol Solut Decimated Destin 275–292
24. Mohanty P, Nair N, Sharma AK (2022) Overcoming overtourism through technology: the case of asian cities. In: Technology application in tourism in Asia: innovations, theories and practices, pp 395–405. Singapore: Springer Nature Singapore, Singapore
25. Gretzel U (2021) Technological solutions to overtourism: potential and limits. In: Mandic A, Petric L (eds) Mediterranean protected areas in the era of overtourism: challenges and solutions, 337–349. Cham, Springer Nature. Cham
26. Li Y, Hu C, Huang C, Duan L (2017) The concept of smart tourism in the context of tourism information services. Tour Manage 58:293–300
27. Zhang Y (2021) Tourist experience and influencing factors in the context of smart tourism. In: Proceedings of the 2021 4th international conference on e-business, information management and computer science, pp 409–417. Cham, Springer, Cham
28. Hassan A, Sharma A (eds) Overtourism, technology solutions and decimated destinations. Springer, Cham
29. Gretzel U, Sigala M, Xiang Z, Koo C (2015) Smart tourism: foundations and developments. Electron Mark 25:179–188
30. Coca-Stefaniak JA (2019) Marketing smart tourism cities–a strategic dilemma. Int J Tour Cities 5(4):513–518
31. Koens K, Melissen F, Mayer I, Aall C (2021) The smart city hospitality framework: creating a foundation for collaborative reflections on overtourism that support destination design. J Destin Mark Manag 19:100376
32. Tham A, Abidin HZ (2023) Getting smart? a research note into smart tourism curriculum and implications on generation alpha and beta. J Smart Tour 3(1):35–39
33. Gong Y, Schroeder A (2022) A systematic literature review of data privacy and security research on smart tourism. Tour Manag Perspect 44:101019
34. Tiwari V, Mishra A, Tiwari S (2023) Role of data safety and perceived privacy for acceptance of IoT-enabled technologies at smart tourism destinations. Curr Issues Tour, 1–16
35. Femenia-Serra F, Ioannou A, Tussyadiah IP (2022) Is smart scary? a mixed-methods study on privacy in smart tourism. Curr Issue Tour 25(14):2212–2238

36. Ritzer G, Liska A (2002) 'McDysneyization and Post Tourism: Complementary perspectives of contemporary tourism In: Rojek C, Urry J (eds) Touring cultures, pp 106–120. Abingdon, Routledge
37. Tribe J, Mkono M (2017) Not such smart tourism? the concept of e-lienation. Ann Tour Res 66:105–115
38. Weaver DB, Moyle BD (2019) 'Tourist stupidity'as a basic characteristic of 'smart tourism': challenges for destination planning and management. Tour Recreat Res 44(3):387–391
39. Bourliataux-Lajoinie S, Dosquet F, del Olmo Arriaga JL (2019) The dark side of digital technology to overtourism: the case of Barcelona. Worldw Hosp Tour Themes 11(5):582–593
40. Baggio R, Micera R, Del Chiappa G (2020) Smart tourism destinations: a critical reflection. J Hosp Tour Technol 11(3):407–423
41. Ivanov S, Webster C (2020) Robots in tourism: a research agenda for tourism economics. Tour Econ 26(7):1065–1085
42. Korstanje ME (2024) AI and robots in science fiction movies: why should we trust in AI? In: Scribano A, Korstanje M (ed) AI and emotions in digital society, pp 141–153. IGI Global, Hershey
43. Sunstein CR (2002) Risk and reason: safety, law, and the environment. Cambridge University Press, Cambridge
44. Ivanov S, Webster C, Berezina K (2022) Robotics in tourism and hospitality. Handb e-Tour 1873–1899
45. Lukita C, Pangilinan GA, Chakim MHR, Saputra DB (2023) Examining the impact of artificial intelligence and internet of things on smart tourism destinations: a comprehensive study. Aptisi Trans Technopreneurship (ATT) 5(2sp):135–145
46. Korstanje ME (2024) Sustainable urban tourism and smart destinations in South America. In: Maxim C, Morrisson A, Day J, Coca-Stefaniak JA (eds) Handbook on sustainable urban tourism, pp 484–495. Cheltenham, Edward Elgar Publishing. Cheltenham

Use of Artificial Intelligence in Controlling Overtourism: A Case of Nainital City, Uttarakhand

Vanshika Bajaj, Zeeshan Ibrar, Sobia Ahsan, and Nisar Khan

Abstract Tourism has become a significant but problematic industry, providing socio-economic benefits while also leading to environmental degradation and infrastructural strain. In the Himalayan region, Nainital, a popular lake town in Uttarakhand, exemplifies the challenges of overtourism, with 3,000–4,000 tourist vehicles entering daily during peak seasons. This influx surpasses the town's carrying capacity and puts immense pressure on city's infrastructure. Current management strategies, such as vehicle barricades and "Nainital Housefull" boards, have proven inadequate. To address these challenges, this research paper proposes the integration of Artificial Intelligence (AI) and smart design strategies for Sustainable Tourism Management. It aims to identify the causes, impacts, and indicators of overtourism in Nainital while analyzing international best practices using AI. The methodology of the study includes primary survey that was carried out with local residents and Key Informant Interviews to assess social, environmental, and economic impacts. The research proposes an AI-based monitoring framework featuring solutions that includes sensor-based crowd control, Tourist Cards, AI-enabled parking systems, and mobile applications. This approach aims to better manage tourist flow and mitigate the adverse effects of overtourism, laying the groundwork for future research in the area.

Keywords Artificial intelligence (AI) · Overtourism · Overcrowding · Artificial intelligence (AI) · Technology · Congestion · Nainital

V. Bajaj · S. Ahsan · N. Khan
Department of Architecture, Jamia Millia Islamia, New Delhi, India

Z. Ibrar (✉)
Department of Architecture and Planning, Netaji Subhas University of Technology (NSUT), New Delhi, India
e-mail: zeeshan@nsut.ac.in

V. C. Sharma et al. (eds.), *Architectural Vision of Overtourism*,
SpringerBriefs in Architectural Design and Technology,
https://doi.org/10.1007/978-981-96-5794-0_8

1 Introduction

Tourism, also known as the "sunrise industry," has gradually developed into one of the world's largest and most influential industries. The rapid development of tourism has impacted the environment while providing economic benefits [1]. The overexploitation of tourism resources has caused problems such as ecological damage, extensive mismanagement of tourism spots, and lack of idle tourism infrastructure [2]. The World Tourism Organization [3, 4] defines overtourism as "the impact of tourism on a destination, or parts thereof, that excessively influences the perceived quality of life of citizens and quality of visitor experiences negatively." Overtourism leads to increased living costs for residents and businesses, congestion, and damage to the natural or built environment [5]. One such case is Nainital, which is a lake town in the Indian Himalayan zone situated in Uttarakhand that is facing the challenge of overtourism.

Nainital is the headquarters of the district in Uttarakhand, India (Fig. 1). Nainital is situated at an elevation of 2,084 m and covers an area of 11.73 km square, with a population of 41,377. Nainital is derived from the Naini Temple, located at the northern end of Naini Lake. The city expanded around the main lake and is known as the "lake district of India." When Uttarakhand was carved out from Uttar Pradesh in 2000, Nainital became one of Northern India's most popular and easily accessible tourist destinations. Tourism has also become an essential source of employment, increasing the region's economic potential [6].

The Uttarakhand State Government created and promoted several programs and policies to boost tourism in the state, with a particular emphasis on Nainital. As a result, the city has become a popular destination for adventure, wildlife, and leisure travelers [7]. With a strong cultural identity and ideal climatic conditions, Nainital has drawn many tourists over the last two decades, resulting in overcrowding. Overtourism is evident in Nainital, where tourist inflows outnumber residents regardless of shared or private locations, causing concern and worry among locals. About 3,000–4,000 tourist vehicles enter the town every day during the peak season, and the numbers go further up during the weekend when tourists from neighboring states like UP and Delhi visit the town [8].

Overtourism in Nainital leads to problems such as waste disposal, traffic congestion, lack of parking spaces, and increased burden on the water supply. These challenges align with global concerns outlined in the United Nations Sustainable Development Goals (SDGs). SDG 11 (Sustainable Cities and Communities) emphasizes the need to make cities inclusive, safe, resilient, and sustainable, which includes addressing overcrowding and ensuring equitable access to resources. SDG 13 (Climate Action) and SDG 15 (Life on Land) underline the importance of protecting ecosystems and combating environmental degradation caused by unsustainable practices like overtourism. Additionally, SDG 8 (Decent Work and Economic Growth) and SDG 12 (Responsible Consumption and Production) call for fostering economic growth while ensuring sustainable resource use and minimizing negative environmental impacts.

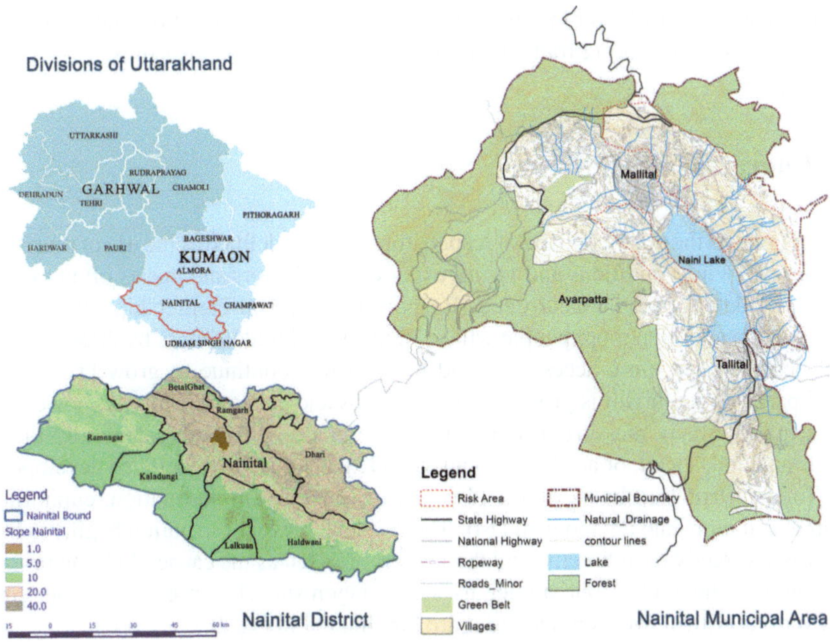

Fig. 1 Nainital—study area. *Source* By Authors

This research focuses on developing AI-based measures to address overtourism in Nainital. AI measures refers to the artificial intelligence-based actions that align with these SDGs to achieve a balance between economic development, environmental sustainability, and quality of life for residents. Artificial Intelligence (AI) offers promising solutions for managing overtourism by analyzing visitor behavior, optimizing resource allocation, and mitigating negative impacts on the environment and local communities [9]. AI-supported measures such as real-time crowd control, smart parking systems, georeferenced databases, and environmental monitoring have been successfully implemented in cities like Amsterdam, London, Venice, and Switzerland [10].

The AI-supported measures have been successfully implemented and have reduced overtourism in these cities. The issue of overtourism in India has impacted the environment and the residents. However, the use of AI has not been explored to tackle the issue in the Indian context [11]. Hence, there is a need to use AI-based solutions to control overtourism in Nainital as it could create an adverse impact and pose a threat to the ecology. The research paper aims to address the problem of overtourism in Nainital City by incorporating Artificial Intelligence (AI)-based measures for Sustainable Tourism Management in line with smart city proposals. The research objectives are to identify the onset, causes, and impacts of overtourism in Nainital based on indicators of overtourism and to analyze the use of AI and technological solutions applied to overtourism in an international context. The paper also assesses the impacts of overtourism in terms of overcrowding and congestion and develops an overtourism monitoring framework incorporating AI-based measures.

2 Literature Review

Overtourism is the polar opposite of responsible tourism, defined as the use of tourism to improve living conditions and visitor experiences. Visitors and guests often experience the deterioration concurrently (Responsible Tourism Partnership). It is estimated that one billion more people will be in the global middle class by 2030, travel will become even more accessible, and tourism will continue to grow [12]. The good part is that tourism is growing, creating jobs and economic activity; however, this expansion may pressure destinations. More places will be threatened by their environmental, social, or aesthetic popularity [13]. George Doxey (1975) developed the Tourism Irritation Index on the four phases of a tourism destination: euphoria, apathy, irritation, and antagonism [14]. The residents' level of irritation begins to be expressed directly or indirectly, and the visitors are seen as the cause of all the problems around them. Overtourism begins with antagonism. The onset of overtourism and various critical events related to it are highlighted in Fig. 2.

In 2016, Skift coined the term overtourism [15]. In 2017, oppression started in Barcelona, Venice, and Berlin. In 2018, the term overtourism was recognized by

Fig. 2 Addressing overtourism through years. *Source* By Authors

the UNWTO; in 2019, London was the first city to use AI-based solutions for overtourism. As Bhakuni [8] mentioned, around two billion tourists arrive annually globally, growing at a steady rate of 6 and 7%. Overtourism does not always deal with numbers; even a small number of people at an ecologically fragile location or a rural location beyond its carrying capacity may lead to more significant complexities than scores of additional people landing up in a city [16]. A systematic literature review has been conducted to identify the parameters and indicators for the onset, causes, and impacts of overtourism, solutions to control overtourism, and applications of AI to control overtourism in the international context. The research areas included in the literature review are AI application for crowd control, AI in smart tourism, the impact of overtourism, the need for and application of AI to control overtourism, the mobility of tourists using AI, and policy regulations (Fig. 3).

Figure 3 depicts the literature review on the impact of overtourism and the need for and application of AI to control overtourism. To analyze the impact of overtourism, various researchers have applied qualitative and quantitative methods. The qualitative

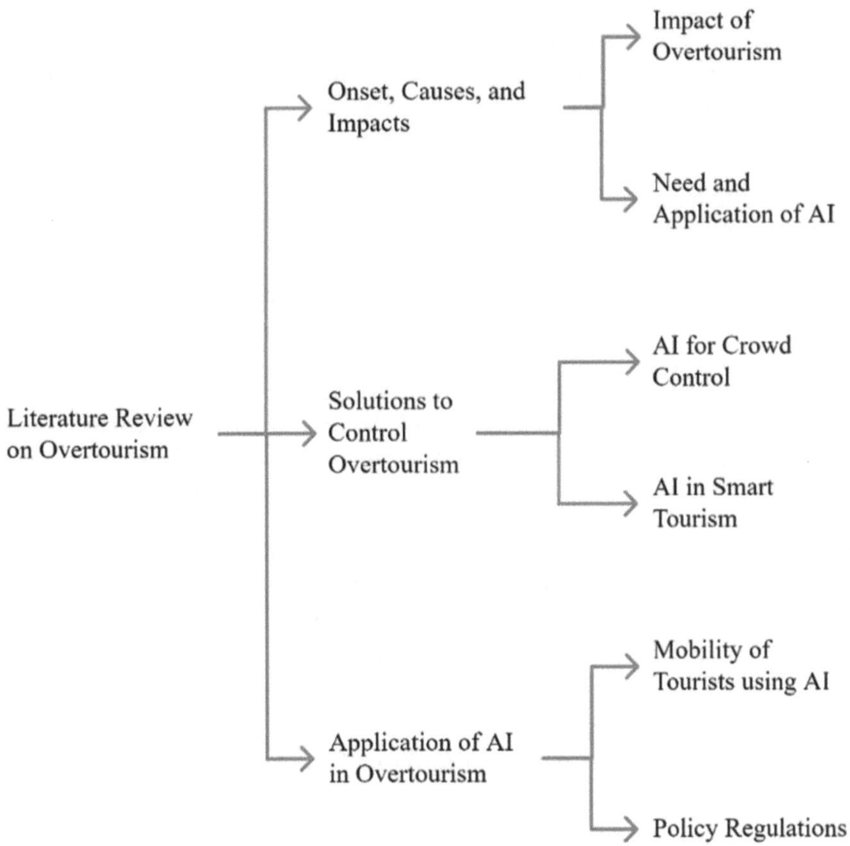

Fig. 3 Research areas in literature. *Source* By Authors

Table 1 Indicators of overtourism

Indicators	Definition	Description
Tourism density	Bed-nights/km2	Annual number of bed-nights per km2
Tourism intensity	Bed-nights/residents	Annual number of bed-nights per resident in the destination
Sharing economy: Airbnb	Number of Airbnb offers	Number of Airbnb offers in a destination
Air transport intensity	Air passengers/Bed-nights	Ratio of number of air passengers to the number of bed-nights
Closeness to airports		Arrival within 50 km
Closeness to cruise ports		Number within 10 km
Closeness to world heritage sites		Number within 30 km

Source Pérez-Garrido [18]

methods include primary variables such as Tourism Density, Tourism Infrastructure, Geographical Setting, and Affordability. The Quantitative Method includes Scientific Models, AI Algorithms, Big Data Analytics, and IOT Tools [17].

In the Qualitative Analysis, Pérez and Flores [16] have analyzed the impact of overtourism based on indicators such as Tourism Density, Tourism Intensity, Number of Airbnb's, Distance from Airports, Share of Tourism in the economy, and Number of destinations in proximity as mentioned in Table 1. Weber [17] has analyzed the impact of overtourism based on indicators viz. number of restaurants, number of cars entering the city, revenue of toll tax, number of tourist destination within a 5 km radius, closeness to other tourist destinations, number of travel agencies, and Air Quality Index (AQI).

The problem of overtourism requires a systematic approach to have sustainable solutions without affecting the economy of the local people. Hence, researchers have worked to formulate strategies based on frameworks to regulate policies. The framework provided by Weber [17] includes three stages, i.e., identification of drivers and impacts and then providing solutions based on the type of driver and impact created by overtourism (Fig. 4). The overtourism framework described above in Fig. 4 includes Drivers, further divided into Supply and Demand areas. The impact stages are identified based on the economic, environmental, and social impact parameters. The solutions are provided to the previous stages.

2.1 *Drivers, Impacts, and Solutions of Overtourism*

The drivers and factors responsible for overtourism are increased affordability, better connectivity, urbanization, bucket list tourism, and the influence of social media and

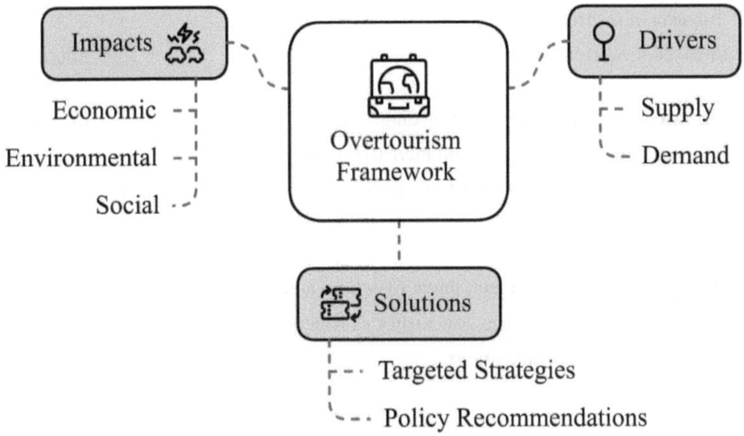

Fig. 4 Framework of overtourism. [17] *Source* Weber

the government. Schemes, marketing strategies, branding, lack of tourist infrastructure and bad governance, and honeypots (i.e., successfully marketed, established destinations) attract more tourists, as well as high dependency and seasonality in tourism. Table 2 describes various factors and drivers responsible for overtourism. High dependency and seasonality in tourism and increased affordability are the major factors contributing to overtourism, as authors have mentioned these drivers.

The impacts of overtourism are increased congestion and overcrowding, infrastructure under pressure and increased energy demand, degradation of quality of life of residents, gentrification, privatization and migration, environmental deterioration, damage to built environment, threat to culture and heritage, degraded tourist experience, low visitor satisfaction, and loss of identity and authenticity. Table 3 describes the impact of overtourism, as mentioned by various authors. Increased congestion and overcrowding, as well as loss of identity and infrastructure under pressure, have been identified as the main impacts of overtourism.

Table 2 Drivers/factors of overtourism

S. no.	Drivers and factors	Source
1	Increased affordability	[2, 7, 19, 20]
2	Better connectivity and urbanization	[17, 20, 21]
3	Bucket list tourism and influence of social media	[17, 19, 20]
4	Govt. schemes, strategies, marketing, and branding	[19, 20]
5	Lack of tourist infrastructure and bad governance	[7, 17, 20, 21]
6	Honeypots (i.e., successfully marketed, established destinations) attract more tourist	[5, 7, 12, 17, 22]
7	High dependency and seasonality in tourism	[5, 17]

Source By Authors

Table 3 Impacts of overtourism

S. no.	Impacts of overtourism	Source
1	Increased congestion and overcrowding	[5, 7, 12, 17]
2	Infrastructure under pressure and increased energy demand	[17, 19, 21]
3	Degradation of quality of life of residents, gentrification, privatization, and migration	[17, 19, 20, 23]
4	Environmental deterioration	[19, 20]
5	Damage to built environment, threat to culture and heritage	[7, 17, 20, 21]
6	Degraded tourist experience, low visitor satisfaction	[6, 7, 12, 19]
7	Loss of identity and authenticity	[2, 7, 19]

Source By Authors

The solutions provided by various authors for overtourism are Implementation of Taxes, Caps and Limitations, Strategy Formulation, City Planning and Zoning, Smart Marketing and Branding, On-ground Visitor Management, Technological Solutions, Community-Based Sustainable Tourism, Review and Adapt Regulations, Creating Itineraries, Spreading Visitors, Review and Adapt Regulations, Spreading Visitors, and Public Awareness Programs using Social Media. Table 4 provides a compiled list of solutions and strategies suggested by various authors to control overtourism.

Technological solutions, implementation of taxes, caps, and limitations, strategy formulation, city planning and zoning, innovative marketing, and branding are the most widely mentioned by authors.

Table 4 Solutions to control overtourism

S. no.	Solutions/Strategies	Authors
1	Taxes, caps, and limitations	[6, 7, 12, 17]
2	Strategy formulation, city planning and zoning, smart marketing and branding	[17, 20, 21]
3	On-ground visitor management	[2, 7, 19]
4	Technological solutions	[17, 20, 24]
5	Community-based sustainable tourism	[7, 17, 20, 21]
6	Review and adapt regulations, creating itineraries, spreading visitors	[19, 20]
7	Public awareness programs using social media	[6, 7, 12, 19, 23]

Source By Authors

2.2 *AI-Based Technological Solutions to Control Overtourism*

Advanced AI-based technological solutions are increasingly deployed to manage overtourism, with applications spanning from basic sensors to sophisticated Virtual Reality (VR) platforms and smart city infrastructure. Cities like London, Barcelona, Helsinki, Paris, Switzerland, and Venice have notably implemented these measures. For instance, sensors (cameras or other devices) are used to detect congestion, providing real-time information to suggest alternative routes or attractions, as seen in Dubrovnik. Amsterdam leverages Virtual Reality videos on its "I Amsterdam" website to highlight attractions and disperse visitors. Similarly, mobile applications, such as "Play London with Mr. Bean," offer new experiences while open-source crowd-controlling systems like Barcelona's "Public Eye" actively manage dense areas.

To achieve temporary dispersion, Smart Tourist cards in destinations like Florence allow visitors to bypass long queues at popular sites. Mobile applications promoting alternative attractions, as in Prague, encourage tourists to explore lesser known areas. Demos marketing efforts are also facilitated by technology; Amsterdam utilizes georeferencing techniques to send targeted messages to tourists in congested areas via digital platforms. Helsinki's mobile applications foster interaction with locals and promote responsible tourist behavior. Furthermore, these technologies optimize urban infrastructure. Real-time information on public transport networks aids in efficient travel, seen in Barcelona's transport management. Amsterdam employs Veovo Blip Track Guest Predictability, using sensors and Big Data to anticipate crowd movements. Smart parking systems, traffic light sensors in Madrid, and smart building technologies for public facilities like museums are also prevalent. Environmental monitoring through sensors, such as those used in Japan and Maryland, US, leverages AI for real-time big data analytics to assess air and water quality.

For follow-up and response to tourism evolution, monitoring tourist flows via mobile device localization, Wi-Fi networks, tourist cards, social media, and credit card payment distribution provides valuable data. Venice and Lisbon, in collaboration with telecommunication companies, use big data analytics to measure tourism pressure and inform strategic decisions. While India has adopted technological solutions for increasing tourist footfall and convenience (e.g., advance booking systems, QR codes at historical sites like Qutub Minar, VR at Statue of Unity, chat boxes, and virtual guides), its direct overtourism control measures are more manual, including physical barricades, warning advisories, tourist taxes, and increased parking charges, in addition to promoting alternate destinations and regulating accommodation.

2.3 Overtourism in Nainital

In the case of Nainital the population has increased to 60,000 in 2024 from 41,000 in 2011 while the floating population is presently 1,40,000 which is more than twice of the city's resident population (Census 2011). According to a 2001 survey by RITES, the projected volume of vehicular traffic for 2011 and 2021, only for Mall Road, was 20,900 and 26,000, respectively. Similarly, based on tourist arrival growth, the projected pedestrian traffic on Mall Road for 2011 and 2021 was estimated at 30,955 and 35,360, respectively. RITES study indicates the Parking demand in peak season for 2011 and 2021 is estimated at 1500 and 1925, respectively [8].

Increased affordability, better connectivity, various activities, and the gateway to the Kumaon region cause overtourism in Nainital. The lack of parking spots due to rugged terrain, seasonality, and reliance on tourism are the reasons that create issues in the city. The impacts are traffic congestion, overcrowding, stress on basic services such as water, degradation of Naini Lake, low AQI, and increased waste generation. The solutions that the authors have proposed are the development of new parking lots in the foothills and the provision of shuttle bus service, tourist tax, optimization of tourist activities, price differentiation, community-based tourism, and spreading visitors over time and space [7, 8]. There is no regulation or monitoring of the carrying capacity of the roads. As a result, any number of vehicles enter the city till the roads are choked, and the traffic spillover results in long trails of stranded vehicles on all the roads entering the town, many times stretching over 5–8 km [7]. The existing solutions include providing temporary parking spaces in peak seasons designated above the hill, barricades, and posters highlighting Nainital Housefull, which are flanked by the entry of Nainital (Fig. 5). The research on overtourism is mainly conducted in European countries, focusing on AI-based solutions. In India, many cities face the problem of overtourism, but little research has been done. In India, technological advancements in tourism have been employed to enhance the tourist experience and tap tourists into unexplored places instead of controlling overtourism. India's national tourism policy promotes technology for sustainable tourism, but no research has been done to indicate technological advancement. To overcome problems such as traffic congestion and overcrowding, manual crowd control and the use of taxes and penalties have been implemented. However, no technology to control overtourism has been employed so far. Hence, there is a need to explore AI-based technological solutions to control overtourism in Nainital.

3 Methodology

The methodology for the research paper consists of three steps—data collection, data analysis, and the development of a framework for controlling overtourism in Nainital (Fig. 6). The data collection includes primary and secondary data. The primary data is collected through surveys and Knowledge Informant Interviews (KII),

Fig. 5 Parking at Nainital and warning banner at the entry of Nainital. *Source* Hindustan Times 2019

reconnaissance survey along the mall road. The primary surveys led to the mapping of tourist attractions and activities, and identification of congestion points and nodes. The secondary data includes tourist footfall statistics by the Tourism Department, Uttarakhand Tourism Management Plan (2007–2022), and case studies to explore and analyze applications of AI-based solutions for overtourism. The second stage, data analysis, includes analysis of the onset, causes, and effects of overtourism based on multiple indicators (Fig. 6).

The indicators for the onset of tourism are Annual Tourist Footfall in comparison to the resident population, floating population in comparison to the permanent population, and tourism density. The indicators for drivers of overtourism are nearness to major urban centers compared to other tourist destinations, difference in temperature in the peak summer, number of destinations in five kilometer (Km) radius, and analysis of the availability of infrastructure as well as carrying capacity of the city. The impact of overtourism in Nainital is analyzed on the basis of social, environmental, and economic impacts. The case studies are analyzed based on Similarity, Feasibility, and Success Rate. Then, they are incorporated into the overtourism framework along with drivers based on demand and supply effects.

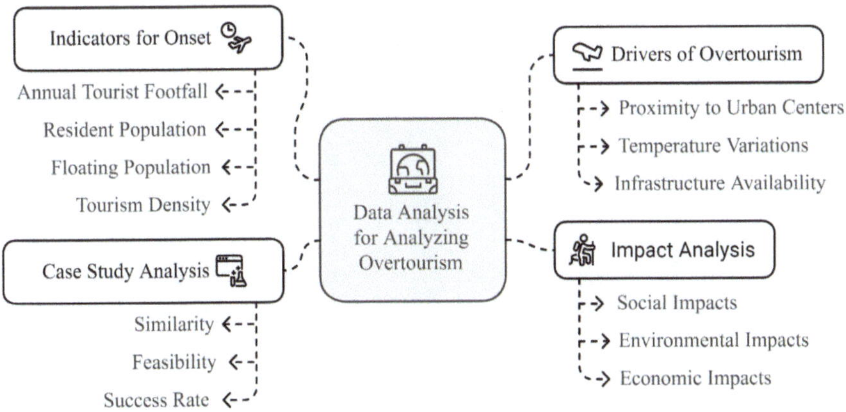

Fig. 6 Methodology flowchart. *Source* By Authors

4 Results and Discussions

The first sub-section assess the population variation that led to the onset of overtourism. The second sub-section analyzes the drivers of overtourism in which proximity, temperature, and availability of infrastructure are assessed. The third sub-section analyzes the impact of overtourism in terms of social, environmental, and economic aspects. The fourth and fifth sub-sections proposed a AI-based solution and overtourism framework for Nainital city.

4.1 Onset of Overtourism

The onset of overtourism is analyzed based on the Annual Tourist Population compared with the resident population and the permanent population compared with the floating population. The tourism density is calculated based on the area of the city and the number of tourists entering the city during peak season. Figure 7 describes annual tourist footfall from 1981 to 2021. It has increased from 4,70,000 in 1981 to 570,000 in 1991 and then decreased to 3,60,000 in 2001. A sudden rise was seen in 2011 and then again dipped to 520,000 from 9,00,000 in 2021 due to COVID-19. The resident population, however, has always been less than 60% of the annual tourist footfall in all years. Figure 8 depicts the floating population in contrast to the permanent population. The city's floating population includes tourists, institutional population, and everyday visitors. The floating population has always been more than twice the city's permanent population from 1981 to 2021.

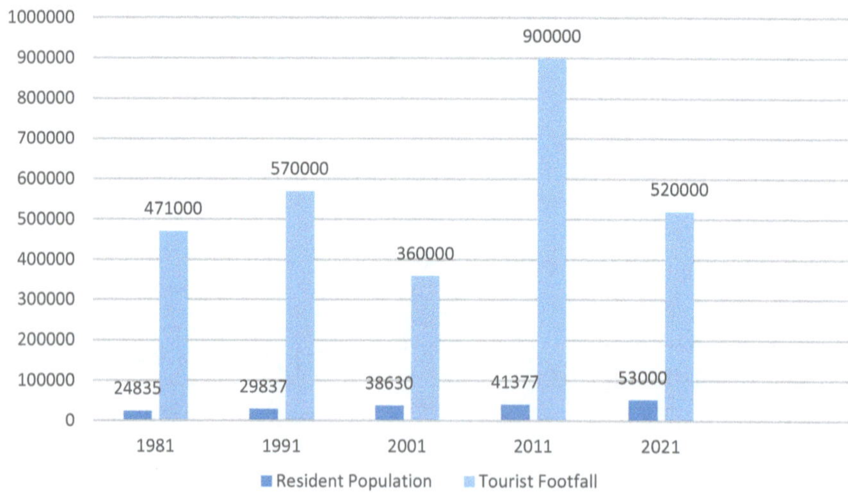

Fig. 7 Annual tourist footfall and resident population. *Source* IEE, UUSDA 2021 (Left)

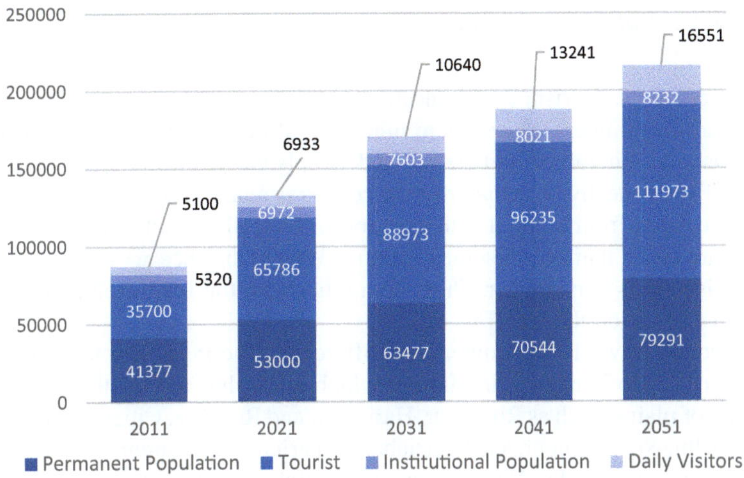

Fig. 8 Current and projected floating population in Nainital. *Source* CSR, 2021 (Right)

The tourist footfall for the year 2021 of four major tourist destinations, i.e., Dehradun, Mussoorie, Tehri, and Nainital cities are presented in Fig. 9. These cities have been selected due to the highest footfall numbers in the year 2021. The tourism density in Nainital is the second highest among the top four cities, with the highest number of footfalls in Uttarakhand.

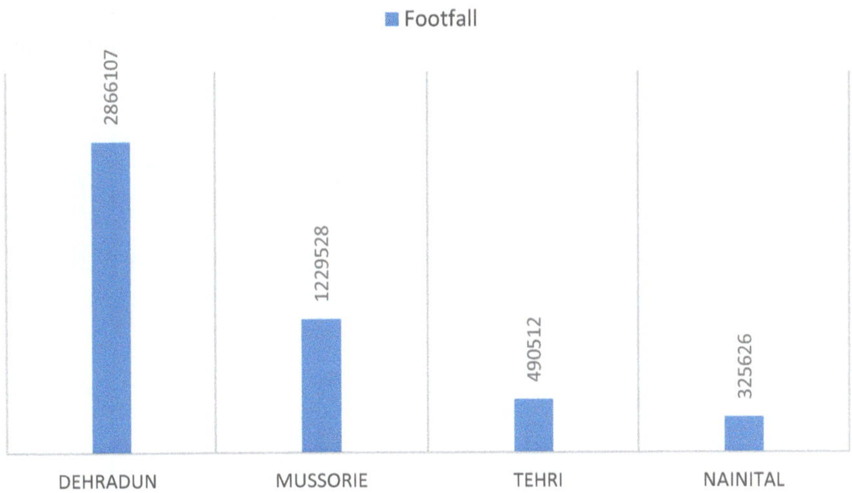

Fig. 9 Tourism footfall. *Source* TSR 2021

4.2 Drivers of Overtourism

The drivers of overtourism are connectivity to major urban centers, difference in temperature in summers, closeness to other tourist destinations, number of tourist attractions in a 5 km radius, and the demand–supply gap in the availability of infrastructure. The connectivity to major urban centers is analyzed by categorizing cities based on distance from Nainital with more tourist footfall. The cities within a 174 km distance are ten in number, the cities within a 314 km distance are thirteen, and the 454 km distance is 9 in number. Thirty-four cities with tourists traveling to Nainital lie within 454 km of Nainital (Fig. 10).

The second driver is the temperature difference. The maximum temperature in Nainital goes up to 26 degrees in summer, which is less than that in other cities in the proximity of plains, such as Rudrapur, Haridwar, and Roorkee. Cities in other states, such as Delhi, Noida, Jaipur, and Ghaziabad, experience high summer temperatures; hence, they are attracted to Nainital, and the footfall rises in summer in Nainital.

Nainital is well connected to other cities through the National Highway, which connects Nainital to major cities such as Delhi and Dehradun (Fig. 11). As part of NH, Mall Road experiences heavy traffic flow during peak seasons, leading to overcrowding and congestion. The other tourist destinations Connected to Nainital are Bhimtal, Bhowali, Mukteshwar, Almora, Ranikhet, Lohaghat, Champawat, etc.

Tourist destinations in the 5 km radius of the city include major tourist destinations such as Naina Devi Temple, Snow View, Neem Karoli Mandir, Zoo, Sankat Mochan Mandir, Tiffin Top, Lovers Point, Himalayan Botanical Garden, Khurpatal, Eco Cave Park, Woodland Waterfall, Kilbury, and Pangot. The lack of supply and demand for infrastructure is analyzed based on the city's carrying capacity (Fig. 12). The vehicular traffic in peak summers is 3600 vehicles entering the lake toll bridge while

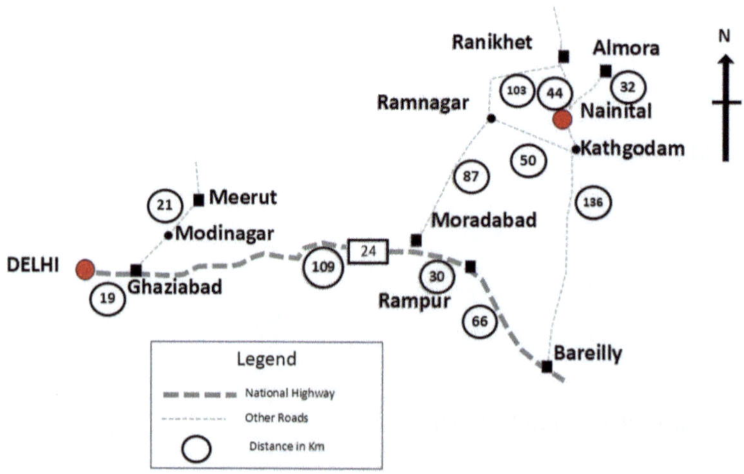

Fig. 10 Connectivity to major urban centers. *Source* NHAI 2018

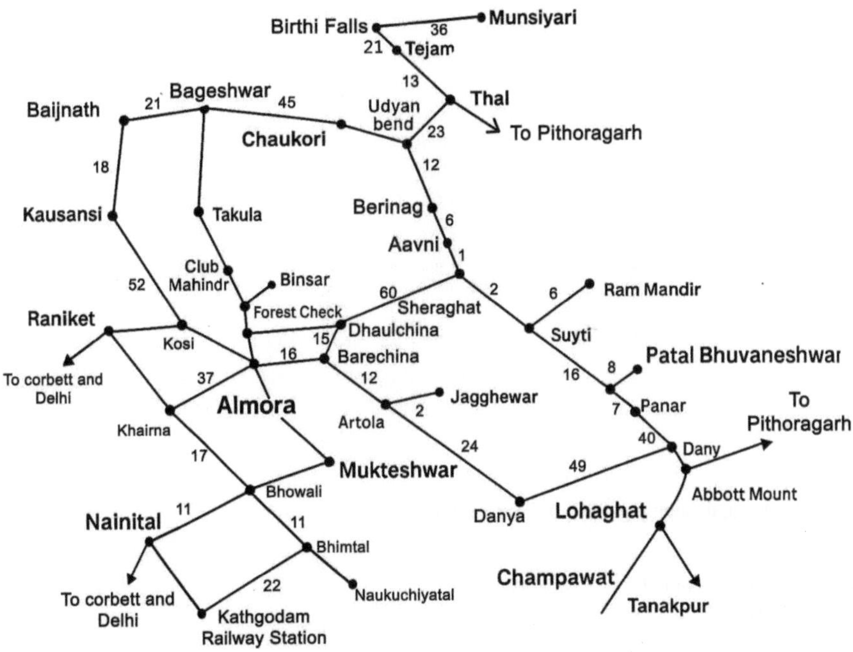

Fig. 11 Connectivity to other tourist places. *Source* Nainital Online

the parking available is 2000. The number of accommodation units with no beds is 6488 while the demand is 8000. The restaurants are 350 in number while the daily footfall is more than 10,600. The travel agencies are 120 in number while the footfall is 10600 in number.

4.3 Impact of Overtourism in Nainital

The impacts of overtourism in Nainital are analyzed using environmental, social, and economic parameters. The environmental impacts are based on the depletion of Naini Lake, the ambient air quality, and the increase in solid waste generation (Figs. 13 and 14).

The physical character of the lake has been analyzed based on area and depth. The area decreased by 7%, while the depth decreased by 55% as the water demand increased to 64% more in the peak season. The water quality has also deteriorated as the Alkalinity and Hardness have increased by 25.8% and 7.9% respectively, and the Transparency has decreased by 37.3%.

The social impact has been analyzed based on changes in built use and the carrying capacity of mall roads. The built use in the 200 m buffer of Naini Lake has changed

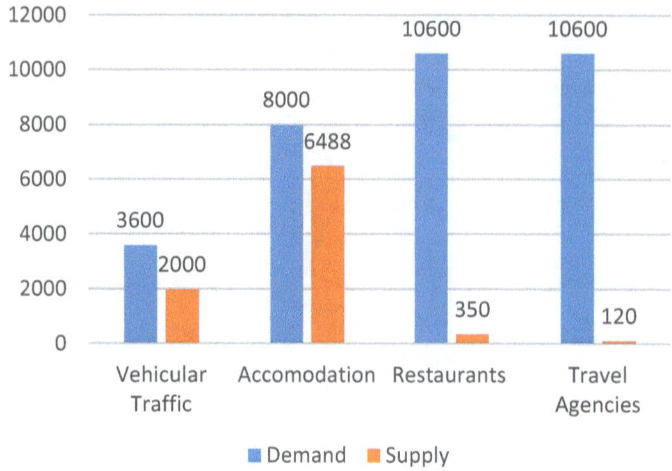

Fig. 12 Demand–supply gap. *Source* By Authors

Fig. 13 AQI of Nainital. *Source* AQImet

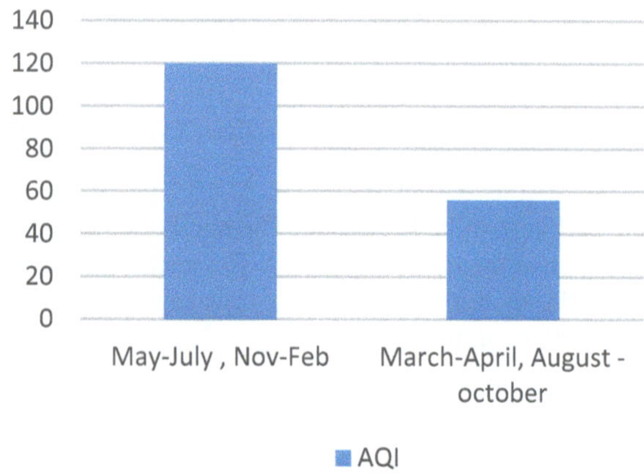

significantly as the commercial use has increased compared to the residential and other uses as shown in Fig. 15.

The carrying capacity is calculated using Eq. 1:

$$CarryingCapacity = NumberofPeople * AreaofMallRoad * AreaperPerson \quad (1)$$

The number of pedestrians on Mall Road in peak season is 35,360 while the carrying capacity calculated is 13500, which leads to overcrowding and congestion on Mall Road, disturbing the social fabric and creating problems for local people.

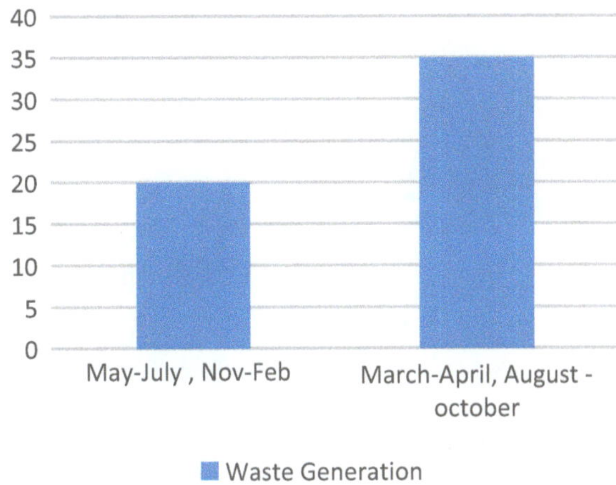

Fig. 14 Waste generation. *Source* IWD 2019

4.4 AI-Based Solution for Overtourism in Nainital

The solution analyzed based on similarity, feasibility, and success rate depicts that smart parking with real-time vacant space information and open-source crowd controlling have the highest scores, i.e., 1 in all three aspects of similarity, success rate, and feasibility. The open-source crowd-controlling system with sensors has a high similarity and success rate score. However, it is low in feasibility as the open-source big data needs to be controlled at various levels, followed by VR, mobile applications, tourist cards, and AI-based water quality detection.

4.5 Proposed Overtourism Framework for Nainital

The overtourism framework for Nainital has been developed based on components including Drivers, Effects, and Solutions. In this framework, the solutions mentioned are based on technological advancements in artificial intelligence (AI) deployed in other cities worldwide. The drivers have been identified based on different indicators. The effects have been classified as environmental, social, and economic. The solutions mentioned in the framework have been connected with the effects which can be controlled through the solutions. Figure 16 depicts the AI-based overtourism framework for Nainital. The solutions mentioned, such as open-source crowd-controlling systems with sensors, smart parking with real-time vacant space information, tourist cards for tickets and vouchers, mobile applications, VR videos, and AI-based water quality detection devices, have been linked with the problems that have been analyzed through the drivers and effects. The open-source crowd-controlling system can solve the issues of overcrowding, congestion, stress on infrastructure, and lack of parking

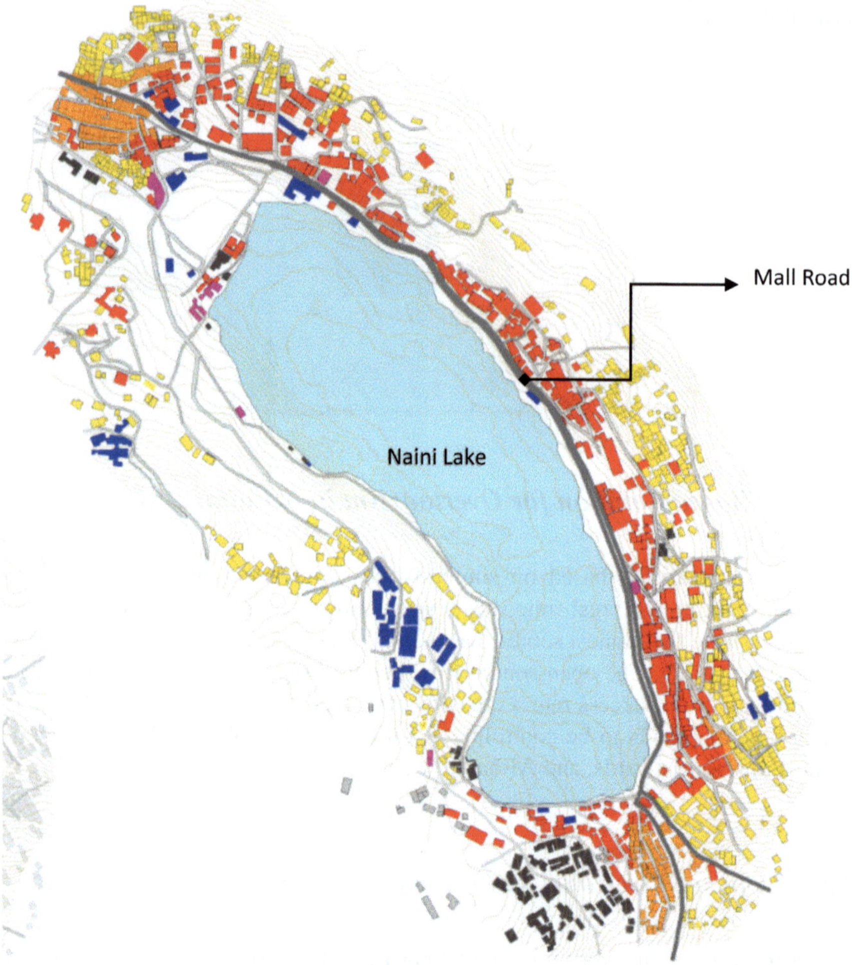

Fig. 15 Change in built use. *Source* By Authors

spaces. The smart car park with real-time vacant spaces can solve the entry issue of more cars than the parking spaces in Nainital.

The tourist card and mobile application will help to avoid overcrowding and congestion. The VR videos and social media marketing focusing the dispersal of tourists to other Taals, i.e., Sattal, Bhimtal, and Naukuchiyatal, under the tagline "Visit the Nine Taals." The AI-based real-time water quality detection of Naini Lake will detect the water quality more frequently based on set parameters.

The literature reveals that overtourism stems from proximity to urban centers, accessibility, and favorable climate, exacerbated by insufficient infrastructure and poor carrying capacity planning. This leads to environmental degradation (e.g.,

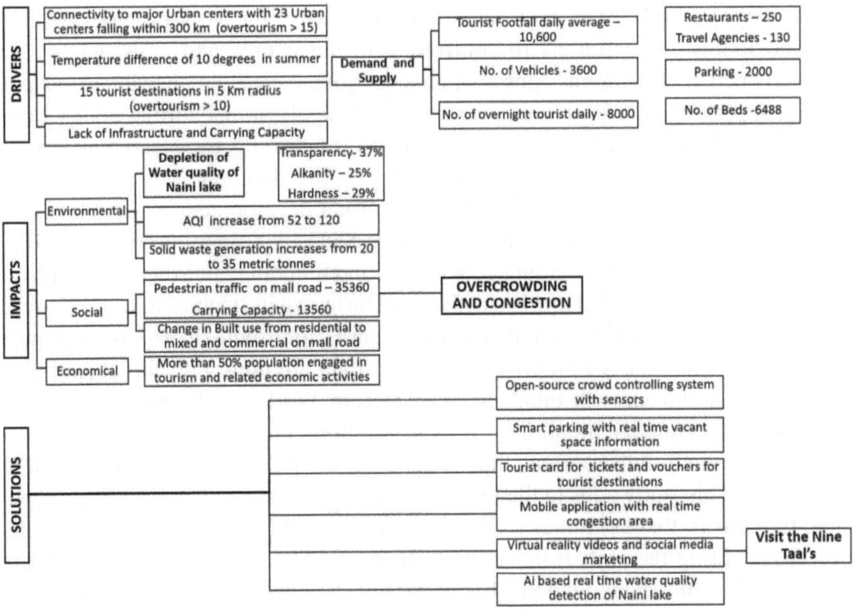

Fig. 16 AI-based overtourism framework. *Source* By Authors

reduced water quality), social disruption (congestion, displacement), and economic over-reliance. Solutions involve AI-based technologies like crowd control and smart parking (successful in Amsterdam, Venice, Barcelona), alongside redistributing tourist flows spatially and temporally. Policy measures such as visitor caps, entry taxes, and community-based sustainable tourism are crucial for effective management. These integrated strategies, combining technology, proactive policy, and sustainable practices, are vital for tourism-dependent regions.

5 Conclusion

This research addresses the considerable issues caused by overtourism in Nainital and puts forward a creative framework that aligns with the United Nations Sustainable Development Goals (SDGs) to alleviate its adverse effects. Overtourism in Nainital is influenced by factors such as its connectivity to 23 urban centers located within a 200-km range, a temperature difference of 10 degrees compared to the plains, and the presence of 15 tourist attractions within a 5-km area. Along with inadequate infrastructure and poor planning regarding carrying capacity, these factors have led to overcrowding, traffic congestion, and harm to the environment, especially affecting Naini Lake, an essential natural and cultural asset in the region.

The suggested AI-driven overtourism management framework provides sustainable and innovative solutions to confront both the causes and consequences of overtourism. These solutions consist of an open-source crowd management system, intelligent parking solutions, tourist cards, mobile applications, virtual reality (VR) videos, and AI-enabled real-time water quality monitoring for Naini Lake. Such initiatives are directly associated with SDG 11 (Sustainable Cities and Communities) by fostering inclusive and sustainable urban development while alleviating congestion in overloaded areas. They also tackle SDG 13 (Climate Action) and SDG 15 (Life on Land) by safeguarding ecosystems and reducing environmental damage resulting from uncontrolled tourism activities. During visits, technologies like crowd management systems, VR content, and tourist cards contribute to dispersing crowds and ensuring fair access to tourism resources, thus aligning with the objective of inclusive economic growth under SDG 8 (Decent Work and Economic Growth). AI-driven water quality monitoring directly supports environmental sustainability by allowing timely interventions that help conserve vital natural resources such as Naini Lake. This study adds to the field of sustainable tourism by presenting a replicable framework that combines technology, policy, and community-focused strategies, highlighting the shift from reactive strategies, like limiting tourist access, to proactive, data-informed methods. By integrating AI-based solutions into Nainital's tourism strategy, the city can secure long-term economic and environmental sustainability, in accordance with the SDGs. Future studies should aim to test this framework in Nainital and other Indian locales to assess its effectiveness, scalability, and role in promoting sustainable tourism development across various contexts.

References

1. Dignesh P, Derashri P (2018) Tourismsector in India growth challenges and opportunities. ImpAl J Interdiscip Res (IJIR) 3(10):201–204
2. Čumlievski N, Brkić Bakarić M, Matetić M (2022) A smart tourism case study: classification of accommodation using machine learning models based on accommodation characteristics and online guest reviews. Electronics (Switzerland) 11. https://doi.org/10.3390/electronics11060913
3. UNWTO (2019) 'Overtourism'? understanding and managing urban tourism growth beyond perceptions volume 2: Case Studies. World Tourism Organization (UNWTO)
4. UNWTO (2018) 'Overtourism'?—understanding and managing urban tourism growth beyond perceptions, executive summary. World Tourism Organization (UNWTO)
5. Żemła M, Szromek AR (2021) Influence of the residents' perception of overtourism on the selection of innovative anti-overtourism solutions. J Open Innov: Technol, Mark, Complex 7. https://doi.org/10.3390/joitmc7030202
6. Kumar D, Ginwal R, Lohani JK, et al (2023) Tourism and its evaluation: A case study of district Nainital. Int J Multidiscip Res Dev
7. Mukherjee T, Mazumder A (2018) An empirical study on induced tourism in Nainital town
8. Bhakuni K (2018) Overtourism in Nainital: tourism at peril. Int J Innov Soc Sci & HumIties Res. ISSN:2454–1826
9. Gretzel U, Sigala M, Xiang Z, Koo C (2015) Smart tourism: foundations and developments. Electron Mark 25:179–188. https://doi.org/10.1007/s12525-015-0196-8

10. Gajdošík T, Marciš M (2019) Artificial intelligence tools for smart tourism development. In: Advances in intelligent systems and computing. Springer Verlag, pp 392–402
11. Krishnan C, Goel R, Garg V (2021) Role of artificial intelligence and its impact on the tourism industry of India. pp 291–302
12. Li D, Du P, He H (2022) Artificial intelligence-based sustainable development of smart heritage tourism. Wirel Commun Mob Comput. https://doi.org/10.1155/2022/5441170
13. Manzo GG, Brise O, Turner R (2017) Coping with success managing overcrowding in tourism destinations
14. Wang yasong, Pfister R, Morais D (2006) Residents attitude towards tourism development, a case of Washington, NC. In: Proceedings of the 2006 northeastern recreation research symposium
15. Ali R (2018) The Genesis of overtourism_ why we came up with the term and what's happened since. Skift
16. Flores-Crespo P, Bermudez-Edo M, Garrido JL (2022) Smart tourism in villages: challenges and the Alpujarra case study. In: Procedia computer science. Elsevier B.V., pp 663–670
17. Weber F (2020) Overtourism. An analysis of contextual factors contributing to negative developments in overcrowded tourism destinations. BEST EN Think Tank XVII
18. Pérez Garrido B, Sebrek SS, Semenova V, et al (2022) Addressing the phenomenon of overtourism in Budapest from multiple angles using unconventional methodologies and data. Sustainability (Switzerland) 14. https://doi.org/10.3390/su14042268
19. Estalella R, Barrin C (2019) HOTREC Position paper on overtourism Putting sustainable tourism on top of the EU policy agenda for the benefits of society overtourism: a pressing issue of concern for society
20. Goodwin H (2017) Responsible tourism partnership working paper 4. October 2017
21. Jordan P, Patras P, Pssaros M (2018) Managing tourism growth in Europe
22. Majid GM, Tussyadiah I, Kim YR, Pal A (2023) Intelligent automation for sustainable tourism: a systematic review. J Sustain Tour 31:2421–2440. https://doi.org/10.1080/09669582.2023.2246681
23. García-Hernández M, Ivars-Baidal J, Mendoza de Miguel S (2019) Overtourism in urban destinations: The myth of smart solutions. Boletin de la Asociacion de Geografos Espanoles. https://doi.org/10.21138/bage.2830
24. Basumatary M, Sarma G (2024) The use of artificial intelligence in tourism industry of India: a critical insight. Migr Lett 21:111–119

Smart Tourism Infrastructure: The Role of Technology in Destination Development

Tomáš Gajdošík

Abstract Tourism infrastructure refers to the set of facilities and services necessary for the tourist to be able to stay in a destination and engage in various recreational and cultural activities typical for tourism. With the advent of smart tourism phenomenon, tourism destinations have possibilities to use the state-of-the-art technologies in infrastructure planning. The chapter explains how technology, such as geographic information system, Internet of Things, blockchain, gamification, and augmented and virtual reality, helps to plan the tourism infrastructure in a smart way. The use of these technologies is illustrated on the mature tourism destination—the High Tatras, Slovakia. The semi-structured interviews with the most important destination stakeholders are used to evaluate the attitudes toward designing and building the infrastructure. The chapter presents that the use of the technologies in a smart way can prevent the negative effects of overtourism and tackle the problems of tourism infrastructure planning.

Keywords Mature destination · Overtourism · Planning · Smart technology · Tourism infrastructure

1 Introduction

In a destination, the final stage of tourism development is realized when physical objects are planned and constructed on-site. This phase falls within the expertise of architects and landscape architects, who collaborate to create effective and appealing tourism environments [1]. These assumptions stress the role of tourism infrastructure and its planning in a tourism destination as it is regarded as an indispensable factor for tourism development. In a broad sense, infrastructure encompasses physical,

T. Gajdošík (✉)
Faculty of Economics, Matej Bel University in Basnká Bystrica, Banská Bystrica, Slovakia
e-mail: tomas.gajdosik@umb.sk

V. C. Sharma et al. (eds.), *Architectural Vision of Overtourism*,
SpringerBriefs in Architectural Design and Technology,
https://doi.org/10.1007/978-981-96-5794-0_9

legal, environmental, and psychological elements that enhance the tourist experience, ensuring that the product is enjoyable, reliable, and sustainable [2]. It encompasses the facilities and services essential for tourists to stay at a destination and participate in various recreational and cultural activities characteristic for tourism. Primary tourism facilities and services include accommodation, restaurants, travel and tour providers, and car rental companies. Secondary facilities cover shopping and convention centers, recreational assets, entertainment venues, and visitor information centers. Lastly, tertiary facilities and services involve health care, emergency and safety services, financial and personal services, and law enforcement and the legal system [3].

To protect unique natural attractions and minimize the negative social, economic, and environmental impacts of tourism, effective tourism infrastructure planning is essential. This involves the coordinated development of attractions (natural, cultural, and man-made), services (such as accommodation, shops, restaurants, visitor information centers, tour and travel operators, currency exchange, medical facilities, postal services, and entry/exit facilitation), and transportation infrastructure, including physical features like trails and transport services. In recent years, the planning of tourism infrastructure has become increasingly complex, due to the changes in the tourism ecosystem.

The tourism ecosystem faces a highly complex and dynamic environment, presenting numerous challenges. Advances in technology, climate change, and increasing uncertainty are reshaping destination planning and development, making it progressively harder to navigate the development of destinations within such a multifaceted socio-technical system. These ongoing changes in the destination landscape have prompted the exploration of new models of destination planning and governance [4]. In response to tackling the challenges in various sectors and industries, the smart approach has emerged. This approach brings value as it enables to sensitively analyze and quickly respond to the changing situation. The smart concept has emerged as a result of the rise of information technology and the need for sustainability [5].

As tourism is highly dependent on information technologies [6, 7] and sustainable development of tourism is a prerequisite of its development [8], the smart phenomenon has also penetrated into the tourism sector [9]. Smart tourism is derived from smart city agenda, paying special attention to tourists and their experiences [10]. Moreover, the concept of a smart destination is closely linked to the concept of the smart city, which stresses the urban planning and management [11] and which strategy is based on the use of technologies in key areas such as the economy, environment, mobility, and governance to transform infrastructures and services [12].

However, so far, the studies on smart tourism have explored the tourist perspective (e.g., [13, 14]), the business level (e.g., [15, 16]), and the destination level (e.g., [10, 17]) and not adequately address the infrastructure planning perspective. Although Boers and Cottrell [18] call for more research on tourism infrastructure planning using the potential of information technologies, there has been a lack of research since their paper on sustainable tourism infrastructure planning using the GIS-supported approach. Thus, this study responds to their call and proposes a novel approach to tourism infrastructure planning building on theories of smart tourism.

2 Smart Tourism Infrastructure Planning

The planning theory highlights the importance of interdisciplinary approaches to tourism planning in destinations. Architecture and urbanism prioritize physically oriented planning, while geography emphasizes the location theory and the tourism system approach. Business and economics advocate for a demand-driven perspective, whereas sociology and psychology underpin participatory and collaborative planning. Environmental studies stress the significance of sustainable development, with a particular focus on carrying capacities [19].

Historically, since the 1940s, tourism planning has primarily emphasized physical and economic aspects, focusing on constructing facilities for tourists, including hotels, restaurants, communication networks, and transportation systems. During this period, tourism growth was largely driven by marketing these facilities in both established and emerging destinations [19]. The negative impacts of tourism on the physical aspects of the destination have pushed forward a broader approach to planning focusing more on capacity issues and sustainability [20] and community engagement [21]. As Gunn [1] explains: *"The tourism environment must accommodate the many activities and movements of people and vehicles. A quantifiable amount of space is necessary, for example, to park a given number of cars, to provide food for a given number of people, or to enable a given number of people to camp. Although the limits vary, rough spatial standards have been set for a great many kinds of sports, games, and recreational areas. So-called oversaturation of destinations* (today referred to as overtourism) *and deterioration of environments is not caused as much by large number of visitors as by inadequate design and poor management of crowds."* The World Tourism Organization [22] outlines that this is above all a matter of planning, regulations, and infrastructure. This assumption stresses the need for planning and taking into account the basic principles of infrastructure development, when tackling the overtourism issues.

Taking into account the principles of infrastructure planning, Baud-Bovy and Lawson [23] focus on standards, environmental integration, contact with nature, increasing the value of resources, grouping of activities, landscaping, distribution of buildings, and focuses of interest, as well as separation of traffic and quality of construction as principles of destination development. Hall [24] adds sustainability and authenticity as prerequisites of sustainable design. Overall, it can be concluded that the most important principles of tourism infrastructure planning are (1) sustainability/environmental, social and economic principles, (2) authenticity, and (3) respecting the standards.

However, nowadays infrastructure planning in tourism destinations faces several challenges. Insufficient or outdated data on tourism development in a destination [25] hinders decision-making processes. Moreover, balancing infrastructure development with environmental protection is critical [18]. Effective planning requires collaboration among various stakeholders, including government agencies, private sector players, and local communities [26]. Conflicting interests can complicate decision-making processes. Infrastructure development must respect and preserve local cultures and heritage, while insensitive planning can lead to community backlash and loss of authenticity [27].

Therefore, trends in infrastructure planning are evolving rapidly with a focus on sustainability and the integration of smart technologies to reduce these challenges [28, 29]. Moreover, the emphasis on sustainability in tourism underscores the need for smart infrastructure planning that minimizes environmental impacts while promoting social equity and economic efficiency with advanced decision-making possibilities. Smart tourism infrastructure planning is closely aligned with the UN Sustainable Development Goals (SDGs) as it integrates technology, sustainability, and inclusiveness to address global challenges. It is linked to SDG 6: Clean Water and Sanitation as well as SDG 7: Affordable and Clean Energy as technologies ensure efficient water management and energy-efficient infrastructure allows reliable and affordable energy access. Smart tourism infrastructure fosters innovation (SDG 9: Industry, Innovation and Infrastructure) and helps in planning and efficient resource management in cities (SDG 11: Sustainable Cities and Communities). Through real-time monitoring it optimizes use of resources (SDG 12: Responsible Consumption and Production) and aids in mitigating the climate change impacts (SDG 13: Climate Action). Using smart information systems enables sustainable interaction with natural ecosystems (SDG 14: Life Below Water and SDG 15: Life on Land) and community participation (SDG 17: Partnerships for the Goals). Integration of smart technologies to tourism infrastructure planning should facilitate informed decision-making processes for planners, allowing for the identification of key trends and the efficient management of tourism-related resources. Smart infrastructure planning uses smart technologies and data analytics to optimize the design and construction of tourism infrastructure in line with the SDGs.

Smart technologies serve as a broad and often ambiguous "umbrella term" encompassing various advanced innovations that elevate connectivity to a new level. They combine hardware, software, and networks to deliver real-time data, enabling more informed and efficient decision-making and include a variety of technologies, such as the Internet of Things (IoT), blockchain, cloud computing, artificial intelligence, mobile applications, virtual reality (VR) and augmented reality (AR), near field communication (NFC), radio frequency identification (RFID), Bluetooth low energy (BLE) beacons, and the latest tourism information systems (e.g., property management systems, geographic information systems, business intelligence systems) [30, 31]. However, so far, the role of smart technologies in tourism infrastructure planning with the focus on overtourism has not been complexly examined.

3 Context and Methods

The aim of this study is to analyze how smart technologies can help in smart(er) planning of tourism infrastructure. The use of these technologies is illustrated on the mature tourism destination—the High Tatras, Slovakia. The High Tatras represents a mature tourism destination located in the Carpathian Mountains in Slovakia. The

High Tatras are a protected area and a national park, which puts pressure on the preservation of their natural beauty and ecosystem. The destination is known for sports and recreational tourism, specifically skiing in winter and hiking in summer, offering a wide range of trails that cater to all levels of tourist experience. The High Tatras are also known for their climatic spas, where visitors can relax and rejuvenate. In recent years, the infrastructure and tourism services in this area have been developing due to the constantly growing demands of tourists, making the High Tatras a leading tourism destination in Slovakia, while the overtourism problems have been rising.

The beginning of the tourism development in the destination started in the nineteenth century when the first Grand Tours took place. Significant development of tourism in the High Tatras has been observed since the beginning of the twentieth century. This period was characterized by the construction of grand hotels as well as lower class accommodation facilities. The development of tourism in the High Tatras was mainly oriented toward high-altitude hiking and spa tourism. The dominant architectural style was the so-called "Swiss style."

Later, the organization of international competitions and sports events became significant. In 1935, the first World Ski Championship was held here, with the second, more significant championship in 1970. The 1970s were criticized for the noise, hustle, and overcrowding of the High Tatras. The dominant architectural style was functionalism. The Tatra settlements began to acquire an increasingly urban character, which is why in the 1980s construction was limited.

Since 2000, a shift away from the traditional product focused on sports tourism and spa tourism has been observed in the High Tatras. Tourism products focused mainly on entertainment and leisure have come to the forefront. Since 2009, several innovations have been implemented in the High Tatras, mainly aimed at the renovation and construction of infrastructure, innovations in organizational structures, and marketing communication, which have contributed to the gradual increase in the number of guests in the High Tatras and reaching the carrying capacities of the destination [32]. Recently, several stylistic trends in architecture can be observed in the destination. For example, there is the so-called "new historicism," which in the High Tatras relies on the Swiss style with an effort for representativeness and decorativeness as a counterbalance to the former monotony and uniformity of functionalism, often, unfortunately, without a sense of taste and without the presence of basic aesthetic values. This creates pressure on the authenticity of the destination. The destination is nowadays facing several problems in infrastructure planning.

To identify these problems, qualitative research was used. The strength of this kind of research is its ability to dive deeply within the problem, and it provides a holistic view to the problematics in its natural settings. The semi-structured interviews with the most important destination stakeholders are used to evaluate the attitudes toward designing and building the tourism infrastructure and analyzing the infrastructure problems that have arisen. Together six interviews were made, namely with the former director of the Tourism Association of the Hight Tatras, manager of DMO High Tatras Foothills, manager of the non-profit organization for culture and art "Vietor Tatranský," marketing manager of the TMR, Inc.—the major developer in

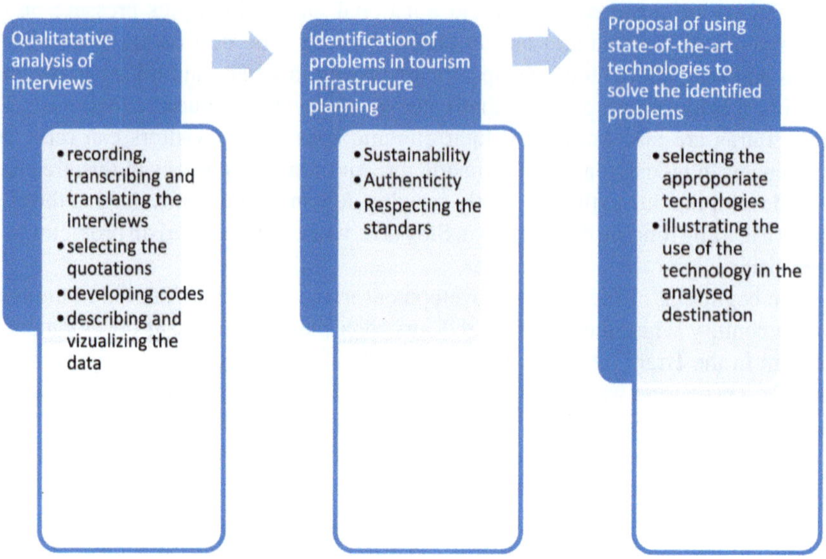

Fig. 1 Research methodology flowchart. *Source* By Authors

the destination, and two architects specializing on the territory. The selection of the respondents was based on the purposive sampling with the aim of obtaining opinions from the tourism sector, profit and non-profit sector, as well as the architectural domain. The interviews were recorded, transcribed, and translated to English for further analysis. After initial reading of data, the most important quotations were selected, and codes were developed using Atlas.ti 24 software. Finally, the results were described and visualized. The aim of the interviews was to identify the problems of tourism infrastructure planning in a mature tourism destination focusing on sustainability, authenticity, and respecting the standards, as the basic principles of tourism infrastructure planning. Based on the identified problems, smart solutions using state-of-the-art technologies are presented, being applicable to similar mature destinations tackling the overtourism issues (Fig. 1).

4 Results and Discussion

Tourism infrastructure planning is critical for the sustainable development of destinations, especially in light of the rapid growth of the tourism sector, which can result in overtourism. While well-resourced destinations attract increasing numbers of visitors, this influx often leads to pressures on local ecosystems and residents of a destination. The need for sustainable tourism practices has prompted discussions around balancing economic growth with environmental and social responsibility.

Effective infrastructure planning must address the complexities of accommodating visitor demands while preserving natural resources, enhancing the residents' quality of life, economic stability, as well as authenticity, while respecting the standards. In the light of these issues, problems of tourism infrastructure planning are analyzed.

4.1 Problems of Tourism Infrastructure Planning

The High Tatras, once a pristine "jewel" of the Slovak Republic, are grappling with severe sustainability challenges, primarily due to unconrolled tourism-related construction. Historically, a balanced approach to development prevailed. Until 1989, a "council" of six key stakeholders—spa operators, Tatra National Park, tourism entrepreneurs, scientists, transport companies, and residents—was governed by consensus, ensuring minimum requirements for each group and strict building restrictions, including limits on bed numbers. If a new construction was desired, beds had to be "sold," maintaining a controlled growth. This respect for carrying capacity and unified planning has drastically eroded. Currently, there are no set physical carrying capacities, and a crucial landscape-ecological plan is absent. This void allows investors to build indiscriminately, with no understanding or regulation of the environmental impact. The density of accommodation facilities now stands at 2,558.92 beds per 100 km^2, vastly exceeding the past recommended value of 886–1,300 beds per 100 km^2. This excessive construction is driven by a surging demand for apartment accommodation, despite the current bed occupancy being a mere 32.92%, indicating significant underutilization of existing capacity. From an economic perspective, new construction is illogical; as a marketing manager notes, external developers, blinded by the "good brand" of the destination, inflate the market, leading to often-empty new apartments and unfulfilled investor promises due to a lack of stakeholder cooperation and economic discussion.

The uncontrolled development also severely impacts residents' quality of life. While locals recognize tourism's economic importance, social carrying capacities are frequently breached, particularly during peak seasons. As a director of non-profit organization highlights, uncoordinated recreational development at the expense of local living conditions leads to increasing problems with transportation and limited resident engagement in urban planning, fostering fear of consequences for those willing to speak out.

A critical loss is the diminishing authenticity of the High Tatras. New buildings often feature designs that completely clash with the traditional Tatra architecture, spoiling the area's aesthetic. An architect points out the prevalence of "new historicism"—an imitative style using inappropriate materials like wooden slats imitating half-timbering, which holds little art-historical value. Organic architecture, which harmoniously integrates with regional cultural heritage, is rare. This architectural degradation stems from a fundamental lack of a unified concept and vision. Paradoxically, the past had a "chief architect department" that harmonized standards, land use, and protected zones; this crucial oversight body was abolished due to cost

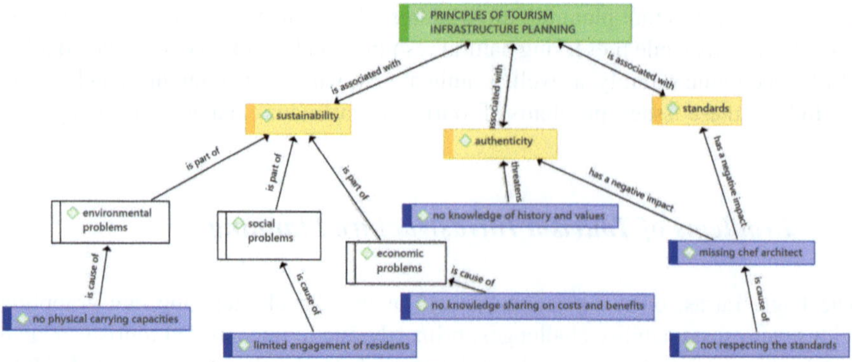

Fig. 2 Problems of tourism infrastructure planning in the destination High Tatras. *Source* By Authors

concerns. While a zoning plan exists and building authorities approve constructions, they lack oversight on design, which is vital for maintaining authenticity. The absence of architectural competitions further exacerbates this issue.

Historical architectural standards, such as ensuring mountain views from every location and strict height restrictions (e.g., maximum two stories in Tatranská Lomnica with no pensions or apartment complexes), are now routinely disregarded. A former DMO director laments that buildings in Tatranská Lomnica exceed height limits, and land coverage, supposedly capped at 40%, is often built up to 90% with concrete fences. This disrespect for standards deprives visitors, who seek tranquility and magnificent views, of the very essence of the High Tatras. The use of non-original materials, like andesite on renovated villas, further demonstrates this disregard for authentic design. In core, the High Tatras face significant problems in tourism infrastructure planning: absence of setting the physical carrying capacities, unknown impacts of new construction, limited resident engagement, an economically irrational real estate market, architectural disrespect for authenticity, and a critical lack of a governing chief architect. This aggressive overdevelopment has permanently altered the natural scenery and diminished the destination's unique values (Fig. 2).

4.2 Using Smart Technologies to Overcome the Problems of Tourism Infrastructure Planning

Using smart technologies represents an opportunity for tourism destinations to face the challenges of infrastructure planning. They can change how architects, engineers, and destination managers work and how they make decisions. This shift toward constant access to real-time information enables to determine what actions to take, when, and where, regarding the identification of infrastructure requirements. The use of the state-of-the-art technologies is illustrated on the identified problems in the destination High Tatras; however, it can be used in other tourism destinations facing similar problems.

In terms of *environmental* sustainability, GIS can aid in the selection and evaluation of locations by utilizing geographic, resource, and environmental data collected through various technologies. It helps visualize the landscape and strategically determine the optimal placement of facilities, ensuring a balance between the built environment and natural surroundings. GIS enables the analysis of terrain and environmental factors, facilitating the identification of the most efficient and eco-friendly options. Furthermore, it supports infrastructure development by identifying suitable land, flood-prone areas, traffic patterns, weather trends, and regions sensitive to environmental impacts [18].

In the conditions where physical carrying capacities are not set, the combination of GIS with IoT sensors can help to estimate the capacities. An interactive dashboard using data from IoT counters can visualize the places where there is an excessive number of beds, number of tourists, degrading the natural environment. Sensors and IoT devices play a crucial role in gathering data for infrastructure planning. They can deliver real-time, detailed information to destination managers. This real-time data can aid in optimizing infrastructure management. Moreover, by integrating capabilities of GIS, IoT sensors, and technologies like AR and VR, it enables real-time geospatial analysis for more informed and accurate decision-making (Fig. 3).

The *social pillar of sustainability* takes into account the interests of various stakeholders. In a destination, where there is a limited engagement of residents in infrastructure planning, blockchain technology can be used. Until now, blockchain technology has primarily been used in infrastructure planning for contracting purposes, as it has the potential to eliminate the multiple layers of contracts and intermediaries

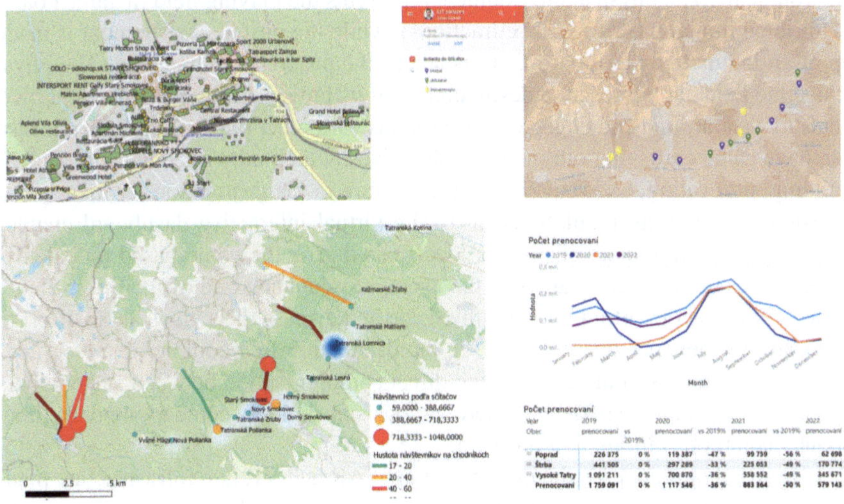

Fig. 3 Proposed dashboards using data form GIS and IoT sensors in the destination High Tatras. *Source* By Authors

involved in the development and completion of infrastructure projects. Blockchain automates the management of contracts and subcontracts, simplifies agreements, and minimizes confusion. However, there is a growing trend to apply blockchain technology to enhance stakeholder participation in infrastructure planning, an area that has been identified as a challenge in destination development [33].

The main characteristics of a blockchain are decentralization, traceability, immutability, anonymity, and auditability. Using a blockchain application can enhance the involvement of residents and other stakeholders in the planning process. The platform can serve as a communication hub, allowing participants to publish content and access smart contracts. Since the records are visible to all, information can be traced and must be approved by the network, enabling stakeholders to monitor and verify the data. This promotes transparency and helps prevent corruption. As planning processes evolve, these changes need to be communicated effectively. Blockchain ensures traceability, as all transactions are permanently visible to anyone at any time. Additionally, blockchain removes the need for trust, as the information is stored in a distributed ledger without a central authority. Furthermore, stakeholders can monitor processes to ensure accountability and ensure that all parties adhere to the plans. Conflicts can be resolved in a fair and just manner. Therefore, by using a blockchain application, stakeholders can better understand their decisions by leveraging transparency, traceability, trust, and accountability in many times controversial infrastructure development projects. Moreover, it can serve as a substitution for the chief architect (Fig. 4).

The *economic pillar,* mainly the knowledge sharing, can be supported by gamification. This approach uses the principles of gaming to achieve goals by fostering learning and collaboration in a safe environment that simplifies reality. To encourage knowledge sharing, it is beneficial to use a game that combines physical and virtual environments. The physical environment represents the playing field (a game board, 3D-printed miniatures of tourist attractions), while the virtual environment is made up of an interactive dashboard that visualizes the progress of the game. The physical game field can represent a destination, and its physical presence allows participants to walk around and engage in discussions with other players. Participants also can have game cards that outline their objectives, which may include tasks such as building, renovating, and removing infrastructure in the destination, or managerial decisions like issuing regulations [34]. The virtual interactive dashboard enables simulations of the impact of players' decisions during the game based on calculations. For example, it can calculate the average occupancy rate of facilities. Stakeholders have the opportunity to communicate and discuss their views in an effort to advocate for their perspectives and see the destination in a complex manner. This innovative method stimulates open and direct communication between stakeholders by creating a unique experience where various stakeholders come together to think about a common solution to a problem.

In terms of *authenticity*, the incorporation of AR and VR into infrastructure planning is transforming the design and experience of tourism destinations. These technologies enhance immersion, enabling professionals and tourists to engage with both real and virtual environments in more meaningful ways. AR can overlay digital

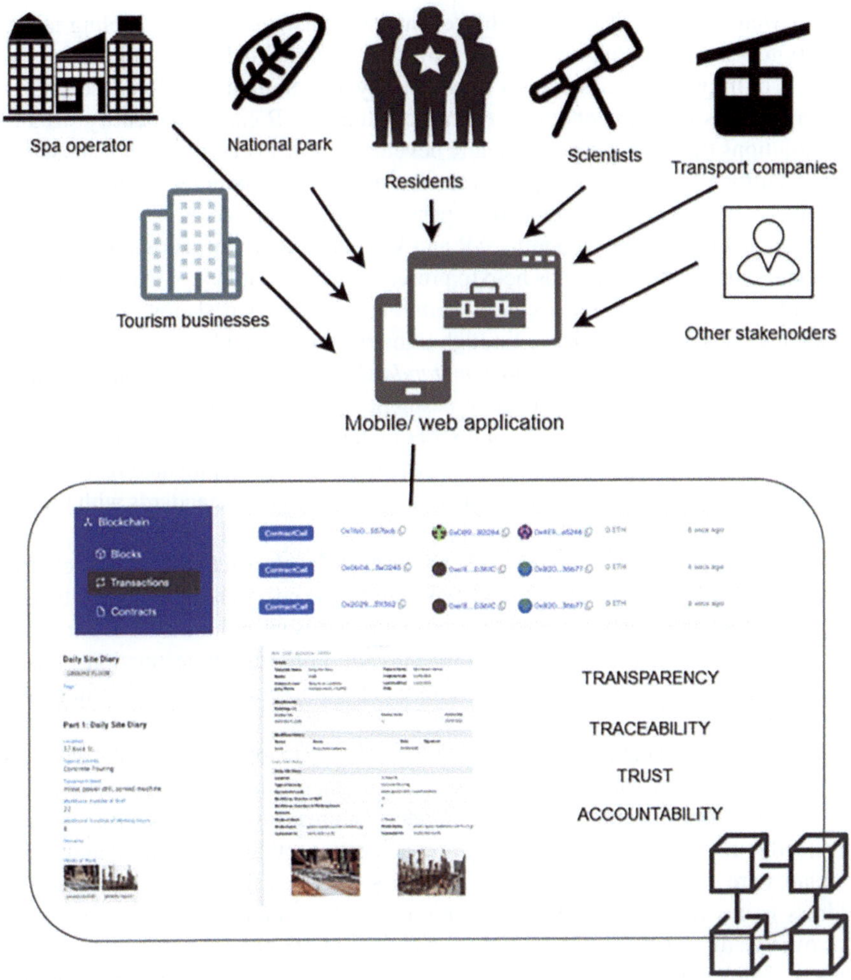

Fig. 4 Structure of a blockchain application for tourism stakeholders. *Source* Elaborated based on [33], 2023

content onto real-world views, providing planners and tourists with real-time additional layers of information [35]. In solving infrastructure problems, AR can enhance the authenticity of a destination by allowing architects and engineers to visualize how new developments will blend with the natural landscape. For instance, AR allows spatial data to be projected onto a physical site, helping planners visualize how new buildings or infrastructure will integrate with the surrounding environment. This enables the preservation of cultural and visual integrity by simulating potential environmental impacts before construction starts.

Similarly, VR provides immersive simulations of upcoming projects, allowing engineers to virtually walk through a planned infrastructure, experiencing the space as if it were finished. This capability enhances decision-making, enabling stakeholders to assess the long-term impact of their designs and make adjustments to preserve or improve the destination's authenticity. By simulating different scenarios, such as various weather conditions or traffic patterns, VR can help identify sustainable solutions that balance infrastructure development with environmental conservation. Additionally, this technology lets tourists preview destinations before their visit, offering a more engaging and realistic experience, which can encourage more sustainable tourism choices. Incorporating AR and VR into tourism infrastructure planning in a mature destination can thus help to preserve the authenticity of a destination by ensuring that new developments are harmonious with the natural environment while enhancing the visitor experience through immersive pre-trip planning.

Respecting *architectural and urban standards* in tourism destinations is crucial to maintaining a destination's integrity, aesthetic appeal, and environmental sustainability. To tackle the problems of not respecting the standards, Building Information Modeling (BIM) and its integration with virtual reality and augmented reality technologies can help tourism planners and architects to adhere to standards with greater precision and collaboration.

BIM software allows architects and engineers to develop detailed 3D models that showcase the physical and functional aspects of a project. These models integrate geometric data and attributes, enabling designers to visualize how the infrastructure will blend with the surrounding environment [36]. For tourism destinations, this is essential because it ensures that new developments complement the local architecture and urban fabric.

One of the key benefits of BIM is its ability to bring together various stakeholders in the design process. Engineers, architects, and urban planners can collaborate effectively from remote locations, reviewing models and making decisions without being physically present. This collaborative approach ensures that architectural and urban standards are consistently upheld, as all parties can view the same 3D models and address issues in real time. The integration of AR technologies, such as headsets that overlay digital models onto real-world sites, can further enhance this process. On-site teams can visualize how a new building will fit into the existing environment, ensuring compliance with local regulations, height or FAR standards, and architectural norms.

The use of VR and AR technologies in conjunction with BIM can further enhance the design and construction process. VR enables immersive experience and helps to ensure that the project meets the aesthetic and functional needs of the tourism destination, providing a realistic preview of how the final structure will fit within the existing landscape of a destination (Fig. 5).

Fig. 5 Using BIM, VR, and AR to preview the fit with existing landscape. *Source*: By Authors

5 Conclusion

Technological advancements have the ability to revolutionize tourism infrastructure planning. Using smart technologies can bring notable improvements in sustainability, authenticity, and respecting standards through the integration of data-driven decision-making, location intelligence, and immersive reality. Smart technologies have laid the foundation for smarter infrastructure planning, which not only boosts efficiency but also enhances resilience by addressing risks and uncertainties. Data-rich environments, empowered by smart technologies, offer a clearer understanding of real-time conditions and trends, leading to more informed and responsive decision-making, thus tackling the overtourism issues.

The integration of smart technologies like GIS, IoT, blockchain, AR, VR, and BIM into tourism infrastructure planning offers powerful tools to address the effects of overtourism while enhancing the sustainability and authenticity of destinations. By enabling real-time data collection, geospatial analysis, and immersive simulations, these technologies allow for more informed decision-making that can help maintain a balance between development and environmental preservation (Table 1).

Architectural and urban design interventions become more precise and collaborative with these tools, ensuring that tourism infrastructure adheres to local standards while minimizing the impact on natural and cultural resources. Blockchain technologies also offer promising solutions for improving stakeholder engagement and transparency, fostering a participatory approach to tourism development.

Table 1 The proposed smart technologies used for solving tourism infrastructure planning problems

Problem of infrastructure planning	Solution using the smart technologies
No physical carrying capacities set	GIS, IoT, AR, VR
Limited engagement of residents in planning	Blockchain app
No knowledge sharing on cost and benefits of new construction among stakeholders	Serious gaming (gamification)
No knowledge of history and values affecting authenticity	AR, VR
Not respecting architectural standards	BIM, AR, VR

Source Own research, 2024

Moreover, the knowledge gained from employing these innovations empowers tourism destinations to handle growth more sustainably, avoiding the pitfalls of overtourism by ensuring that infrastructure is well-planned, environmentally conscious, and respectful of the destination's unique characteristics. This technology approach helps safeguard both the integrity of the destination and the quality of life of residents, fostering long-term resilience.

However, the implementation of the proposed technologies in a destination requires huge financial resources. This challenge can be tackled by international or national grand schemes. The smart tourism development is supported many international projects, e.g., in Europe. National government policies are also starting to focus on this issue.

In the conditions of High Tatras, Slovakia, a new fund on tourism development is being implemented since 2025. The sustainability of tourism and its infrastructure will be financed from the fund. Applicants will be able to use the funds for projects focused on digitalization and modernization of existing facilities, or the construction of new attractions and infrastructure in line with the sustainable development of tourism.

However, findings from the analyzed destination are transferable to other geographical regions, mainly mature destinations struggling with overtourism, where tourism infrastructure planning struggles with similar problems. Tourism destinations can benefit from the proposed technological solutions to tackle their problems.

Acknowledgements The research was supported by the research projects VEGA 1/0136/23 From resilience to sustainability. The impact of data on sustainable and competitive development of tourism and VEGA 1/0360/23: Tourism of the New generation—responsible and competitive development of tourism destinations in Slovakia in the post-COVID era.

References

1. Gunn C (1972) Vacationscape: designing tourist regions. Bureau of Business Research, University of Texas at Austine, Austin
2. Boopenand S, Khadaroo J (2020) Infrastructure—tourism. In: Jafari Jafar H, Xiao (Ed), Encyclopedia of tourism (pp 1–2). Cham: Springer International Publishing. https://doi.org/10.1007/978-3-319-01669-6_447-2
3. Dwyer L, Forsyth P, Dwyer W (2020) Tourism economics and policy. (2, Ed.) (Vol 2nd edition). Bristol: Channel View Publications
4. Gajdošík T (2022) Smart tourism destination governance: technology and design-based approach. Routledge, Smart Tourism Destination Governance. https://doi.org/10.4324/9781003269342
5. Joss S, Sengers F, Schraven D, Caprotti F, Dayot Y (2019) The smart city as global discourse: storylines and critical junctures across 27 cities. J Urban Technol 26(1):3–34. https://doi.org/10.1080/10630732.2018.1558387
6. Buhalis D (2003) eTourism. Information technology for strategic tourism management. Edinburgh: Pearson Education Limited
7. Benckendorff P, Xiang Z, Sheldon P (2019) Tourism information technology (3rd editio.). Wallington: CABI
8. Hall CM (2019) Constructing sustainable tourism development: The 2030 agenda and the managerial ecology of sustainable tourism. J Sustain Tour 27(7):1044–1060. https://doi.org/10.1080/09669582.2018.1560456
9. Gajdošík T (2018) Smart tourism: concepts and insights from Central Europe. Czech J Tourism 7(1):25–44. https://doi.org/10.1515/cjot-2018-0002
10. Buhalis D, Amaranggana A (2014) Smart tourism destinations. In: Xiang Z, Tussyadiah L (Eds), Information and communication technologies in Tourism 2014 (pp 553–564). Springer International Publishing. https://doi.org/10.1007/978-3-319-03973-2
11. Ivars-Baidal JA, Celdrán-Bernabeu MA, Femenia-Serra F, Perles-Ribes JF, Vera-Rebollo JF (2023) Smart city and smart destination planning: examining instruments and perceived impacts in Spain. Cities 137. https://doi.org/10.1016/j.cities.2023.104266
12. Bakıcı T, Almirall E, Wareham J (2013) A smart city initiative: the case of Barcelona. J Knowl Econ 4(2):135–148. https://doi.org/10.1007/s13132-012-0084-9
13. Femenia-Serra F, Neuhofer B, Ivars-Baidal JA (2019) Towards a conceptualisation of smart tourists and their role within the smart destination scenario. Serv Ind J 39(2):109–133. https://doi.org/10.1080/02642069.2018.1508458
14. Gajdošík T (2020) Smart tourists as a profiling market segment: Implications for DMOs. Tour Econ 26(6):1042–1062. https://doi.org/10.1177/1354816619844368
15. Ballina FJ (2022) Smart business: the element of delay in the future of smart tourism. J Tourism Futures 8(1):37–54. https://doi.org/10.1108/JTF-02-2020-0018
16. Buhalis D, Leung R (2018) Smart hospitality—Interconnectivity and interoperability towards an ecosystem. Int J Hosp Manag 71(April):41–50. https://doi.org/10.1016/j.ijhm.2017.11.011
17. Gretzel U, Sigala M, Xiang Z, Koo C (2015) Smart tourism: foundations and developments. Electron Mark 25(3):179–188. https://doi.org/10.1007/s12525-015-0196-8
18. Boers B, Cottrell S (2007) Sustainable tourism infrastructure planning: a GIS-supported approach. Tour Geogr 9(1):1–21. https://doi.org/10.1080/14616680601092824
19. Rahmafitria F, Pearce PL, Oktadiana H, Putro HPH (2020) Tourism planning and planning theory: historical roots and contemporary alignment. Tourism Manag Perspect 35. https://doi.org/10.1016/j.tmp.2020.100703
20. Tosun C, Jenkins CL (1998) The evolution of tourism planning in Third-World countries: a critique. Progress Tourism Hospitality Res 4(2):101–114. https://doi.org/10.1002/(SICI)1099-1603(199806)4:2<101::AID-PTH100>3.0.CO;2-Z
21. Saarinen J, Rogerson CM, Hall CM (2017) Geographies of tourism development and planning. Tour Geogr 19(3):307–317. https://doi.org/10.1080/14616688.2017.1307442

22. UNWTO (2018) 'Overtourism'?—understanding and managing urban tourism growth beyond perceptions, executive summary. 'Overtourism'?—Understanding and Managing Urban Tourism Growth beyond Perceptions, Executive Summary. Madrid: World Tourism Organization (UNWTO). https://doi.org/10.18111/9789284420070
23. Baud-Bovy M, Lawson F (1998) Tourism and recreation handbook of planning and design. Architectural Press, Oxford
24. Hall CM (2008) Tourism planning: policies, processes and relationships (2 nd editi.). Prentice Hall
25. Wang L (2024) Enhancing tourism management through big data: design and implementation of an integrated information system. Heliyon 10(20):e38256. https://doi.org/10.1016/j.heliyon.2024.e38256
26. Roxas FMY, Rivera JPR, Gutierrez ELM (2020) Mapping stakeholders' roles in governing sustainable tourism destinations. J Hosp Tour Manag 45:387–398. https://doi.org/10.1016/j.jhtm.2020.09.005
27. Rickly JM, McCabe S (2017) Authenticity for tourism design and experience. In: Fesenmaier DR, Xiang Z (Eds), Design science in tourism: foundations of destination management (pp 55–68). Cham: Springer International Publishing. https://doi.org/10.1007/978-3-319-42773-7_5
28. Mandic A, Mrnjavac Ž, Kordic L (2018) Tourism infrastructure, recreational facilities and tourism development. Tourism Hosp Manag 24(1):41–62. https://doi.org/10.20867/thm.24.1.12
29. Pasquinelli C, Trunfio M (2023) Sustainability-oriented innovation in smart tourism: challenges and pitfalls of technology deployment for sustainable destinations (Tourism on the Verge.). Cham: Springer Nature Switzerland
30. Gajdošík T, Orelová A (2020) Smart technologies for smart tourism development. In: Silhavy R (Ed), Advances in intelligent systems and computing (Vol 1225 AISC, pp 333–343). Cham: Springer. https://doi.org/10.1007/978-3-030-51971-1_27
31. Sustacha I, Baños-Pino JF, Del Valle E (2023) The role of technology in enhancing the tourism experience in smart destinations: a meta-analysis. J Destin Mark Manag 30. https://doi.org/10.1016/j.jdmm.2023.100817
32. Gajdošíková Z, Gajdošík T, Kučerová J (2019) Slovak winter tourism destinations: future playground for tourists in the Carpathians. Tourism Hosp Manag 25(1):161–178. https://doi.org/10.20867/thm.25.1.7
33. Ietto B, Rabe J, Muth R, Pascucci F (2023) Blockchain for citizens' participation in urban planning: the case of the city of Berlin. A value sensitive design approach. Cities 140. https://doi.org/10.1016/j.cities.2023.104382
34. Koens K, Klijs J, Weber-Sabil J, Melissen F, Lalicic L, Mayer I, … Aall C (2022). Serious gaming to stimulate participatory urban tourism planning. J Sustain Tourism 30(9):2167–2186.https://doi.org/10.1080/09669582.2020.1819301
35. Ismael D (2024) Immersive visualization in infrastructure planning: Enhancing long-term resilience and sustainability. Energy Efficiency 17(7). https://doi.org/10.1007/s12053-024-10270-5
36. Banfi F (2021) The evolution of interactivity, immersion and interoperability in HBIM: digital model uses, VR and AR for built cultural heritage. ISPRS Int J Geo-Inform 10(10). https://doi.org/10.3390/ijgi10100685

Tomáš Gajdošík is an associate professor at the Department of Tourism at the Faculty of Economics of Matej Bel University in Banská Bystrica. In his scientific research and teaching activities he deals with smart tourism development, information technologies, data analysis, management and governance of tourism destinations. He is an expert of the European Commission in the projects Smart Tourism Destinations, European Capital of Smart Tourism, Crisis Management and Governance in Tourism. He is the author of scientific and professional monographs, university textbooks and teaching texts.

The Role of Technology in Managing Overtourism: The Case of the Historic Centre of Florence

Martina Franco, Marta Conte, Carlo Francini, Gianluca Belli, and Giuseppe De Luca

Abstract The purpose of this research is to provide a detailed picture of the development of the tourism phenomenon within a World Heritage Site, specifically the Historic Centre of Florence. Understanding the current state of affairs is essential for creating strategies that can sustainably relaunch tourism, expanding the focus to a broader territorial context and other sites that on the contrary are experiencing undertourism, such as the Medici Villas and Gardens. In this context, the Municipal Administration has adopted a restrictive policy aimed at improving the livability of the site, through an amendment to the urban planning instrumentation. Simultaneously, an analytical approach based on innovative technologies has been developed within the research activities of the Heritage Research Joint Laboratory to reconstruct the growth, development, and scenarios of overtourism. The phenomenon of overtourism has been analyzed using a database system structured with tourist density indicators and geolocation data within a Geographic Information System. Currently, the collected data is used only partially, lacking a comprehensive, systemic approach to its analysis and utilization. Efforts are now being made to integrate and expand this information through an integrated Information System for the unified management of the data collected on the territory. This system will support and provide crucial assistance for policy makers in their decision-making processes. In line with the principles of the International Charter for Cultural Tourism and supporting the

M. Franco (✉) · M. Conte · G. Belli · G. De Luca
University of Florence, Florence, Italy
e-mail: martina.franco@unifi.it

M. Conte
e-mail: marta.conte@unifi.it

G. Belli
e-mail: gianluca.belli@unifi.it

G. De Luca
e-mail: giuseppe.deluca@unifi.it

C. Francini
Municipality of Florence, Florence, Italy
e-mail: carlo.francini@comune.fi.it

V. C. Sharma et al. (eds.), *Architectural Vision of Overtourism*,
SpringerBriefs in Architectural Design and Technology,
https://doi.org/10.1007/978-981-96-5794-0_10

Sustainable Development Goals, the paper aims to illustrate strategies in the area of land governance that focus on diversified management of tourism activities.

Keywords World heritage site · Data analysis · Overtourism impacts · Managing change · Tourism management · Sustainability · Management plan · Governance

1 Introduction

1.1 Tourism, Sustainability, and Territorial Governance in the UNESCO Context

The rapid growth of the global tourism industry has, over the years, placed increasing pressure on destinations, transforming what was once a primary source of income for local economies into a sector that risks causing irreversible damage. According to the first World Tourism Barometer published by the UN World Tourism Organization (UNWTO) [1], nearly 790 million tourists traveled internationally in the first seven months of 2024—an 11% increase compared to 2023 and only 4% below 2019 levels. This brings us to 96% of pre-pandemic numbers, largely driven by strong demand in Europe and the reopening of markets in Asia and the Pacific. Despite ongoing debate and uncertainty surrounding its definition, overtourism can be described as "the impact of tourism on a destination, or parts of it, that excessively and negatively affects the quality of life of residents and/or the visitor experience" [2].

This study specifically examines the effects of overtourism on "cultural tourism" destinations, a phenomenon that gained prominence in the 1990s, with many of these sites' appeal as tourist destinations being reinforced by their World Heritage Site status. Currently, the World Heritage List (WHL) includes 1,223 sites, with the majority located in Europe (about 573 sites, or 46.85% when including North America) [3]. Italy offers a particularly telling example of overtourism, especially in cities like Florence and Venice, where UNESCO designation remains a powerful driver of socio-economic changes that affect residents' well-being [4–6], while also posing a potential threat to the authenticity of these sites.

At the international level, the World Heritage Centre initially took a cautious stance toward tourism. As early as 1976, the ICOMOS Cultural Tourism Charter, concerned about the growing scale of the phenomenon, stated: "cultural tourism cannot be considered separately from the negative, despoiling and destructive effects which the massive and uncontrolled use of monuments and sites entails" [7]. Today, growing concerns about the excessive number of visitors increasingly prompt site managers to adopt containment policies and focus on responsible tourism management [8–10]. This shift offers opportunities to improve social resilience and well-being. With the International Cultural Tourism Charter [11], which redefines the principles of the Brussels Charter, the International Council on Monuments and Sites (ICOMOS) has recognized both the importance of balancing the differing needs of visitors and residents, as supported by various studies [12–14], and the potential of the tourism sector as a source of community growth and, more broadly, of regional development. The underlying principle of this approach is sustainability, a concept introduced in 2015 with the 2030 Agenda for Sustainable Development, which has increasingly gained prominence in the global market.

To support the adoption and implementation of sustainable and effective tourism practices, the World Heritage Centre has recently introduced a series of tools and methodologies. For example, the UNESCO Sustainable Tourism Toolkit (2021) [15] helps site managers develop long-term strategies for preserving Outstanding Universal Value (OUV), while maximizing tourism benefits and sharing proven best practices from around the world. The Enhancing Our Heritage Toolkit 2.0 [16], published in 2023, offers a "multicriteria" approach for evaluating the effectiveness of management strategies in minimizing the impacts of key threats to the integrity and authenticity of heritage sites. Specifically, the "Management Planning Framework" highlights the importance of aligning territorial planning procedures with integrated programming and site management strategies. This synergy can be further enhanced through collaboration among the various stakeholders involved in conservation and enhancement efforts, enabling them to explore future transformation scenarios and address challenges such as high tourist pressure.

Within this theoretical framework, various perspectives emerge on the impact that UNESCO designation can have on the performance of tourist destinations, which in turn depends on multiple factors [17]. While the World Heritage Site (WHS) designation can enhance tourism appeal [18–20] by increasing awareness of the site's exceptional values [21], several studies indicate that in some cases, World Heritage status does not significantly affect tourism demand [22–24], but may instead play a more substantial role in safeguarding the site. Other studies [25, 26] suggest a negative relationship between UNESCO sites and destination performance, resulting from overly high expectations from both tourists and service providers, leading to potential disappointment. These contrasting outcomes become even more relevant when considering the necessary balance between good conservation practices, heritage enhancement, urban planning, and sustainable tourism [27–29], which serves as the foundation for the work presented in this article.

1.2 The "Historic Centre of Florence" and the "Medici Villas and Gardens in Tuscany": A Brief Contextual Analysis

This study aims to advance the understanding of the future management of two contrasting phenomena—overtourism and undertourism—affecting the UNESCO World Heritage Sites of the "Historic Centre of Florence" and the "Medici Villas and Gardens in Tuscany." The analysis begins by examining the effects of the tourism market on the city of Florence and aims to outline new potential research directions and management proposals that could support local governance tools (Fig. 1).

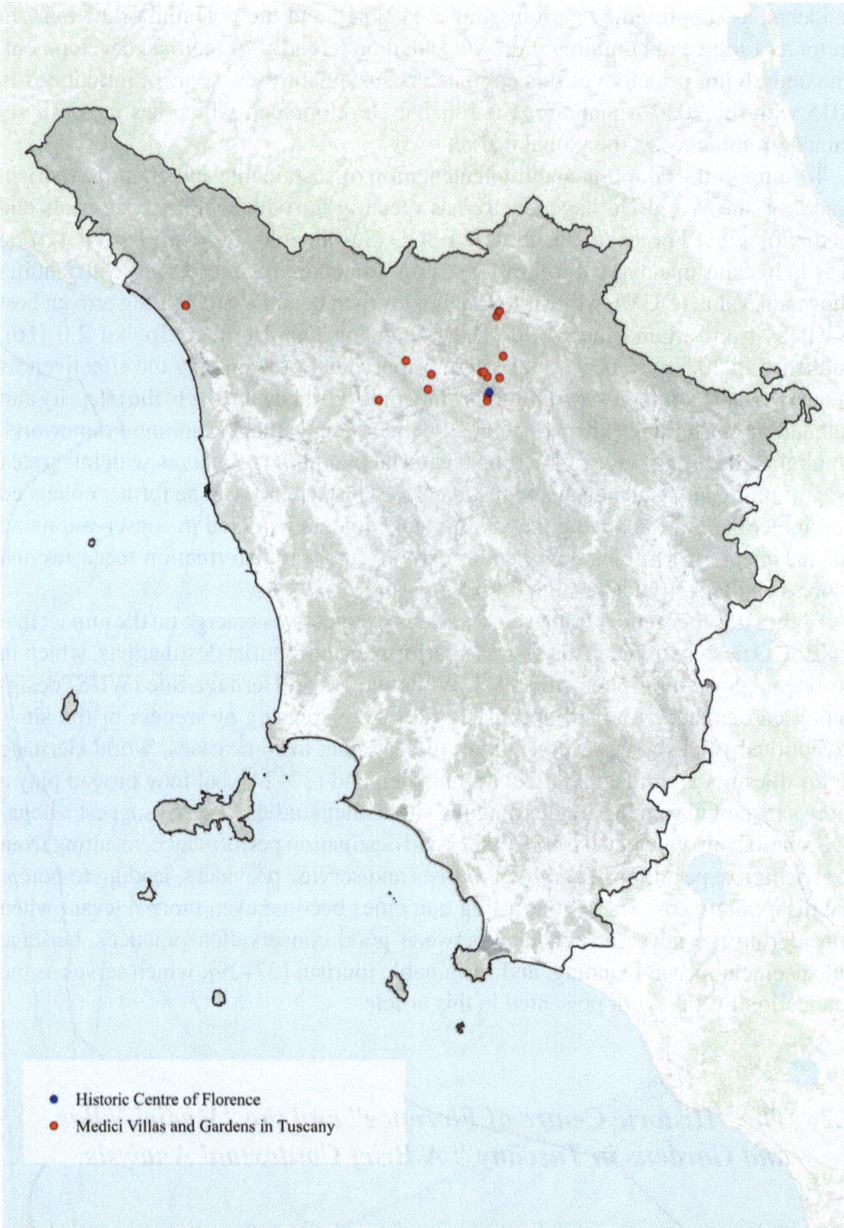

Fig. 1 Geographical distribution of the Historic Centre of Florence and the Medici Villas and Gardens in Tuscany Sites (Cartographic base and Geophysical localization of UNESCO sites—Tuscany Region Geoscope. Cartographic processing by the authors)

Regional map showing the spatial distribution of the two sites:

The Historic Centre of Florence

The Historic Centre of Florence was inscribed on the World Heritage List on December 17, 1982. Since 2006, the issue of overtourism has remained one of the most significant challenges for the site, as highlighted in the latest Management Plan (PdG) [30], developed according to the Historic Urban Landscape approach [31]. The rapid post-pandemic increase in tourist arrivals has created *saturation areas*, causing a series of negative impacts that seriously threaten not only the site's Outstanding Universal Value (OUV) but also the livability of the city and the well-being of its residents. Besides the gentrification and consequent depopulation of the Historic Centre, the negative effects of mass tourism include the deterioration of the site's attributes, pollution, environmental pressure, and the increasing homogenization of commercial activities, which primarily cater to visitors' needs.

The Florence World Heritage and Relations with UNESCO Office of the Municipality of Florence, in collaboration with the joint research laboratory HeRe Lab—Heritage Research (University of Florence and Municipality of Florence), continues to identify strategies to address the issues arising from tourism growth and to mitigate its possible impacts. Among the projects outlined in the Management Plan are the Firenze Greenway (developed in collaboration with the Medici Villas and Gardens) and the Feel Florence web portal, both aimed at alleviating congestion in the most crowded circuits by promoting alternative areas that are less frequented by visitors.

Medici Villas and Gardens in Tuscany

In 2013, the serial site "Medici Villas and Gardens in Tuscany" was inscribed on the World Heritage List. The process of updating its latest Management Plan (MP) [32] included data collection and stakeholder consultation to assess the current state of the fourteen components, the factors generally affecting the OUV, and the severity of these threats. The importance of tourism for the site development led the Tuscany Region to commission a quantitative analysis of visitor flows, which was completed in September 2023. The study, conducted by IRPET, highlighted that the tourism competitiveness of the components is closely tied to the "competitiveness" of the territories they are located in (Florence, Mugello, Empolese-Montalbano, and Versilia). Other critical issues identified in the Management Plan include the need to improve physical accessibility (through dedicated transport or soft mobility routes) and digital accessibility. Several initiatives have been launched in this regard, while others are still in the planning phase.

2 Literature Review

2.1 State of the Art and Contextualization of the Topic

Although tourism is vital for the economy of many destinations, the phenomenon of overtourism, exacerbated by inadequate management, continues to have a significant impact on destinations, particularly if the cities of art are concerned [33]. Given the complexity of the issue, numerous studies have explored methodologies for managing tourist flows, using tools such as carrying capacity indicators [34, 35] and Big Data [36], as well as examining regulations implemented internationally [37, 38].

In the case of Florence, studies on tourist carrying capacity have been essential in defining new scenarios for sustainable development that aim to preserve the city's social fabric, economy, and cultural identity in the long term. As will be discussed in the following sections, such systematical analysis requires the availability of official and up-to-date data.

As previously mentioned, one of the most critical aspects of tourism management is the establishment of a clear legislative framework for all stakeholders and policymakers. This requires a preliminary study of the evolution of both national and regional regulations, which developed alongside the growth of the tourism sector itself.

The key legislative measures adopted at the national level can be summarized as follows:

- From the early twentieth century to the 1930s, laws were enacted to promote national tourism, introduce a tourist tax, set health and safety standards, establish tourism management bodies, and classify accommodation facilities.
- In the post-war period, more comprehensive regulations were introduced, making tourism and the hotel industry matters of regional legislative competence, followed by the establishment of the Ministry of Tourism in 1956.
- From the 1980s, driven by the need for regulatory restructuring, Italy began aligning its tourism legislation with that of other European countries, setting out fundamental principles for managing tourism and the hospitality industry. Notable measures include.
  - Law No. 217/1983, which set out general principles and allocated competencies between the State and Regions, though inconsistencies persisted in regional legislation.
  - Law No. 135/2001, which replaced the previous law, redistributed competencies between the State and Regions, defined the Ministry of Tourism as the governing body for the tourism sector, and established the National Tourism Conference, local tourism systems, and funds to support tourism services.

- Further legislative changes were introduced in the early 2000s, starting with the reform of Title V of the Constitution, which reshaped regional powers and functions, as referenced in Article 117. In 2011, the Tourism Code (Legislative Decree No. 79) was developed to revive the sector.
- In the past eight years, multiple regulations have been enacted, including Law No. 198/2023, which introduced.
  - the National Identification Code (NIC), effective from November 2, issued by the Ministry of Tourism to monitor all tourism-related businesses.
  - safety regulations for equipment installation, mandatory for obtaining the NIC.
  - a national database for accommodation facilities and short-term rentals.
  - penalties, starting in 2025, for failing to display the NIC on all listing platforms, including Airbnb.

Despite various attempts to regulate the management of tourist flows, a clear national regulatory framework has yet to be fully established, and the regional bureaucratic systems remain diverse and complex.

2.2 The Regional Legislative Framework

In Tuscany, tourism regulations are set by Regional Law no. 86/2016 and subsequent amendments, which replaced the earlier Regional Law no. 42/2000. This law covers various aspects of the tourism sector and accommodation facilities [39].

The key elements of the regional legislative framework include

- classification of accommodation facilities with defined minimum standards;
- specific regulations for short-term rentals and tourist accommodations, in line with national laws;
- promotion of sustainable tourism by encouraging eco-friendly practices, such as the use of renewable energy in accommodations and the preservation of the region's natural and cultural heritage;
- implementation of a tourist tax, with rates and rules varying by municipality;
- initiatives to promote tourism, including efforts to enhance the region's artistic, cultural, and landscape heritage through partnerships with both public and private entities.

In summary, the main objectives of these updates are

- restructuring governance with a focus on empowering municipalities;
- improving the competitiveness of hotel accommodations through simplified regulations and an expansion of services;
- clarifying the roles and responsibilities of non-hotel accommodations, particularly for guesthouses and B&Bs;
- introducing specific guidelines for short-term rentals in response to the rise of the sharing economy;

- updating regulations concerning travel agencies.

An analysis of the current situation shows that, while regional legislation does not fully address the gaps in national law, it assigns local administrations the task of developing their own strategies for managing tourism accommodations.

2.3 *Local Management: The Case of the Historic Centre of Florence*

The shortcomings in both national and regional legislation have prompted the City of Florence to implement local initiatives that, although limited in scope, seek to regulate urban land use in accordance with Article 23-ter of DPR 380/2001 and Articles 98 and 99 of Regional Law no. 65/2014. The attempt to regulate tourism at the local level is part of the broader territorial governance framework for local authorities. This framework regards urban planning measures necessary for guiding urban development not only from a building and infrastructure perspective but also in terms of social, economic, and cultural growth [40]. In this context, several urban planning actions have been undertaken to protect the Historic Centre, recognized as "invariant" in the Structural Plan since 2010. This designation reinforces the goal of focusing future interventions on maintaining the site's Outstanding Universal Value (OUV).

Key actions taken by the City of Florence include

- The introduction of a tourist tax and the approval of regulations for short-term rentals;
- A memorandum of understanding signed with Airbnb Ireland in 2016 to facilitate the collection of the tourist tax, followed by a 2018 agreement for Airbnb to directly collect the tax;
- The inclusion of a ban on new retail, food, and beverage establishments in the "Regulation for the Protection of the Cultural Heritage of the Historic Centre";
- The creation of specific rules to preserve historic and/or traditional economic activities within the Historic Centre;
- A second memorandum of understanding to promote Florence as a tourist destination, while also developing strategies to decentralize tourist flows.

In line with these efforts, in 2023, the Florence administration amended the Urban Planning Regulation (UPR) to manage properties rented for short-term tourism purposes, introducing restrictions that go beyond those in higher level legislation. At the regional level, the Consolidated Text on the Regional Tourism System (LR no. 86/2016) refers to the provisions of Legislative Decree no. 79/2011 (Article 54), identifying non-hotel accommodations, such as guesthouses, B&Bs, vacation homes, and historic residences, that can be established in residential properties. These businesses can start operating by submitting a certified notice to the administration, without requiring prior authorization. Local urban planning is governed by the previously mentioned laws, as well as Regional Law no. 65/2014 (Articles 98 and 99) and DPR 380/2001 (Article 23-ter). These regulations give local authorities the power to define the functional characteristics of the building categories they oversee, allowing them to introduce restrictions on potential uses. In response, the City of Florence introduced the functional category of "temporary residence," which includes non-hotel accommodations and short-term rentals. At the same time, restrictions were imposed in areas most heavily affected by tourism-related businesses. The urban area most impacted by tourism is the World Heritage Site, known in urban planning terms as Zone A1—Area of the Historical Nucleus. In terms of territorial extension, Zone A1 coincided with the boundaries of the Historic Centre of Florence until 2021, when the perimeter of the World Heritage Site was expanded to include the San Miniato al Monte complex and the surrounding area, as approved in decision 44 COM 8B.56 by the World Heritage Committee (Fig. 2).

The relevant restrictions in the Technical Implementation Rules (TIR) refer to

- Article 19—Classification of Uses, which introduces the category of temporary residence, encompassing short-term rentals and non-hotel accommodations;
- Article 65—Area of the Historical Nucleus (Zone A), which explicitly prohibits the use of housing as temporary residences [41].

In the long term, these measures are expected not only to reduce the impact of mass tourism but also to gradually restore the Historic Centre to the resident community. However, these non-retroactive rules have drawn considerable criticism from industry stakeholders, leading to legal appeals in the Regional Administrative Court (RAC). Following the court's ruling on July 10, 2024, and the inauguration of the new City Council, on July 30, 2024, an amendment to the Structural Plan and Operational Plan, two key urban planning instruments, was adopted. This amendment officially incorporated the category of temporary residence and enforced its prohibition within the Historic Centre (Table 1).

The amendment approval process is still ongoing; however, its initiation clearly reflects a political commitment to addressing the impacts of unregulated tourism development within the historic city.

3 Research Methodology

3.1 The Historic Centre of Florence Between Gentrification and Touristification: Spatial Analysis and Digital Mapping

Urban processes like gentrification and touristification are often analyzed independently, even though they are interrelated and often difficult to distinguish [42]. "Gentrification" usually describes changes in land use aimed at increasing profitability,

Fig. 2 Discrepancies between the areas of the Historic Centre of Florence World Heritage Site and the "Historic Centre" or Zone A1 (Topographic map, Boundaries of the Municipality of Florence, Boundaries of the Property of the Historic Centre of Florence—Tuscany Region Geoscope; Boundaries of the of neighborhood Q1 and Zone A1—Open Data of the Municipality of Florence. Cartographic processing by the authors)

while "touristification" refers to transformations in a given area driven by tourism [43]. Indicators of touristification include the rising number of short-term rentals to meet visitor demand, tourism's impact on GDP, and job creation within the sector. Recent literature reveals different perspectives on the relationship between these two phenomena [44–47]. However, in the case of the Historic Centre of Florence, the

Table 1 Chronology of phases for amendments to the Urban Planning Regulation (UPR). Data source Municipality of Florence Del. CC DC/2024/00057

Urban planning tools	Date and regulatory reference	Procedure status and details of amendments
Structural plan	06.22.2011	Approved
	Del. CC n.2011/C/00036	
Urban planning regulation	04.02.2015	Approved
	Del. CC n.2015/C/00025	
New structural plan and operational plan	03.13.2023	Adoption of New Urban Planning Instruments
	Del. CC DC/2023/00006	
Amendment to urban planning regulation	10.02.2023	
	Del. CC DC/2023/00039	Adopted
		Amendment to Articles 19 and 65 of the Urban Planning Regulation concerning "short-term tourist rentals"
Adoption of the UPR amendment challenged by 15 appeals to the Regional Administrative Court (RAC)		
New structural plan and operational plan	12.29.2023	Approval of the counterarguments. The approved text of the TIR has been aligned with the ongoing amendment to the UPR
Municipal council measure	03.27.2024	Through a mayoral amendment, the text of the Technical Implementation Rules of the new Operational Plan is updated by eliminating the changes introduced by the amendment to the UPR
10.07.2024—Issuance of a ruling by the RAC deeming the effects of the UPR amendment ineffective due to the City Council's amendment, following the approval of the counterarguments to the new Operational Plan on 27.03.2024		
June 2024—Municipal elections		
	07.30.2024	Adoption of the amendment "Short-Term Tourist Rentals. Amendment to the Technical Implementation Rules of the Operational Plan"
Amendment to new Structural Plan and Operational Plan	Del. CC DC/2024/00057	
08.14.2024–Deadline for observations on the amendment		

interplay between gentrification and touristification has resulted in a growing imbalance among city users: residents, temporary inhabitants, and occasional visitors. This study adopts a spatial approach to exploring the relationship between gentrification and touristification, aiming to build a preliminary knowledge framework that could inform future territorial governance strategies. The current landscape consists of areas with diverse resources and varying tourism potential, marked by their proximity to the UNESCO serial site "Medici Villas and Gardens" (see image below) (Fig. 3).

A defining aspect of this research has been the review of recent developments in the tourism sector. The analysis not only examines the impact of gentrification in Florence but also compares data on accommodation supply with demand-side insights, seeking to offer a comprehensive view of the progression of both phenomena. The research extends from the Historic Centre core area to economically

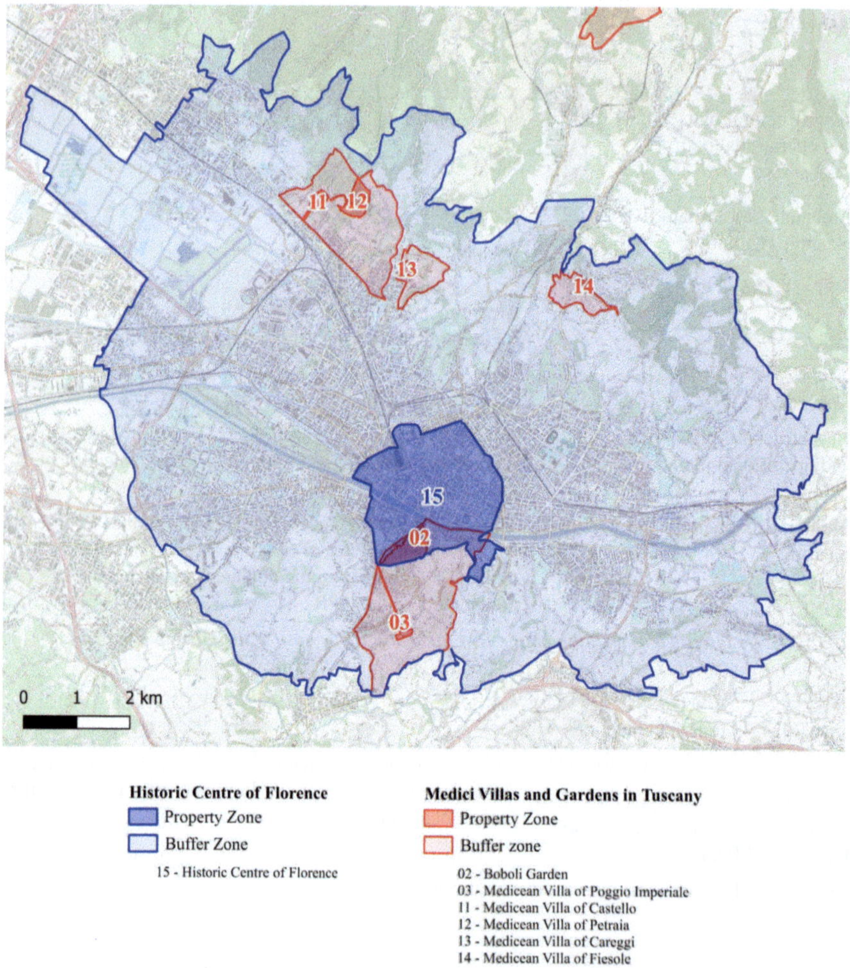

Fig. 3 Historic Centre of Florence and Medici Villas and Gardens in Tuscany: Properties and Buffer Zones (Cartographic base and Geophysical localization of UNESCO sites—Tuscany Region Geoscope. Cartographic processing by the authors)

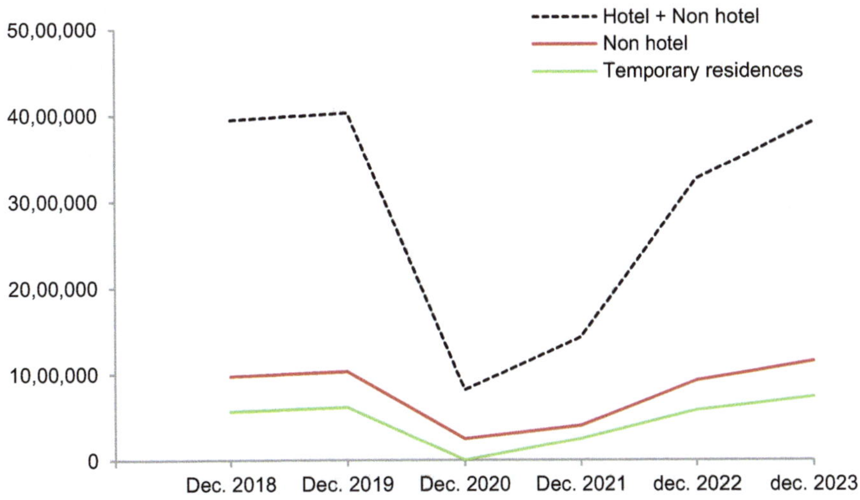

Fig. 4 Tourist arrivals by accommodation type in the Municipality of Florence (Tourism database—Region of Tuscany)

marginal zones surrounding the World Heritage Site, aiming to estimate the distribution of non-hotel accommodations, specifically "temporary residences," defined as residential properties rented for periods not exceeding 30 days. This is complemented by an analysis of demographic and socio-economic changes within the same spatial–temporal framework. The following images illustrate tourist flow trends by accommodation type, with a particular focus on temporary residences (Figs. 4 and 5).

For the descriptive statistical analysis, open institutional data were consulted, along with platforms that provide additional insights into the short-term rental market (Table 2).

In this context, cartography plays a crucial role, given the increasing need to implement digital tools that enhance territorial understanding, support decision-making processes, and monitor the effectiveness of planned interventions. The City of Florence has developed a Territorial Information System (TIS) to manage a municipal database, aimed at providing up-to-date information to internal departments, external public and private entities, and the general public. The TIS is composed of several platforms:

- MapStore, which systematizes university research results related to the Historic Centre [48];
- OpenData, which facilitates data sharing and shapefile downloads [49];
- WebRU and WebPO, which enable interactive publication of documents related to urban planning tools [50, 51].

The HeRe Lab—Heritage Research Laboratory has led several initiatives to enhance the cartographic capabilities of the municipal database, particularly for

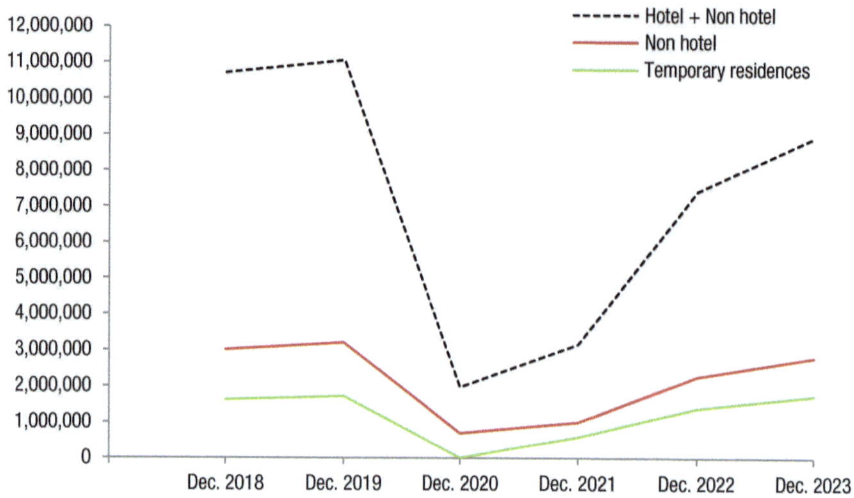

Fig. 5 Tourist stays by accommodation type in the Municipality of Florence (Tourism database—Region of Tuscany)

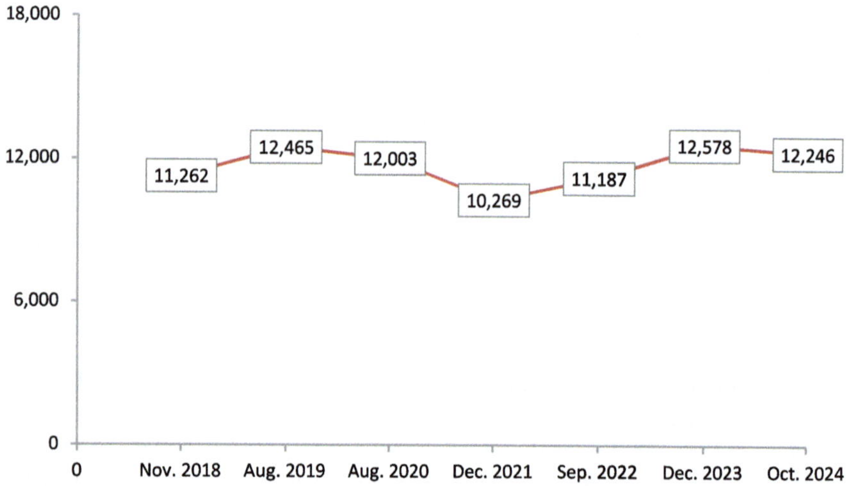

Fig. 6 Number of housing units listed on Inside Airbnb for Florence (Inside Airbnb; Romano A., 2024)

the WebGIS MapStore. This has been made possible through Ministry funding, including the Florence Heritage Data System project [52]. In line with the administration's objectives, the project involved gathering data (in CSV format) from multiple sources and indexing them. To map the information for inclusion in the TIS, Geographic Information System (GIS) software was required, specifically the open-source Quantum GIS. This approach allowed for the management of various

Table 2 Summary table of analyzed and processed data, list of sources consulted

Data	Sources	Processing
Resident population	Statistical data—City of Florence	Comparative table
		GIS cartography
Tourist flows	Statistical data—Region of Tuscany	Cartesian diagram
	Open data—Region of Tuscany	
	Tourism database—Region of Tuscany	
Sharing economy	Literature review	Cartesian diagram
	Database—Inside Airbnb	GIS cartography

data types, including vector data, satellite images, and georeferenced raster data. The interpolation process involved linking alphanumeric data (e.g., demographic data) with geographic data (e.g., polygons or points on a map).

The data processing workflow was structured into the following phases:

- collection of tabular data;
- collection of geographic data, including:
  - Properties and Buffer Zones of UNESCO sites;
  - delimitation of Florence's neighborhoods and municipal OpenData;
  - administrative boundaries of Florence and GEOscopio of the Tuscan Regional Authority;
  - cartographic bases from the Web Map Service (WMS) and GEOscopio of the Tuscan Regional Authority;
- importation of polygon vector data (shapefiles) and CSV tables into QGIS;
- identification of a common field between the vector data and the CSV table for data association;
- verification of the results;
- cartographic rendering.

The project's completion will allow for the expansion of the MapStore TIS, integrating the processed data with the municipal property registry (ACI) and aligning it with the UTM WGS84 reference system [53].

4 Discussion and Results

The rapid growth of the sharing economy, identified in this study as a major driver of gentrification, has led to an unchecked expansion of touristification, particularly in the city's central areas.

Analysis of a dataset spanning seven years (2018–2024) in the neighborhood Q1 (Historic Centre) of Florence reveals a 9.42% increase in short-term rental listings, based on data from Inside Airbnb and various studies on the subject [54]. This increase has fueled inflation in the housing market, with rental prices rising by 38.89% and property sale values increasing by 12.45% (Fig. 6).

The inflationary pressure created by the spread of temporary residences has also directly contributed to the phenomenon of displacement, observed in two forms: the displacement of the resident population and of the real estate market dynamics [55] (Table 3).

To assess the intensity of gentrification and touristification separately in Florence, heat maps were created using QGIS tools based on previously cited data (see Figs. 7, 8, and 9). Each map uses a sequential color gradient, with darker tones indicating higher values [56]. This approach provides insight into the current state of the phenomena and potential future developments.

The maps indicate that certain parts of the city have become increasingly specialized in tourism-related services, generating negative externalities that affect the livability and identity of these transformed urban spaces.

Preliminary analysis shows that the density of tourist accommodations (see Fig. 7) is significantly concentrated in the city center (73.3%), with a strong cluster within the UNESCO World Heritage Site, which serves as the phenomenon's core and radiates outward. One notable expansion area is Q2 (Campo di Marte) to the northeast, with 1,287 short-term rental listings (10.5% of the total), followed by Q5 (Rifredi) with 1,007 listings (8.2%) [57]. Listings of entire properties (10,180 recorded on Inside Airbnb in October 2024) dominate, accounting for 63.1% in Q1, followed by Q2 (Campo di Marte) with 7.8%.

The spread of short-term rentals beyond the city center is expected to continue, driven by the newly approved Urban Planning Regulation amendment. This shift may be attributed to the saturation of the Historic Centre and a growing demand for more "authentic" experiences, prompting visitors to choose less crowded areas with services they can share with residents, such as cultural centers, local markets, and, most importantly, public transportation networks that make tourist attractions more accessible. Further research on "digital reputation," through user-generated reviews, could offer insights into how demand shapes urban tourism practices, the spatial distribution of short-term rentals, and broader market trends [58].

Table 3 Number of residents by district from 2018 to 2024 (Statistical data—City of Florence)

Year	Q1	Q2	Q3	Q4	Q5
2018	66.532	91.389	41.211	68.188	109.209
2019	65.813	90.789	40.961	67.658	107.684
2020	64.213	89.415	40.191	66.297	105.321
2021	65.203	89.312	40.301	66.036	105.094
2022	64.483	89.787	40.371	66.626	106.233
2023	64.158	90.031	40.300	66.636	106.651
2024	63.442	89.749	40.235	66.670	106.856

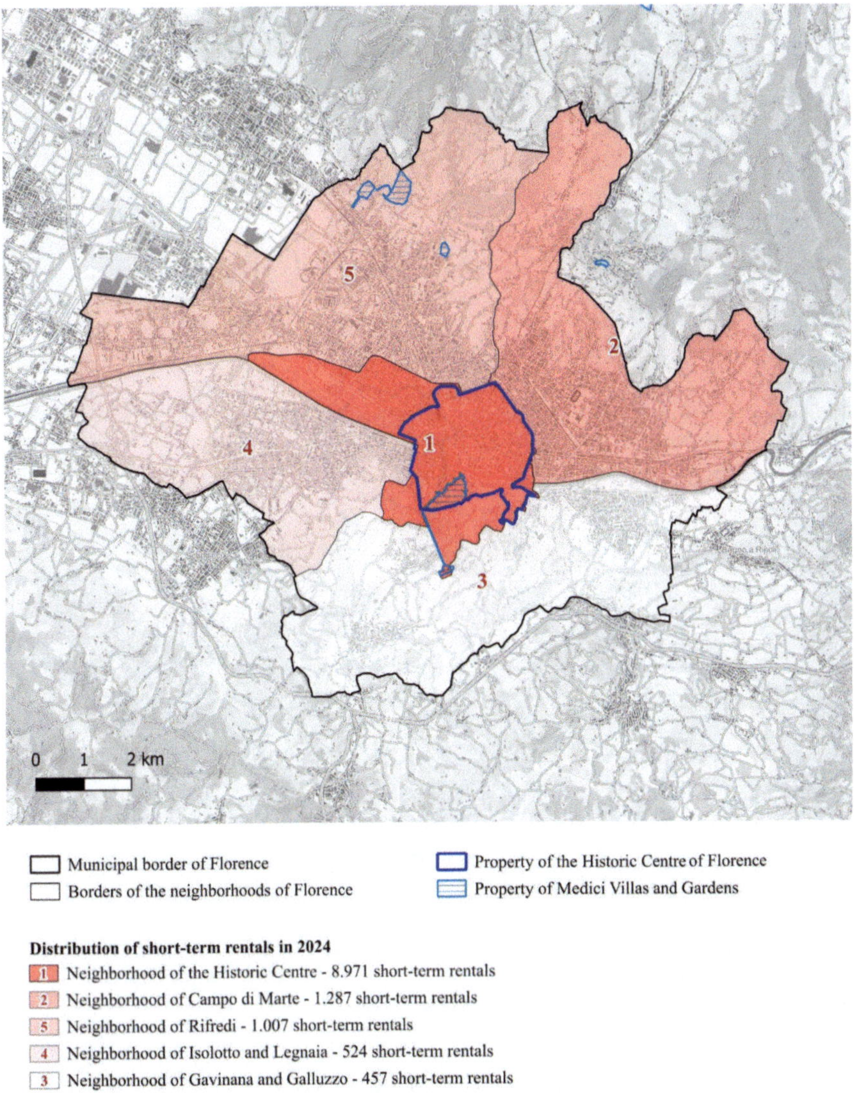

Fig. 7 Distribution of short-term rentals in 2024 (Topographic map, Boundaries of the Municipality of Florence, Boundaries of UNESCO Site Properties—Tuscany Region Geoscope, distribution of short-term rentals in 2024 Inside Airbnb. Cartographic processing by the authors)

Another notable trend is the lower concentration of tourist accommodations in Q4 (Isolotto and Legnaia) and Q3 (Gavinana and Galluzzo), the latter encompassing Villa of Poggio Imperiale, an area characterized by limited infrastructure (Table 4).

The natural expansion of the accommodation sector into more peripheral urban areas, provided it is guided by specific regulations that ensure a more balanced

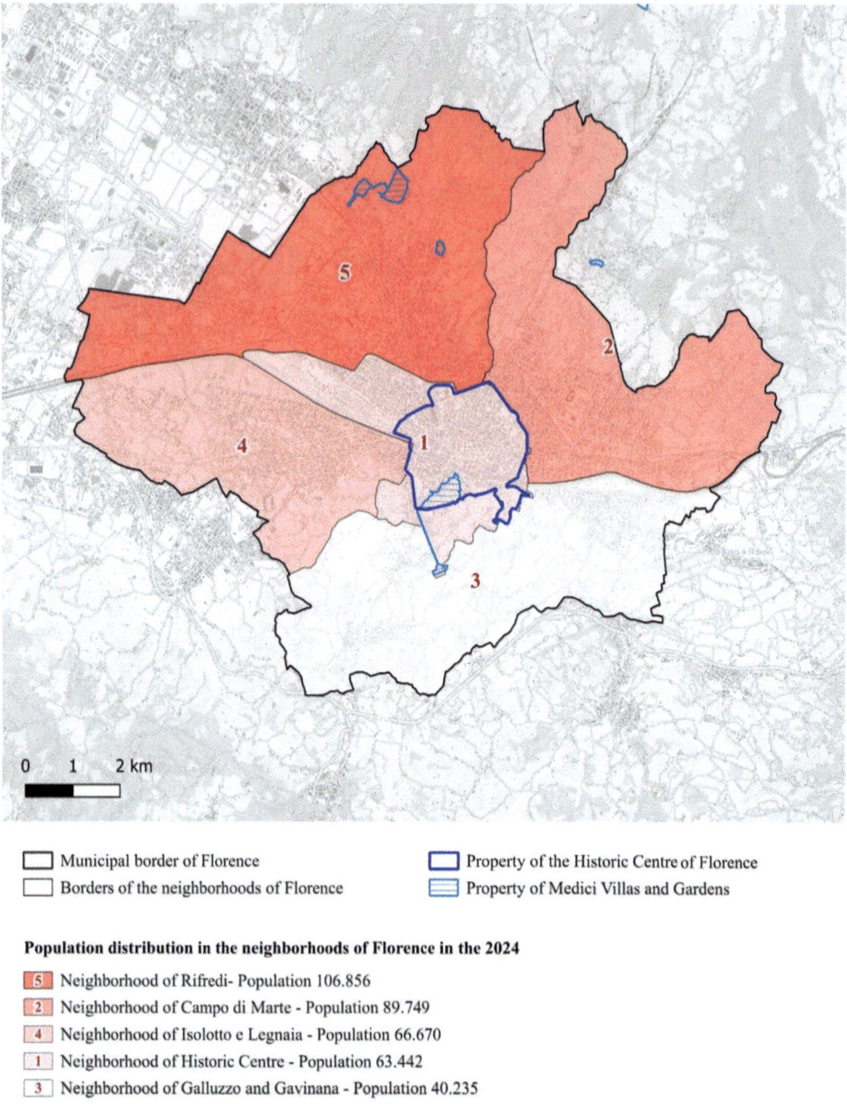

Fig. 8 Population distribution in the neighborhoods of Florence in 2024 (Topographic map, Boundaries of the Municipality of Florence, Boundaries of UNESCO Site Properties—Tuscany Region Geoscope, population distribution in the neighborhood of Florence in 2024 Municipality of Florence. Cartographic processing by the authors)

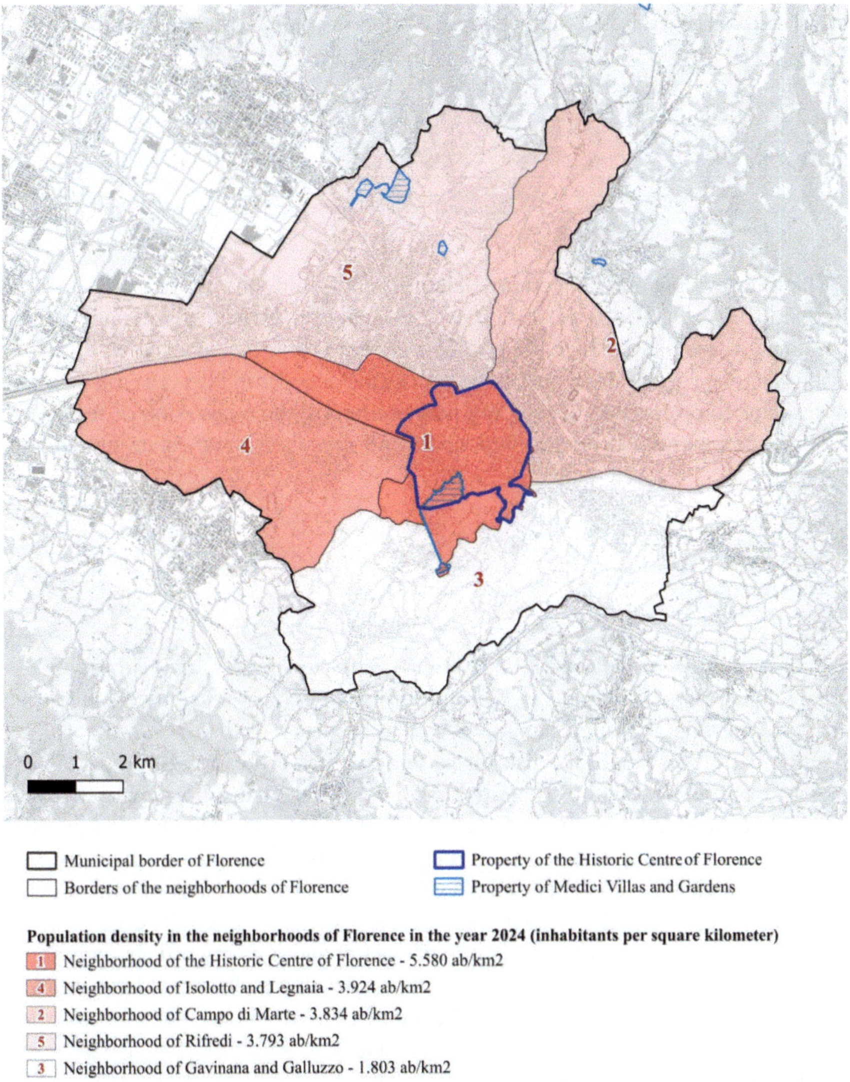

Fig. 9 Population density in the neighborhoods of Florence in the year 2024 (Topographic map, Boundaries of the Municipality of Florence, Boundaries of UNESCO Site Properties—Tuscany Region Geoscope, population density in the neighborhoods of Florence in 2024 by the Municipality of Florence. Cartographic processing by the authors)

and controlled distribution, could play a strategic role. It would facilitate a more widespread distribution of tourist flows across the territory and encourage a gradual shift toward currently less attractive destinations, potentially transforming these areas into key hubs for inclusive and sustainable tourism.

Table 4 Percentage change in short-term rental distribution in Florence by district 2018–2024 (Inside Airbnb)

Year	Q1	Q2	Q3	Q4	Q5
2018	8.198	1.276	451	498	839
2024	8.971	1.287	457	524	1.007
	+ 9,42%	+ 0,86%	+ 1,33%	+ 5,22%	+ 20,02%

The most evident and concerning outcome, confirming the above analysis, is the direct link between the spread of short-term rentals and depopulation dynamics. This phenomenon, persisting since the post-World War II era, continues to be driven by various factors, such as the reduction in residential occupancy rates, the relocation of the resident community to developing neighborhoods, the conversion of residential units into commercial or tertiary functions, and speculative real estate investments [59].

5 Conclusion

The comparison between the estimated number of short-term rental listings and the actual number of active rentals in Florence brings to light significant considerations for managing this growing phenomenon. The accuracy of data is a critical issue, affecting both the availability of information and the verification of data from key databases, particularly at the municipal level. For instance, the discrepancy between the 12,411 estimated listings reported by the Centre for Tourism Studies in June 2024 [60] and the 17,282 registered on the tourist tax portal in August of the same year [61] reveals a serious gap in centralized and reliable data. This lack of accurate data poses a major challenge for local administration, especially given the urgent need to manage a phenomenon that could have substantial and ongoing effects on both the city's structure and residents' quality of life. To develop a clearer understanding of the current state of accommodations in the municipality, thematic maps were created. These maps not only help to understand the real impact on the area but also the socio-economic changes affecting the community. The challenges encountered during the research phase underscored the need to refine the methods for data collection and dissemination, primarily at the local level, to improve the effectiveness of future territorial development strategies.

From a legislative perspective, there is a clear absence of specific regulations governing tourist flows and accommodations. This indicates that national legislation should focus on providing a framework for regional statutes, making it easier for smaller administrations to implement sustainable urban development practices. These should include integrated models for the conservation and use of sites under significant human pressure, as well as those considered "marginal" due to low tourist appeal. Greater institutional collaboration is also necessary to introduce new approaches focused on promoting the universally recognized values of World Heritage Sites (WHS), often overlooked, and on positioning tourism as a driver of sustainable growth for both territories and host communities, in line with the principles of the International Cultural Tourism Charter.

Finally, a key issue identified during the research is the inconsistency in the boundaries of the areas most affected by overtourism, including the A1 Historical Nucleus defined by municipal planning tools, the UNESCO World Heritage Site Property, and the neighborhood Q1, also known as the "Historic Centre." The lack of alignment between these boundaries and the regulations governing them suggests a need to resolve this fragmentation and establish more cohesive planning strategies.

Acknowledgements The authors would like to thank Tatiana Rozochkina for her gentle and careful linguistic assistance.

References

1. UNWTO: World tourism barometer data Homepage. https://www.unwto.org/un-tourism-world-tourism-barometer-data. Accessed 3 Oct 2024
2. World Tourism Organization (UNWTO): Overtourism? Understanding and Managing Urban Tourism Growth beyond Perceptions (2018)
3. UNESCO: World Heritage List Statistics Homepage. https://whc.unesco.org/en/list/stat/. Accessed 3 Oct 2024
4. Bertocchi D, Camatti N (2022) Tourism in Venice: mapping overtourism and exploring solutions. In: A research agenda for urban tourism, pp 107–125. Ed. Elgar, KU Leuven
5. Del Bianco C, Montedoro L (2023) Cultural tourism pressure on historic centres: Its impact on public space and intervention strategies for its mitigation. Florence, Italy, as a case study. In: Housing environment, pp 80–86. Ed. Wacław Seruga, Cracow
6. Cheer J. M: Travel without leaving a trace Homepage. https://courier.unesco.org/en/articles/travel-without-leaving-trace (2024)
7. ICOMOS: International Cultural Tourism Charter Homepage. https://www.icomosictc.org/p/1976-icomos-cultural-tourism-charter.html#:~:text=It%20can%20be%20considered%20the,economic%20benefit%20and%20cultural%20education. (1976)
8. De Luca G, Dastgerdi AS, Francini C, Liberatore G (2020) Sustainable cultural heritage planning and management of overtourism in art cities: lessons from atlas world heritage. In: Sustainability, by MDPI
9. De la Calle-Vaquero M, García-Hernández M, Mendoza de Miguel S (2020) Urban planning regulations for tourism in the context of overtourism. Applications in historic centres. sustainability. In: Sustainability, pp 1–22 by MDPI
10. Menegus G (2024) Notes for a tourist city government. In: Italian papers on federalism, pp 79–101. Ed. ISSiRFA—CNR, Rome
11. ICOMOS: International Cultural Tourism Charter Homepage. https://www.icomosictc.org/p/2022-icomos-international-cultural.html (2022)
12. Li M, Wu B, Cai L (2008) Tourism development of World Heritage Sites in China: A geographic perspective. In: Tourism management, pp 308–319. Ed. Elsevier, Amsterdam
13. Zhuang X, Yao Y, Li J (2019) Sociocultural impacts of tourism on residents of world cultural heritage sites in China. In: Sustainability by MDPI
14. García-Buades ME, García-Sastre MA, Alemany-Hormaeche M (2022) Effects of overtourism, local government, and tourist behavior on residents' perceptions in Alcúdia (Majorca, Spain). In: Journal of outdoor recreation and tourism. Ed. Elsevier, Amsterdam

15. UNESCO: World Heritage Sustainable Tourism Toolkit Homepage. https://whc.unesco.org/en/sustainabletourismtoolkit/ (2021)
16. UNESCO: Enhancing Our Heritage Toolkit 2.0 (EoH 2.0) Homepage. https://whc.unesco.org/en/eoh20/ (2023)
17. Yang Y, Xue L, Jones TE (2018) Tourism-enhancing effect of world heritage sites: panacea or placebo? A meta-analysis. In: Annals of tourism research, pp 29–41. Ed. Elsevier, Amsterdam
18. Yang C, Lin H, Han C (2010) Analysis of international tourist arrivals in China: The role of World Heritage Sites. In: Tourism management, pp 827–837. Ed. Elsevier, Amsterdam
19. Su Y, Lin H (2014) Analysis of international tourist arrivals worldwide: the role of world heritage sites. In: Tourism management, pp 46–58. Ed. Elsevier, Amsterdam
20. Cellini R, Cuccia T (2016) UNESCO sites as public goods: comparative experiences in Italy. In: Journal of contemporary economics (REC), pp 553–569. Ed. UFRJ, Rio de Janeiro
21. UNESCO WHC: World Heritage Education Kit Homepage. https://whc.unesco.org/en/activities/567/ (2008)
22. Cellini R (2011) Is UNESCO recognition effective in fostering tourism? A comment on Yang, Lin and Han. In: Tourism management, pp 452–454. Ed. Elsevier, Amsterdam
23. Huang C, Tsaur J, Yang C (2012) Does world heritage list really induce more tourists? Evidence from Macau. In: Tourism management, pp 1450–1457. Ed. Elsevier, Amsterdam
24. Ribaudo G, Figini P (2017) The puzzle of tourism demand at destinations hosting UNESCO world heritage sites: an analysis of tourism flows for Italy. In: Journal of travel research. Ed. SAGE, New York
25. Cuccia T, Guccio C, Rizzo I (2013) Does UNESCO inscription affect the performance of tourism destinations? a regional perspective. In: ACEI working paper series, New South Wales
26. King LM, Halpenny E (2014) Communicating the World Heritage brand: visitor awareness of UNESCO's World Heritage symbol and the implications for sites, stakeholders and sustainable management. In: Journal of sustainable tourism, pp 768–786. Ed. Routledge, London
27. Bocchio C, De Luca G, Francini C, Liberatore G (2021) World heritage sites in action: experiences in governance, population and tourism management. In: Heritage in action. Proceedings XXIII SIU national conference, Turin 17–18 june (Vol. 06), pp. 15–22. Ed. Planum, Rome
28. Landorf C (2009) Managing for sustainable tourism: a review of six cultural World Heritage Sites. In: Journal of sustainable tourism, pp 53–70. Ed. Routledge, London
29. Somsamone V (2017) Sustainable tourism in the world heritage town Luang Prabang, Lao PDR. In: Journal of appropriate technology, pp 162–175. Ed. Academic Society for Appropriate Technology, Seoul
30. Florence World Heritage and Relations with UNESCO Office Homepage, The Management Plan of the Historic Centre of Florence, UNESCO World Heritage (2022)
31. UNESCO: Recommendation on the Historic Urban Landscape Homepage. https://whc.unesco.org/uploads/activities/documents/activity-638-98.pdf (2011)
32. Tuscan Regional Authority Homepage, The Management Plan for the Medici Villas and Gardens in Tuscany, UNESCO World Heritage (2024)
33. Séraphin H, Gladkikh T, Thanh TV (2020) Overtourism causes, implications and solutions. In: Overtourism: causes, implications and solutions. Palgrave Macmillan, Vol 448
34. Liberatore G, Biagioni P, Ciappei C, Francini C (2021) Preserving UNESCO world heritage sites: assessing and monitoring the carrying capacity of the historic city centre of Florence. Available at SSRN 3993868
35. McKinsey & Company: Coping with success: managing overcrowding in tourism destinations. World Travel and Tourism Council, UNWTO, Londra
36. Meini M, Nocera R, Spinelli G (2013) Spatial analysis of tourist mobility through geotagging tools: an experiment with the Panoramio web community, Trieste
37. Celata F, Tourism, hypertourism and cultural heritage in Rome: dynamics, issues, evidence and research perspectives Homepage. https://riabitiamoroma.it/wp-content/uploads/2024/05/Celata-Turismo-iperturismo-e-patrimonio-culturale-a-Roma.pdf (2023).
38. Bernardi M (2018) The impact of AirBnB on our cities: gentrification and 'disneyfication' 2.0 Homepage, https://labgov.city/theurbanmedialab/the-impact-of-airbnb-on-our-cities-gentrification-and-disneyfication-2-0/

39. Tuscan Regional Authority: Single text of the regional tourism system Homepage. https://www.regione.toscana.it/-/turismo-la-nuova-legge-regionale-testo-unico-sul-sistema-turistico-regionale
40. Municipality of Florence: Urban Planning Regulations Homepage. https://regolamentourbanistico.comune.fi.it/
41. Municipality of Florence: Variation to the Municipal Operational Plan Homepage. https://webpo.comune.fi.it/docs/poc_firenze/documenti/Art_64_Em_var00ter_adot.pdf
42. Castro Noblejas H, Sortino JF, Orellana-Macías JM (2023) Mapping method for the integrated analysis of gentrification and touristification: the case of Málaga (Spain). In: Cuadernos Geograficos, pp 109–129 by ORES
43. Ojeda AB, Kieffer M (2020) Touristification. Empty concept or element of analysis in tourism geography?. In: Geoforum by ScienceDirect
44. Vollmer L (2019) Estrategias contra la gentrificación: Por una ciudad desde abajo. Ed. Katakrak Liburuak, Pamplona
45. Cocola-Gant A (2018) Tourism gentrification. In: Elgar E (ed) Handbook of gentrification studies. Publishing, Cheltenham, pp 281–293
46. Sequera J, Nofre J (2018) Shaken, not stirred: New debates on touristification and the limits of gentrification. City J, 843–855, New York
47. De la Calle-Vaquero M (2019) Touristification of urban centres: an attempt to clarify the debate. In: Bulletin of the Spanish association of geography. Ed. AGE, Madrid
48. GeoSolution: MapStore Homepage. https://www.geosolutionsgroup.com/
49. Municipality of Florence: Municipal Open Data Homepage. https://opendata.comune.fi.it/
50. Municipality of Florence: WebRU Homepage. https://webru.comune.fi.it/webru/pc/index.jsp
51. Municipality of Florence: WebPO Homepage. https://webpo.comune.fi.it/webpo/pc/index.jsp#home
52. University of Florence, Heritage Research Laboratory: Florence Heritage Data System Homepage. https://www.firenzepatrimoniomondiale.it/progetti/florence-heritage-data-system/
53. Capitanio C, Chiesi D, Franco M (2022) Florence heritage data system. A Monitoring and evaluation model for world heritage sites archaeological restoration, Florence
54. Romano A (2024) Airbnb offer trend in Florence 2021–2023 homepage https://www.nocodegeography.com/platforms/andamento-offerta-airbnb-firenze-2021-2023/
55. Lippi G (2020) Airbnb's Impact on the City the conflict between tourist gentrification and the right to the City. Thesis for: undergraduate degree in economic development, international cooperation and conflict management, Florence
56. Brewer CA (1994) Color use guidelines for mapping and visualization. In: Visualization in modern cartography, pp 123–148. Ed. Pergamon, Devon
57. Inside Airbnb: Short-term rental distribution in Florence Homepage. https://insideairbnb.com/florence/. Accessed 13 Oct 2024
58. Celata F, Capineri C, Romano A (2020) A room with a (re)view. Short-term rentals, digital reputation and the uneven spatiality of platform-mediated tourism. In: Geoforum by ScienceDirect
59. Celata F (2023) Hypertourism, short rentals, urban policies and economies: reflections on Rome Homepage. https://riabitiamoroma.it/wp-content/uploads/2024/06/Celata-Iperturismo-affitti-brevi-politiche-ed-economie-urbane-riflessioni-su-Roma.pdf
60. Centre for Tourism Studies: IRPET: Report on Tourism in Tuscany, Florence Homepage. https://www.irpet.it/wp-content/uploads/2024/07/Rapporto-Turismo-in-Toscana-IRPET_31.07.2024.pdf (2023/24)
61. Municipality of Florence: Tourist tax information portal Homepage. https://servizi.comune.fi.it/servizi/portale-ids.

Martina Franco is an urban planning graduate with a combined background in professional practice and academic research. Beginning with collaborations with architectural and urban planning firms during her university studies, she continued this work after graduation, while also conducting research at the Department of Architecture of the University of Florence. Her research interests include both structural planning and environmental assessment, with a recent focus on Spatial Information Systems for World Heritage Sites.

Marta Conte is an architect and research fellow at the Department of Architecture of the University of Florence, after specializing in the field of conservation, enhancement and management of architectural and landscape heritage. Her research interests focus on the management of World Heritage Sites, in line with the principles of UNESCO Conventions and Programs. Since 2022 she has been a member of the research group HeRe Lab—Heritage Research, a Joint Laboratory established between the University of Florence and the Municipality of Florence.

Carlo Francini art historian, is since 2005 the site manager of the UNESCO World Heritage Site "Historic Center of Florence" and the responsible of the Florence World Heritage office of the Municipality of Florence. He is the scientific coordinator of the Italian World Heritage Association and co-director of HeRe Lab—Heritage Research, the Joint Research Laboratory in heritage-related topics between the Municipality of Florence and the University of Florence. Since 2022 he has been a member of the Italian delegation to the UNESCO World Heritage Committee.

Gianluca Belli is a professor in history of architecture carries out teaching and research at the Department of Architecture of the University of Florence. Since 2019 he has held a course in History of Architecture at the Catholic University "Nostra Signora del Buon Consiglio" in Tirana. He also teaches History of Architecture at the School of Specialization in Architectural Heritage and Landscape at the University of Florence, a School of which he has been director since 2021. He was director of the architecture history journal "Opus Incertum" from 2015 to 2020, and is currently director of the FUP series "Essays. Architecture, Planning, Landscape, Design".

Giuseppe De Luca is a professor at the Department of Architecture of the University of Florence, an honorary academician at the Academy of Drawing Arts of Florence and an expert member of territorial and urban planning in the Superior Council of Public Works. He is the co-director of HeRe Lab—Heritage Research, the Joint Research Laboratory in heritage-related topics between the Municipality of Florence and the University of Florence. The main publications revolve around cooperative planning at regional level and cultural heritage at urban level.

MIX
Papier aus verantwortungsvollen Quellen
Paper from responsible sources
FSC® C105338
FSC
www.fsc.org

If you have any concerns about our products,
you can contact us on
ProductSafety@springernature.com

In case Publisher is established outside the EU,
the EU authorized representative is:
Springer Nature Customer Service Center GmbH
Europaplatz 3, 69115 Heidelberg, Germany

Printed by Libri Plureos GmbH
in Hamburg, Germany